Industrial Applications: New Solutions for the New Era

Editors

Marcos de Sales Guerra Tsuzuki
Marcosiris Amorim de Oliveira Pessoa
Alexandre Acácio de Andrade

MDPI • Basel • Beijing • Wuhan • Barcelona • Belgrade • Manchester • Tokyo • Cluj • Tianjin

Editors
Marcos de Sales Guerra Tsuzuki
Universidade de São Paulo
Brazil

Marcosiris Amorim de Oliveira Pessoa
Universidade de São Paulo
Brazil

Alexandre Acácio de Andrade
Universidade Federal do ABC
Brazil

Editorial Office
MDPI
St. Alban-Anlage 66
4052 Basel, Switzerland

This is a reprint of articles from the Special Issue published online in the open access journal *Machines* (ISSN 2075-1702) (available at: https://www.mdpi.com/journal/machines/special_issues/industrial_applications).

For citation purposes, cite each article independently as indicated on the article page online and as indicated below:

LastName, A.A.; LastName, B.B.; LastName, C.C. Article Title. *Journal Name* **Year**, *Volume Number*, Page Range.

ISBN 978-3-0365-3327-8 (Hbk)
ISBN 978-3-0365-3328-5 (PDF)

© 2022 by the authors. Articles in this book are Open Access and distributed under the Creative Commons Attribution (CC BY) license, which allows users to download, copy and build upon published articles, as long as the author and publisher are properly credited, which ensures maximum dissemination and a wider impact of our publications.

The book as a whole is distributed by MDPI under the terms and conditions of the Creative Commons license CC BY-NC-ND.

Contents

About the Editors vii

Preface to "Industrial Applications: New Solutions for the New Era" ix

Ahmad Abubakar, Carlos Frederico Meschini Almeida and Matheus Gemignani
Review of Artificial Intelligence-Based Failure Detection and Diagnosis Methods for Solar Photovoltaic Systems
Reprinted from: *Machines* 2021, 9, 328, doi:10.3390/machines9120328 1

André C. M. Cavalheiro, Diolino J. Santos Filho, Jônatas C. Dias, Aron J. P. Andrade, José R. Cardoso and Marcos S. G. Tsuzuki
Safety Control Architecture for Ventricular Assist Devices
Reprinted from: *Machines* 2022, 10, 5, doi:10.3390/machines10010005 37

Carlos A. B. Reyna, Ediguer E. Franco, Alberto L. Durán, Luiz O. V. Pereira, Marcos S. G. Tsuzuki and Flávio Buiochi
Water Content Monitoring in Water-in-Oil Emulsions Using a Piezoceramic Sensor
Reprinted from: *Machines* 2021, 9, 335, doi:10.3390/machines9120335 53

Cody Berry, Marcos S. G. Tsuzuki and Ahmad Barari
Data Analytics for Noise Reduction in Optical Metrology of Reflective Planar Surfaces
Reprinted from: *Machines* 2022, 10, 25, doi:10.3390/machines10010025 65

Fabio A. A. Andrade, Ihannah P. Guedes, Guilherme F. Carvalho, Alessandro R. L. Zachi, Diego B. Haddad, Luciana F. Almeida, Aurélio G. de Melo and Milena F. Pinto
Unmanned Aerial Vehicles Motion Control with Fuzzy Tuning of Cascaded-PID Gains
Reprinted from: *Machines* 2022, 10, 12, doi:10.3390/machines10010012 83

Joabe R. da Silva, Gustavo M. de Almeida, Marco Antonio de S. L. Cuadros, Hércules L. M. Campos, Reginaldo B. Nunes, Josemar Simão and Pablo R. Muniz
Recognition of Human Face Regions under Adverse Conditions—Face Masks and Glasses—In Thermographic Sanitary Barriers through Learning Transfer from an Object Detector
Reprinted from: *Machines* 2022, 10, 43, doi:10.3390/machines10010043 101

Miguel Angel Orellana, Jose Reinaldo Silva and Eduardo L. Pellini
A Model-Based and Goal-Oriented Approach for the Conceptual Design of Smart Grid Services
Reprinted from: *Machines* 2021, 9, 370, doi:10.3390/machines9120370 117

Paulo Henrique Martinez Piratelo, Rodrigo de Azeredo, Eduardo Yamao, Jose Filho, Gabriel Maidl, Felipe Lisboa, Laercio de Jesus, Renato Neto, Leandro Coelho and Gideon Leandro
Blending Colored and Depth CNN Pipelines in an Ensemble Learning Classification Approach for Warehouse Application Using Synthetic and Real Data
Reprinted from: *Machines* 2022, 10, 28, doi:10.3390/machines10010028 143

Robert R. Gomes, Luiz F. Pugliese, Waner W. A. G. Silva, Clodualdo V. Sousa, Guilherme M. Rezende and Fadul F. Rodor
Speed Control with Indirect Field Orientation for Low Power Three-Phase Induction Machine with Squirrel Cage Rotor
Reprinted from: *Machines* 2021, 9, 320, doi:10.3390/machines9120320 169

Tiago Yukio Fujii, Victor Takashi Hayashi, Reginaldo Arakaki, Wilson Vicente Ruggiero, Romeo Bulla, Jr., Fabio Hirotsugu Hayashi and Khalil Ahmad Khalil
A Digital Twin Architecture Model Applied with MLOps Techniques to Improve Short-Term Energy Consumption Prediction
Reprinted from: *Machines* **2022**, *10*, 23, doi:10.3390/machines10010023 . . . . **193**

Vitor Furlan de Oliveira, Marcosiris Amorim de Oliveira Pessoa, Fabrício Junqueira and Paulo Eigi Miyagi
SQL and NoSQL Databases in the Context of Industry 4.0
Reprinted from: *Machines* **2022**, *10*, 20, doi:10.3390/machines10010020 . . . . **219**

Vitor Neves Hartmann, Décio de Moura Rinaldi, Camila Taira and Arturo Forner-Cordero
Industrial Upper-Limb Exoskeleton Characterization: Paving the Way to New Standards for Benchmarking
Reprinted from: *Machines* **2021**, *9*, 362, doi:10.3390/machines9120362 . . . . **245**

About the Editors

Marcos de Sales Guerra Tsuzuki graduated from Escola Politécnica from USP (1985), received a Master's in Information Processing Engineering at Yokohama National University (1989), a PhD in Mechanical Engineering at Escola Politécnica from USP (1995) and became a Full Professor at Escola Politécnica from USP (2000). He is chair of TC 5.1—Manufacturing Plant Control and Vice-Chair of TC 8.2—Biological and Medical Systems, both at IFAC. He has experience in the field of mechatronics engineering, with an emphasis on CAD/CAM, geometric modeling, and solid modeling. He is currently an associate professor at the Escola Politénica from USP.

Marcosiris Amorim de Oliveira Pessoa holds a degree in Mechanical Engineering (1995), an MS in Electrical Engineering in the field of Computer Engineering (2003), and a D.Sc in Mechanical Engineering in the field of Automation and Control Engineering (2015). He completed his postdoctoral studies at the University of New South Wales (UNSW—Sydney, Australia) at the School of Computer Science and Engineering (CSE—2018). From 2006, he worked as a laboratory specialist engineer at Escola Politécnica of the University of São Paulo and is a researcher in the Automation Systems Laboratory (LSA-PMR-EPUSP); from 2017–2018, was a Research Associate Visitor from the University of New South Wales (UNSW—Sydney, Australia) at CSE. Has experience in Mechanical Engineering, Mechatronic Engineering, and Computer Engineering, emphasizing the following sectors: Industry 4.0, cyber–physical systems, Internet of Things, virtual enterprises, modeling and simulation, time windows, scheduling heuristics, and constraint programming.

Alexandre Acácio de Andrade, also know as Alexandre Acassio, holds a degree in Electrical Engineering with an emphasis on Energy and Automation (1997), a Master's degree in Electrical Engineering (Automation) (2001), and a Ph.D. (2007) in Electrical Engineering (Automation), all from the Polytechnic School at the University of São Paulo. He was a Professor at Faculdade de Engenharia São Paulo from 2002 to 2012. He has 10 years of experience in Industrial Automation projects based on PLCs (Rockwell Automation Agreement—EPUSP), working mainly on the following subjects: artificial intelligence, data mining, supervisory systems, programmable logic controllers, DCS, data networks, low-cost automation, and innovation. He has been a full-time Professor at UFABC since 2012. He has been a Professor of the Master's in Innovation Management Engineering program since 2015.

Preface to "Industrial Applications: New Solutions for the New Era"

The authors from the conference IEEE-INDUSCON 2021, held in São Paulo, Brazil, were invited to submit an expanded version of their papers to a Special Issue of *Machines*: "Industrial Applications: New Solutions for the New Era", https://www.mdpi.com/journal/machines/special_issues/industrial_applications.

The special theme "Innovation in the time of COVID-19" was chosen for IEEE-INDUSCON 2021. The COVID-19 pandemic caused significant challenges that required quick responses from governments and industries. Recent changes have set the scene for an increased awareness of the importance of engineering as a problem solver. Meanwhile, these new challenges are opening up new opportunities for academia and industry innovators to create new solutions to the problems of this new era.

This Book presents twelve selected papers that are related to industrial applications in several major topics of the conference, including: artificial intelligence [1]; a ventricular assist device supervisory control system [2]; an ultrasonic technique to estimate water concentration in water-in-crude oil emulsions [3]; a method to remove noise from a laser scan of planar data [4]; control strategies for unmanned aerial vehicles [5]; infrared thermography fever detection for COVID-19 [6]; microgrid model-based systems engineering [7]; a blend of convolutional neural network pipelines that classifies products from an electrical company's warehouse [8]; the study and implementation of speed control applied to a squirrel cage type of three-phase induction machine [9]; the effective prediction of energy consumption with a smart home digital twin integrated with MLOps [10]; the advantages and disadvantages of relational (SQL) and non-relational (NoSQL) data models for Industry 4.0 [11]; a device and procedure involving static and dynamic tests to assess exoskeletons [12]; machine learning [1,10], and others.

The editors are pleased with the results of this Special Issue and appreciate the authors' valuable contributions and would like to thank the reviewers for their efforts and time spent on improving the submissions, and the publisher for their excellent work and cooperation.

References

1. Abubakar, A.; Almeida, C.F.M.; Gemignani, M. Review of Artificial Intelligence-Based Failure Detection and Diagnosis Methods for Solar Photovoltaic Systems. *Machines* **2021**, *9*, 328.

2. Cavalheiro, A.C.M.; Santos Filho, D.J.; Dias, J.C.; Andrade, A.J.P.; Cardoso, J.R.; Tsuzuki, M.S.G. Safety Control Architecture for Ventricular Assist Devices. *Machines* **2022**, *10*, 5.

3. Reyna, C.A.B.; Franco, E.E.; Durán, A.L.; Pereira, L.O.V.; Tsuzuki, M.S.G.; Buiochi, F. Water Content Monitoring in Water-in-Oil Emulsions Using a Piezoceramic Sensor. *Machines* **2021**, *9*, 335.

4. Berry, C.; Tsuzuki, M.S.G.; Barari, A. Data Analytics for Noise Reduction in Optical Metrology of Reflective Planar Surfaces. *Machines* **2022**, *10*, 25.

5. Andrade, F.A.A.; Guedes, I.P.; Carvalho, G.F.; Zachi, A.R.L.; Haddad, D.B.; Almeida, L.F.; de Melo, A.G.; Pinto, M.F. Unmanned Aerial Vehicles Motion Control with Fuzzy Tuning of Cascaded-PID Gains. *Machines* **2022**, *10*, 12.

6. da Silva, J.R.; de Almeida, G.M.; Cuadros, M.A.d.S.L.; Campos, H.L.M.; Nunes, R.B.; Simão, J.; Muniz, P.R. Recognition of Human Face Regions under Adverse Conditions—Face Masks and Glasses—In Thermographic Sanitary Barriers through Learning Transfer from an Object Detector. *Machines* **2022**, *10*, 43.

7. Orellana, M.A.; Silva, J.R.; Pellini, E.L. A Model-Based and Goal-Oriented Approach for the Conceptual Design of Smart Grid Services. *Machines* **2021**, *9*, 370.

8. Piratelo, P.H.M.; de Azeredo, R.N.; Yamao, E.M.; Bianchi Filho, J.F.; Maidl, G.; Lisboa, F.S.M.; de Jesus, L.P.; Penteado Neto, R.d.A.; Coelho, L.d.S.; Leandro, G.V. Blending Colored and Depth CNN Pipelines in an Ensemble Learning Classification Approach for Warehouse Application Using Synthetic and Real Data. *Machines* **2022**, *10*, 28.

9. Gomes, R.R.; Pugliese, L.F.; Silva, W.W.A.G.; Sousa, C.V.; Rezende, G.M.; Rodor, F.F. Speed Control with Indirect Field Orientation for Low Power Three-Phase Induction Machine with Squirrel Cage Rotor. *Machines* **2021**, *9*, 320.

10. Fujii, T.Y.; Hayashi, V.T.; Arakaki, R.; Ruggiero, W.V.; Bulla, R., Jr.; Hayashi, F.H.; Khalil, K.A. A Digital Twin Architecture Model Applied with MLOps Techniques to Improve Short-Term Energy Consumption Prediction. *Machines* **2022**, *10*, 23.

11. de Oliveira, V.F.; Pessoa, M.A.d.O.; Junqueira, F.; Miyagi, P.E. SQL and NoSQL Databases in the Context of Industry 4.0. *Machines* **2022**, *10*, 20.

12. Hartmann, V.N.; Rinaldi, D.d.M.; Taira, C.; Forner-Cordero, A. Industrial Upper-Limb Exoskeleton Characterization: Paving the Way to New Standards for Benchmarking. *Machines* **2021**, *9*, 362.

Marcos de Sales Guerra Tsuzuki, Marcosiris Amorim de Oliveira Pessoa, Alexandre Acácio de Andrade
Editors

Review

Review of Artificial Intelligence-Based Failure Detection and Diagnosis Methods for Solar Photovoltaic Systems

Ahmad Abubakar *, Carlos Frederico Meschini Almeida and Matheus Gemignani

Department of Electrical Engineering and Automation, Escola Politecnica da Universidade de São Paulo, São Paulo 05508-010, Brazil; cfmalmeida@usp.br (C.F.M.A.); matheusg@alumni.usp.br (M.G.)
* Correspondence: namatoyaa@usp.br; Tel.: +55-11997680626

Abstract: In recent years, the overwhelming growth of solar photovoltaics (PV) energy generation as an alternative to conventional fossil fuel generation has encouraged the search for efficient and more reliable operation and maintenance practices, since PV systems require constant maintenance for consistent generation efficiency. One option, explored recently, is artificial intelligence (AI) to replace conventional maintenance strategies. The growing importance of AI in various real-life applications, especially in solar PV applications, cannot be over-emphasized. This study presents an extensive review of AI-based methods for fault detection and diagnosis in PV systems. It explores various fault types that are common in PV systems and various AI-based fault detection and diagnosis techniques proposed in the literature. Of note, there are currently fewer literatures in this area of PV application as compared to the other areas. This is due to the fact that the topic has just recently been explored, as evident in the oldest paper we could obtain, which dates back to only about 15 years. Furthermore, the study outlines the role of AI in PV operation and maintenance, and the main contributions of the reviewed literatures.

Keywords: artificial intelligence; photovoltaics; fault detection; machine learning; operation and maintenance; renewable energy

1. Introduction

The rapid development of technology and social advancements has led to the skyrocketing of energy demand, which has, in turn, resulted in an increase in fossil fuel generation of energy [1,2] This has raised concerns of high CO_2 emission into the atmosphere due to the combustion of fossil fuels [3,4], which leads to global warming, GHG emissions, climate change, and other environmental issues [5]. Owing to the global commitment to overcome these issues by reducing fossil fuel energy generation to the bare minimum, the renewable industry has experienced an exponential growth and development in recent years. Renewable energy sources, especially solar [6,7], have been increasingly adopted for residential, commercial, and industrial applications [8–10]. The 2020 first quarter (Q1 2020) report of the National Renewable Energy Laboratory (NREL) stated that at the end of 2019, the installed solar PV capacity totaled 627 GW_{DC}, an increase of 115 GW_{DC} from the previous year [11].

Solar PV systems, however, need constant maintenance in order to efficiently operate over time. Therefore, strategies have to be in place to effectively monitor and maintain these systems. Various conventional methods are deployed by experts to carry out preventive, corrective, and predictive maintenance activities [12]. These methods usually equip the PV system with ground fault detection interrupters (GFDI) and overcurrent protection devices (OCPD). However, most of the time, they are not sufficient enough for detecting certain faults due to low irradiance conditions, nonlinear output characteristics, PV inverter maximum power point trackers (MPPT) or high fault impedances [13]. The need for more adequate and intelligent strategies of detecting and diagnosing faults in PV systems has encouraged the adoption of AI-based methods. These methods utilize machine learning to train models to detect and locate various faults, monitor the general health status, and help the maintenance engineers of PV systems to rapidly expedite system recovery [13].

Over the years, AI has been utilized in PV system's design [14], MPPT [15], power prediction [16], as well as failure detection and protection [17–21]. The authors of [22] detailed the key characteristics that distinguish fault detection and diagnosis methods in PV systems. According to the authors, these characteristics include rapidity in detecting defects, input data (climate and electrical data), and selectivity (ability to distinguish between different faults). In addition, the study divided AI-based fault detection and diagnosis methods into two categories. First, visual and thermal methods for detecting discoloration, browning, surface soiling, hotspot, breaking, and delamination. Second, electrical methods for detecting and diagnosing faulty PV modules, strings, and arrays, such as arc faults, grounding faults, and diode faults. In [17], the adaptive neuro-fuzzy inference framework was adopted for the development of a smart fault detection approach for PV modules, while in [18], a dual-channel convolutional neural network model with a feature selection structure was proposed for PV array fault diagnosis. K-NN was used by the authors of [19] for the modeling of PV systems based on experimental data for detection, while in [20], a wavelet-based protection strategy was presented for the detection of a series of arc faults interfered by multicomponent noise signals in grid-tied PV systems. In [21], a PV fault detection algorithm that integrates two bi-directional input parameters based on the artificial neural network was presented. A novel extreme learning machine (ELM)-based modeling method, featuring high training speed and generalization performance was proposed in [23], using current-voltage (I-V) curves measured at different operating conditions, for the characterization of the electrical behavior of PV modules. The authors in [24] attempted to improve the integration of PV systems into the electrical network by controlling the converter and inverter. This is achieved through the introduction of an adaptive reference PI (ARPI) for the inverter aimed at enhancing the system performance by supporting low voltage ride through (LVRT) capability and smoothing of the PV power fluctuations during variable environmental conditions. The authors of [25] proposed a model based on the geographic information system (GIS)-based reinforcement learning, for the optimal planning of rooftop PV system. The model considers the uncertainty of future scenarios across the buildings lifecycle. Another AI application is PV systems, which is utilized in [26] for the optimal dispatch of PV inverters in unbalanced distribution systems using reinforcement learning, while the authors of [27] used AI for the optimal design of a phase change material integrated renewable system with on-site PV, radiative cooling, and hybrid ventilations. In addition, reinforcement learning with the fuzzified reward approach was used in [28] for controlling the MPPT of PV systems. Another MPPT algorithm of the PV system, which is based on irradiance estimation and the multi-Kernel extreme learning machine, was presented in [29] in order to reduce investment costs and improve PV system efficiency, while in [30], the deep reinforcement learning approach was used for MPPT control of partially shaded PV systems in smart grids. Two other literatures on the application of AI for MPPT control are provided in [31,32]. The authors of [31] presented a new combined ELM variable steepest gradient ascent MPPT for the PV system, while the authors of [32] presented a novel meta-heuristic optimization algorithm based MPPT control technique for partially shaded PV systems. As stated earlier, AI was also utilized for PV power prediction as in the case of [33], where the short-term PV power prediction was achieved using a hybrid improved Kmeans-GRA-Elman model based on multivariate meteorological factors and historical power datasets. Moreover, AI was used for a similar application in [34], where the deep learning and wavelet transform integrated approach was used for short-term solar PV power prediction. The authors of [35] developed an intelligent real-time power management system, where an incremental unsupervised neural network algorithm was used to predict the output power and then detect the power fluctuations occurrence of a grid-tied PV system. A comparative study on short-term PV power prediction using the decomposition based ELM algorithm was presented in [36], while in [37], a short-term PV power forecasting model using the hybrid deep learning approach was proposed.

This study presents an extensive review of AI-based methods and techniques of fault detection and diagnosis reported in various literatures. The contribution of the study is in outlining the characteristics of the reviewed AI-based methods and their effectiveness in rapidly and efficiently detecting faults with minimal error, since the effectiveness of a fault detection and diagnosis method depends on the following factors: Its ability to detect a fault and pinpoint its location in the shortest possible time; its relative affordability; and ease of use [38]. The structure of the remaining part of the paper is as follows. Section 2 discusses the various types of faults that occur in PV systems; Section 3 introduces artificial intelligence and machine learning; Section 4 provides a review of the AI-based fault detection and diagnosis methods proposed in various literatures; and Section 5 concludes the present work and discusses its perspective.

2. Types of PV System Faults

Over time, PV systems experience fault occurrences that affect the system's operating efficiency, may cause damage to the system components, and may also lead to dangerous fire threats and safety hazards. PV system faults are classified as physical, environmental or electrical faults [39]. Panel faults, such as PV cell internal damages, cracks in panels, bypass diodes, degradation faults, and broken panels are classified as physical faults [39]. Shade faults due to bird dropping, dust accumulation, cloud movement, and tree shadows are classified as environmental faults [39]. Faults that are classified as electrical faults include MPPT faults, open-circuit faults, ground faults, line-line faults, short-circuit faults, arc faults, and islanding operation [39,40]. This section briefly discusses the different types of faults peculiar to PV systems.

2.1. Shading Faults

Shading occurs when objects, such as trees, neighboring buildings, and overhead power lines, cast shadows on PV modules [41,42]. Shading in PV arrays could be homogeneous, where there is a balanced reduced irradiation across the PV panels or non-homogeneous, where there is an unbalanced reduced irradiation across the panels [39].

2.2. Arc Faults

A frequent high-power discharge of electricity through an air gap between conductors causes this type of fault [43–45]. The two forms of arc faults include first, the series arc faults that usually originate from solder separation, connection corrosion, cell damage, rodent damage or abrasion from numerous sources. Second, parallel arc faults that result from insulation failure in current-carrying conductors [43,44].

2.3. Line-Line Faults

A line-line fault is an unintentional short-circuit between two points with differing voltage potentials [46–48]. These faults are more difficult to detect than other faults and are frequently misinterpreted as short-circuit faults in grounded PV systems, since the fault current is determined by the voltage differential between two fault spots [39]. The most common types of line-line faults are intra string faults, which are short-circuit faults between two locations on the same string, and cross string faults, which are short-circuit faults between two places on separate threads [39].

2.4. Ground Faults

To protect users from a possible electric shock, it is common practice that the metallic parts of the PV array are grounded using earth-grounding conductors (EGC) [39]. The term "ground fault" refers to any unintentional connection between a current-carrying conductor and an EGC that results in a current flow to the ground [45,46,49,50].

2.5. Other Faults

This refers to any fault that cannot be categorized under any of the previously discussed faults. These types of faults include MPPT and inverter faults that mostly occur due to inverter components failure, such as IGBTs, capacitors, and converter switch failure [51–53]; bypass diode faults resulting from a massive reverse current flow during faults, which leads to short-circuits [54,55]; blocking diode faults, also as a result of a reverse current flow [39]; open-circuit faults caused by items falling on PV panels, physical failure of panel-panel cables or joints, and sloppy termination of cables, plugging, and unplugging connectors at junction boxes [56]; faulty connections damage of connecting cables or a wrong connection of panels [57]; battery bank failures due to abnormal charging conditions [39]; and blackouts caused by natural disasters, such as a storm and lightning [58]. Most of these mentioned faults are usually due to an after effect of the other faults [39].

3. Artificial Intelligence and Machine Learning

Machine learning (ML) is a subsection of AI, which, in turn, is one of the most recent branches of science and engineering that emerged shortly after the Second World War, in order to attempt to understand and build intelligent entities [59]. ML thrives on extracting meaning from a large amount of data. Therefore, ML refers to the tools and technology that can be used to answer questions using the currently available data. Philosophy, statistics, biology, computational complexity, information theory, artificial intelligence, cognitive science, and control theory are some of the fields that machine learning draws on concepts and results [60]. In recent years, numerous machine learning algorithms and applications have been successfully created and utilized for different applications from autonomous vehicles, where these algorithms learn to drive on public highways, detect fraudulent transactions using data-mining programs, and are applied to information-filtering systems. This has led to critical advancements in the foundations of this field, which are theory and algorithms.

According to [61–63], machine learning is a branch of artificial intelligence (AI) that allows a system to learn from experience and improve without the need to be explicitly programmed. Its goal is to create computer programs that can access data and learn on their own [61]. Basically, the concept of ML is the use of fed-in data and algorithm to generate artificial knowledge, which is guided by sets of pre-defined analytical rules for pattern recognition in collected data [60]. In simple clear terms, ML is the use of quantitative and qualitative data to answer questions with ease and precision. Using ML, computers and IT systems are able to perform tasks, such as discovering, extracting, and summarizing relevant data, making predictions based on analyzed data, independently adapting to certain developments, calculating probabilities for specific results, and optimizing processes based on recognized patterns [61].

AI and ML are found in our everyday activities from transportation in the form of self-driving and self-parking cars, image recognition, google searches, fraud detection in the banking and finance sector, diagnosis and prescription in medical fields, recommendation systems, as well as social media applications [59]. In PV system applications, ML and AI are used in predicting solar radiation, sunshine duration, and clearness index; mean temperature; and insolation and diffusion fractions [64]. In addition, they are used for the sizing, configuring, modeling, simulation, and control of PV systems, detection and diagnosis of faults, and the forecasting of the output electricity from standalone and grid-connected PV systems [22,64–72].

3.1. Machine Learning Types

3.1.1. Supervised Learning

In this ML type, models are defined in advance and systems learn from the given input and output pairs, i.e., the input data and desired output are labeled [61]. With enough data knowledge, one can help the machine connect the dots with supervised learning using the labeled sample data and correct output.

3.1.2. Unsupervised Learning

Here, the AI learns without the aid of predefined target values, i.e., the model is required to identify patterns in an unlabeled input data [61]. Learning and improving by trial and error is key to unsupervised learning. Unlike supervised learning, here you are not working with labeled data, you are not showing the machine the correct output. You are using different algorithms to let the machine connect the dots by studying and observing data. In unsupervised learning, the chances of the machine to find patterns or classifications that humans can never see is very high.

3.1.3. Semi-Supervised Learning

This is a combination of supervised and unsupervised learning advantages [61]. Here, training starts with a small amount of dataset in order to allow the machine to get familiarized with the data. In addition, the machine studies and observes the data to expand its vocabulary/database using inductive reasoning. Another form of semi-supervised learning is the transductive reasoning, which allows one to narrow down the unlabeled data using unknown knowledge of collected data. Semi-supervised learning is not very common in machine learning applications.

3.1.4. Reinforcement Learning

In this situation, the model is granted autonomy to engage with a dynamic environment that gives feedback based on rewards and punishments, i.e., the model is taught through positive and negative interactions [61]. This method of learning differs significantly from the other three methods. The machine iterates until the outcome is enhanced each time, coming closer and closer to high-quality output.

3.1.5. Multitask Learning

Multitask learning helps several algorithms share their experience with each other, thereby helping them learn concurrently rather than individually [73].

3.1.6. Ensemble Learning

Ensemble learning is a combination of two or more algorithms that form one single algorithm [74]. Here, it has been observed that a collection of algorithms almost always outperforms an individual algorithm when carrying out a particular task [75].

3.1.7. Neural Network Learning

Neural networks, also referred to as artificial neural networks (ANN), are derived from the biological concept of brain cells called neurons. Therefore, to understand ANN, one has to be familiar with how neurons work [74]. An ANN functions on three layers (input layer, hidden layer, and output layer), in the same way as the brain neurons work on four parts (dendrites, nucleus, soma, and axon) [74]. The input layer receives the data, which is then processed by the hidden layer, before it is sent as a calculated output to the output layer [76].

3.1.8. Instance-Based Learning

In this case, the algorithm learns a specific pattern and then applies it to new data [74]. This learning method becomes more sophisticated as the amount of data grows [77].

3.1.9. Evolutionary Computation

Evolutionary computation is currently a distinct branch of artificial intelligence inspired by nature [78,79], with smart methods based on evolutionary algorithms targeted at solving various real life problems through natural processes involving live things [79]. It is based on random processes, data regeneration, and data replacement within a system, such as a personal computer or any other data center. A variety of evolutionary computation approaches are utilized for different applications, including image processing, cloud computing, and grid computing [79].

4. Artificial Intelligence-Based Failure Detection and Diagnosis Methods

Most of the AI-based methods used in failure detection and diagnosis adopt ML models, such as support vector machines, wavelets, neural networks, fuzzy logics, decision trees, graph-based semi-supervised learning (GBSSL), regression, etc. for the development of models and algorithms that are trained to learn the relationships between the input and output parameters of PV systems. The data are obtained from experimentally accurate PV model measurements, and then split into training and test datasets. This section reviews and discusses the AI-based detection and diagnosis techniques proposed in various literatures. Figure 1 shows a distribution of the available literature on fault detection and diagnosis methods for PV systems based on the mentioned ML models. It presents a rough estimate of the number of conference and journal papers published to date. In general, the amount of publications on the subject is relatively low, which is due to the fact that it is a newly explored area of research that dates back to only about 15 years.

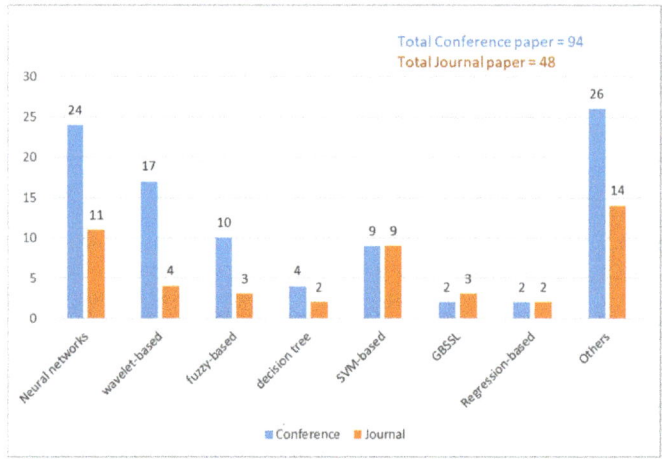

Figure 1. Distribution of AI-based fault detection and diagnosis methods for PV systems found in the available literature.

4.1. Neural Network-Based Methods

The study presents a literature review concerning the AI methods of fault detection and diagnosis that are neural network based. Table 1 presents a summary of the different methods presented in the subsection, pointing out the main contribution of each reviewed literature.

In [80], a simple and effective Bayesian neural network (BNN) model for estimating power losses in PV plants owing to soiling was devised. Four models were built based on the Bayesian neural network (BNN) to assess the performance of two plants for dirty and clean module conditions under standard test conditions (STCs). The loss due to the soiling impact is shown by the difference in the STC power between the two conditions. The study found that utilizing a BNN model rather than a polynomial model for calculating the STC power of a PV system is more successful due to various factors that affect the polynomial model's performance, including the database size. Figure 2 presents a schematic diagram of the BNNs used for calculating the STC power, consisting of an input layer (which has solar irradiance and cell temperature as input), a single hidden layer (estimated during the training process), and an output layer (which provides the STC power output produced by the plant). Bayesian regularization, a process of updating the weight and bias values according to the Levenberg-Marquardt optimization technique, which helps in reducing a combination of squared errors and weights, and determining the correct combination to produce a network that generalizes well, can greatly improve the generalization ability of neural networks [81]. BNN is basically back propagation (BP) with an additional ridge parameter added to the objective function [80]. The study provides an important contribution as it helps the operation and maintenance personnel in decision making between washing cost and losses in energy production.

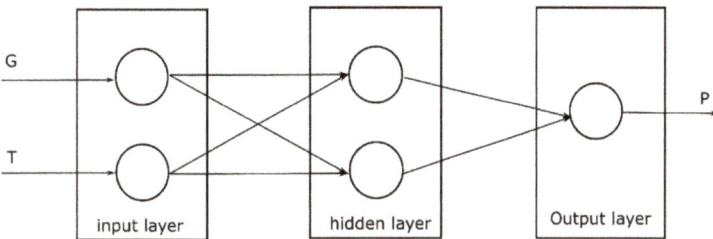

Figure 2. Schematic block diagram of Bayesian regularization [80].

In order to successfully detect and categorize PV array faults, the authors of [82] employed a deep two-dimensional convolutional neural network (CNN) to extract features from two-dimensional scalograms generated from the PV system data. The study took into account five different fault scenarios as well as the use of MPPT. There are two variations of the proposed method. First, the last layers of a pre-trained AlexNet CNN [83] are fine-tuned to generate a six-way classifier in the first configuration. In the second configuration, features are extracted from a specific layer of a pre-trained AlexNet and then combined with a classical classifier. The suggested model's performance is compared to machine learning- and deep learning-based models. The suggested method surpasses the previous methods in terms of detection accuracies for both noisy and noiseless data, according to the authors. They also illustrated the need of representative and discriminative features for categorizing errors (rather than using raw data), especially in noisy environments. Automatic feature extraction based on deep learning has been found to be superior to manual feature extraction. In order to better explain the proposed method, Figure 3 presents a flowchart showing the proposed PV array fault diagnosis method and existing methods. Another neural network-based method of fault detection and diagnosis in solar PV systems, which uses Elman neural network (ENN), is presented in [84]. The study examines the implicit mining link between original data and fault types, develops multiple hypothesis models, and analyzes the mean and variance of diagnostic errors to determine which diagnostic model is optimal. The suggested fault diagnosis approach based on ENN overcomes the problem of PV system multi-source and multi-type defect identification by minimizing the number of sensors, which only collects PV operation data and data from the atmospheric environment.

The main contribution of [66] is the proposition of a technique for isolating and identifying faults that occur in the PV system, and its implementation into a field programmable gate array (FPGA) with real-life application effects. The proposed approach detects and diagnoses faults that occur in PV bypass diodes, cells, modules, and strings. It accomplishes this by examining a set of parameters, such as current, voltage, and the number of peaks in the I-V characteristics that indicate normal and abnormal PV system operation. Two separate algorithms are used in this strategy. The first algorithm isolates defects with different combination attributes using a signal threshold approach. The second technique uses an ANN-based approach to identify errors that have the same mix of features. The approach is low-cost and easily adaptable to large-scale PV systems. The block diagram of the proposed fault detection technique based on the threshold approach and ANN is presented in Figure 4.

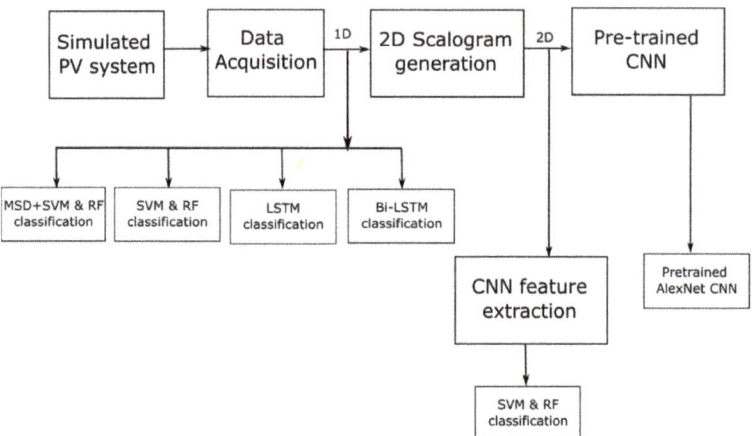

Figure 3. Flowchart showing the proposed method of PV array fault diagnosis and existing methods [82].

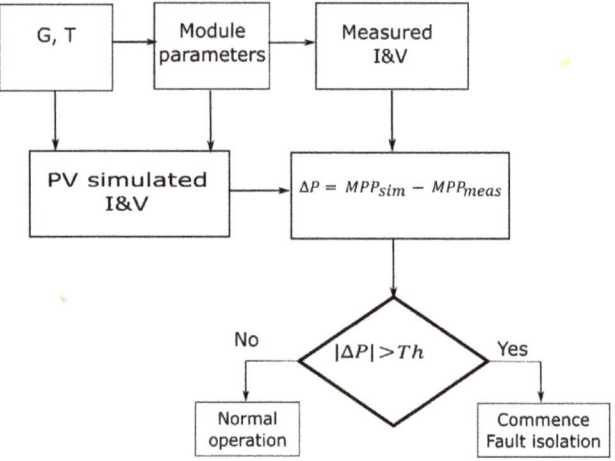

Figure 4. Block diagram of the proposed fault detection technique based on the threshold approach and ANN [66].

A method for detecting the islanding phenomenon in the PV system was introduced in [85]. To detect islanding actions, the method employs a multi-variable method based on an extended neural network that combines passive and active detection modes. The method combines the extension theory's extension distance with a neural network's learning, recalling, generalization, and parallel computing capabilities. The study used an extension neural network to distinguish between power quality interference (voltage swells, voltage dips, power harmonics, and voltage flickers) and actual islanding operations at the grid power end, in order for the islanding phenomenon detection system to cut off the load correctly and promptly when a real islanding operation occurs. The detection algorithm is created and translated using a PSIM software package based on the C language and written in dynamic-link library (DLL) modules. The signals sent by DLL are passed back to the controller to complete the islanding detection control. An enhanced machine learning based approach for the detection and diagnosis of short-circuit faults, and a complete disconnection of the string from an array in the DC side of the grid-connected PV system is presented in [86]. The process uses a probabilistic neural network (PNN) classifier with one diode model (ODM) and a parameter extraction method to create a trustworthy model of a real-world PV system. Two PNN classifiers are used in the proposed method, one is for detecting fault occurrences and the other for diagnosing the type of fault. There are four stages to the method: Array parameter extraction; experimental model validation; elaboration of database of both healthy and problematic operations; and network design, training, and testing based on the best-so-far ABC algorithm. The contribution of this study is highlighted in the model's ability to detect a fault, while also pinpointing its origin. However, for the method to be effective, the high-quality database which is not always readily available, is required to deal with classification problems. To deal with this issue, the authors suggested having in place a trusted simulation model, which is able to mimic the exact healthy and faulty behaviors of a PV system. Another PNN-based intelligent method for PV system health monitoring was proposed in [56], which can detect and categorize short- and open-circuit faults in real time, as well as locate the faulted PV string in a grid-connected PV system. To detect and diagnose faults, the suggested technique uses data obtained from various sensors in PV systems, such as voltage, current, irradiation, and temperature which are used to deliver information on fault occurrence. The PNN used in the method has four layers: The input layer (the number of neurons in the layer represents the number of training and testing samples); the hidden layer (whose pattern units are equal to the training set sample space); the summation layer (the number of neurons is equal to the number of sample space classes); and the output layer or the decision layer (containing one neuron which provides the classification decision). Moreover, it was developed and validated in computer programs utilizing a novel approach to PV system modeling that only requires data from the manufacturer's datasheets provided under normal operating cell temperature conditions (NOCT) and STCs. The modeling approach is an improvement to the previous approaches where STC conditions, I-V characteristics or NOCT conditions are used but never combined together. This systematically builds a relationship between the ideality factor, thermal voltage, and series resistance with the PV module temperature using the manufacturers' datasheet elements. The PV system simulation model is then used to implement and validate the PNN-based detection model and classification method. The authors of [87] explored real time online fault detection for PV modules under partial shading conditions. The approach suggested in the paper is an intelligent method that uses artificial neural networks (ANN) to estimate the output PV current and voltage under varying operating conditions utilizing solar irradiance and cell temperature meteorological factors. Since it performs the real time correlation of estimated performances with the measured performances under variable conditions, the method can also be used to detect the possible anomalies in PV modules. The model proposed is independent of the measured PV module performance which makes its system of fault detection autonomous. Figure 5 depicts the fault detection flowchart of the proposed method. The results of the study show that the proposed method can accurately estimate the output and detect any decrease in output power without requiring any complex calculations or mathematical models. However, it does necessitate that the ANN be trained on a regular basis in order to accurately estimate the output parameters. The approach could also be used in PV arrays or large-scale PV plants, as well as in low-cost microcontrollers for real-time applications.

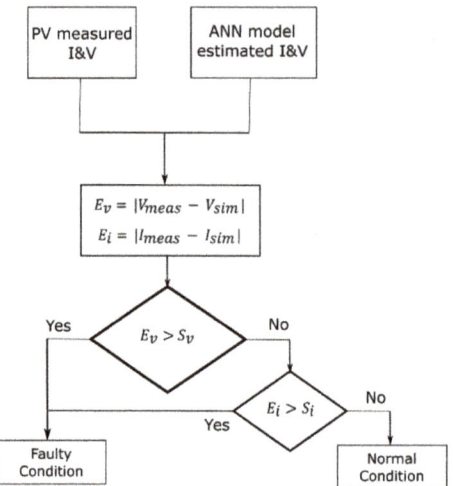

Figure 5. Proposed fault detection flowchart diagram [87].

According to the authors in [88], few data collecting systems in solar power plants focus on intelligent reasoning of the plant's state, despite the fact that the data of these collecting systems offer a wide range of capabilities. To tackle this problem, the authors presented a description of a novel data acquisition system. A Bayesian belief network-based (BNN) fault detection and diagnostic system was then built, which analyzes the acquired data for the existence of faults and intelligent reasons for potential causes of the detected faults. The BNN-based model uses a graphical representation of a problem in the form of a hierarchical network, with nodes representing random variables and directed arcs expressing the probability regarding the dependencies between these variables. Each node has a set of states, each with a probability distribution associated with it. Each arc reflects a conditional probability based on the preceding nodes. A node's state could be as simple as true or false in the most basic scenario. In a more complicated instance, the set of states could include multiple discrete states, such as low, medium, high, and very high. The measurements obtained by a BBN-based inference engine produce a change in the probability values in the respective nodes, which impacts the connected nodes in the network and leads to the automatic derivation of a decision on the likely reason of a failure. By developing a framework for analyzing sensor results and translating them to a Bayesian belief network using the Netica API, the system lays the groundwork for future advancements. The Fault Identification for Nasa Exploration Missions and Navigation (FINEMAN) system was developed as an integrated package with four main components: A connectivity interface for remotely retrieving data from the plant's data acquisition system; a preprocessor for relevant measurement selection; a fault injector for failure testing simulation; and a Netica implemented BBN interface engine. The research adopted in [64] was also ANN-based, and unlike in [87], where ANN was used for the PV module under partial shading conditions, the approach in [64] utilized AI technology for automatic MPPT fault detection and failure type judgement. The method requires five features (solar irradiation, installed capacity, MPPT power, MPPT voltage, and MPPT current) for machine learning. Each inverter's MPPT performance is gathered and modeled using a machine learning algorithm rather than a rule-based programming approach. In addition, an inefficient MPPT is discovered by comparing the real and expected power output. Moreover, faulty equipment is automatically recognized, a knowledge-based system determines the type of failure, and an alarm with the failure diagnostic is communicated to the user by a mobile device or email. The field sites collect data on power production every 5 min. As a result, the entire computation and communication process in the system takes only 5 min to complete. Using this AI method, the authors were able to efficiently manage hundreds of projects at the same time, while also optimizing O&M performance with minimal work and resources. A novel algorithm utilizing the genetic algorithm to optimize the topology of ANN was introduced in [72]. The new algorithm is an online optimized neural network-based fault diagnostic and repairing system, aimed at providing a solution to the problem of complexity and high costs, associated with the other fault detection methods. The method offers the following factors: High speed diagnostic process as it diagnoses multiple faults in parallel; can remotely replace faulty components with good ones; can be used for modern complex PV systems; and has the ability to partition the PV panel into sub-areas. As a result, the diagnostic procedure is divided into two parts. The first is concerned with identifying the failure where the proposed method diagnoses a PV system using a genetic algorithm to optimize the topology structure of the neural network. The system is implemented in five steps: (1) Chromosome representation, which uses binary digits for topology network representation; (2) initial population, which is constructed from random individual sizes; (3) cross operator, which combines the parents to obtain two offspring using the uniform crossover; (4) mutation operation, where random mutation is used for genetic algorithm; and (5) fitness function, which is used to minimize the error value of the neural network representing the fitness function of the genetic algorithm. The second part of the proposed method is solely concerned with determining the cause of the failed area by dividing the structure of the PV system into three modules (the PV panel module, the charger module, and the battery module). Following this step, the ANN is trained with multiple types of faults for the three divided PV system structures and then the ANN begins the diagnostic process. When the proposed technique's results were compared to fuzzy-based and classic neural network-based diagnostic systems, it was discovered that the proposed method produced better results. The authors of [89] presented a neural network-based method for modeling the relationship (MPP) of a shaded PV array and environmental parameters (solar irradiance, sun angle, ambient temperature) in non-uniform settings. Similar to the shading factor, this neural-network-based function can characterize the shadow impacts on a solar PV array over time. As a result, the neural network model is able to eliminate the inaccuracy produced by the shading factor's complex calculation. In contrast to the prior efforts that only address the uniform shadow on the PV array, this method considers the non-uniform shadow and illumination. The proposed method's procedure is as follows:

- The shadow ratio is defined by the solar height and solar azimuth angles, which may be simply calculated from the time of day for a specific geographic location. Therefore, the neural network's inputs are the sun's irradiation levels, angle, and ambient temperature. The neural network's output has the highest solar PV array output power.
- Experimental data are acquired by taking measurements numerous times a day, for several days, while the solar PV array is partially shaded by a nearby object. One set of measured data is used to train the neural network, while the other is used to test it. The neural network's accuracy in estimating the PV array's maximum output power is tested using the test data.
- With low computational effort, the neural network can forecast the output power of solar PV arrays at any solar irradiation level, at any time of day, and at varied ambient temperatures over a long period of time.

The authors of [90] proposed a PV prognostics and health management (PHM) technique. The system was created to track the health of photovoltaic systems, measure degeneration, and provide maintenance recommendations. It employs a system-specific ANN model, which eliminates the need for prior knowledge of system components and design. Two detection techniques were tested in order to better monitor the health of the PV system. The energy loss fault detection system uses a neural network model to compare the sum of power loss over a lengthy period of time. An alarm threshold can be set to detect the long-term effects, such as soiling or material degradation, and alert the user to the need for maintenance. In the event of a catastrophic system failure, such as the loss of a string of modules or an inverter failure, the acute fault detection technique examines the potential of the PV system performing below model expectations and should warn the user. When the two techniques are combined, the short- and long-term PV system defects can be detected. The metrics of the two combined methods can also be used for pre-emptive inspection and maintenance, as they allow the system operator to identify the PV system failure precursors linked to failure modes.

Table 1. Summary of neural network-based methods.

Authors	Reference	Year	Contribution
A.M. Pavan, A. Mellit, D. De Pieri, S.A. Kalogitou	[80]	2013	Evaluation of the effect of soiling in large PV plants using BNN and regression polynomial
F. Aziz, A. Ul Haq, S. Ahmad, Y. Mahmoud, M. Jalal, U. Ali	[82]	2020	Classification of line-line fault, open-circuit fault, partial shading, and arc fault using the convolutional neural net
G. Liu, W. Yu	[84]	2018	Diagnosis of open-circuit fault, overall shading effect, and partial shading effect using the Elman neural net
W. Chine, A. Mellit, V. Lughi, A. Malek, G. Sulligoi, A.M Pavan	[66]	2016	Identification of eight different faults using ANN
K.H. Chao, C.L Chiu, C.J. Li, Y.C. Chang	[85]	2011	Islanding phenomenon detection in PV systems using ANN
E. Garoudja, A. Chouder, K. Kara, S. Silvestre	[86]	2017	Detection and diagnosis of faults in the DC side of PV systems using the probabilistic neural network
M.N. Akram, S. Lotfifard	[56]	2015	Modeling and health monitoring of the DC side of PV systems using the probabilistic neural network
H. Mekki, A. Mellit, H. Salhi	[87]	2016	ANN based modeling and fault detection of partially shaded PV modules
A. Coleman, J. Zalewski	[88]	2011	Intelligent fault detection and diagnosis in the PV system using the Bayesian belief network
M. Chang, C. Hsu	[64]	2019	PV O&M optimization using ANN
A. H. Mohamed, A.M. Nassar	[72]	2015	PV system fault diagnosis using the genetic algorithm-optimized ANN
D.D. Nguyen, B. Lehman, S. Kamarthi	[89]	2009	Performance evaluation and shadow effects in PV arrays using ANN
D. Riley, J. Johnson	[90]	2012	Prognostics and health management of PV systems using ANN

4.2. Regression-Based Methods

In this subsection, PV system fault detection methods that are regression-based are reviewed. Table 2 summarizes these methods, highlighting their main contributions.

For the detection of abnormal situations in the PV system, a novel approach has been employed in [91] based on regression and SVM, to compute the ideal power generation, which takes into account all of the three categories of failures in PV systems, namely failures of PV modules, inverters, and other components. Furthermore, the proposed method makes use of variables that are already present in a small-scale PV system, eliminating the need for the installation of extra expensive sensory equipment. Power, voltage, and current are collected from the power conversion system (PCS). Solar irradiance on the surface of the PV panels is measured with a pyranometer, and ambient and PV cell temperatures are monitored with a thermometer. As a result, the suggested PV abnormal condition detection system can be efficiently used in a small-scale PV system or as an early warning system for the PV operator/owner to undertake further system inspections. Merged regression and support vector machine (SVM) models were used to create the PV abnormal condition detection system. The regression model is used to calculate the expected power generation for each solar irradiation, which is then fed into the SVM model. The SVM model, on the other hand, uses numerous variables, including the expected power generation, which is generated from the regression model to determine the abnormal condition of the PV system. Since the data used as the model's input variable is acquired from the PCS, the proposed model does not require the installation of extra measurement devices and can be constructed at minimal cost. The detection system's accuracy is also increased by taking into account the daylight time and interactions between the independent variables, as well as using the multi-stage k-fold cross-validation technique. The proposed detection system is evaluated using real data from a PV site, and the findings show that it can successfully discern between normal and abnormal PV system conditions using basic measures. The authors in [92] proposed a condition monitoring technique based on an online regression PV array performance model, in which PV array production, POA irradiance, and module temperature/maximum power (MPP) measurements are collected during the system's initial learning phase and used to automatically parameterize the system online using regression modeling. After the model has been automatically parameterized and optimized using the regression modeling methods, the condition monitoring system enters the normal operating phase, in which the performance model is used to anticipate the PV array's power production. The authors claimed that using projected and measured PV array output power values, the condition monitoring system based on the Sandia array performance model (SAPM) [92,93], could detect power losses in the PV array of larger than 5%. When compared to the existing model-based condition monitoring systems, the suggested method is unique in that it can take advantage of the I-V scanning capabilities of a new generation of commercial PV inverters [92], according to the authors. Using the real MPP collected from the I-V curve, as well as the ambient condition sensors, the system can generate an accurate performance model of the PV array in question during field operation. The authors' proposed method has the following advantages: Simple commissioning and operation requirements; potential applicability to a wide range of PV system configurations; and it does not require modeling and testing of the PV array prior to installation. It does, however, have an initial commissioning phase where ambient conditions and array MPP measurements are used for the automatic commissioning of the PV array. Figure 6 presents the condition monitoring systems' learning or commissioning phase.

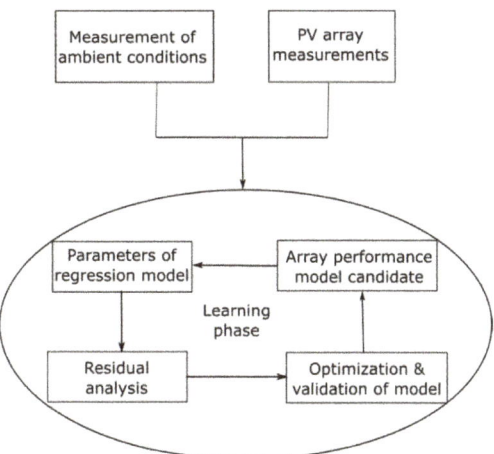

Figure 6. Condition monitoring systems' learning phase [92].

Unlike Reference [92], which only used a regression model, Reference [94] used a combination of linear regression and artificial neural networks, as well as solar irradiation, ambient temperature, and maximum power point (MPP) characteristic variables of PV modules obtained from I-V tracers at the PV installation site, to predict the performance of soiled PV modules. In the study, two methods (linear regression and neural networks) were used in modeling the output of soiled PV panels. In the first method, multiple linear models were created based on the last cleaning cycle time stamp, thus predicting the panel's output as a function of solar irradiance. In the second method, the neural network, date, time, and irradiance, and sometimes, the temperature data were utilized and information was inputted to predict the output of the solar panel. In both methods, PV panel generation predictions were used to have high accuracies. Another regression based fault detection model is presented in [95]. The study proposed a smart algorithm for the diagnosis and prognosis of reversed polarity fault in PV generators, utilizing a hybrid optimization of a support vector regression (SVR) technique by a k-NN regression tool (k-NNR). The main contribution of the study included the development of a smart prognosis algorithm that detects, locates, and characterizes a reversed polarity fault at the cell, bypass, and string levels. For data processing in a linear space, the SVR requires a kernel function in order for the data to be transformed in a new greater dimension space. To overcome the SVR output indetermination, the study utilizes k-NNR for predicting the approximate value of the SVR undetermined output. The method was validated using 50 samples of a typical generator and the results showed homogeneous and reliable predictions.

Table 2. Summary of regression-based methods.

Authors	Reference	Year	Contribution
F.H. Jufri, S. Oh, J. Jung	[91]	2019	Detection of abnormal conditions in PV systems using the combined regression and support vector machine
S. Spataru, D. Sera, T. Kerekes, R. Teodorescu	[92]	2013	Condition monitoring of PV array based on the online regression model
S. Shapsough, R. Dhaouadi, I. Zualkernan	[94]	2018	Predicting the performance of soiled PV modules using the linear regression and BP neural network
W. Rezgui, N.K. Mouss, M.D. Mouss, M. Benbouzhid	[95]	2014	Reversed polarity fault prognosis in PV generators using the regression algorithm

4.3. Decision Tree-Based Methods

The study also presents a review of the AI methods of fault detection and diagnosis that are decision tree-based. Table 3 shows a summary of the methods presented in the subsection, with an outline of the main contribution of each literature.

A number of fault detection and diagnosis methods that are decision tree-based are provided in [67,96,97]. In [67], failure detection routines (FDRs), that use obtained datasets of grid-tied PV systems to accurately obtain and classify exhibited faults, were created. The FDRs are made of two stages, namely the fault detection stage, which uses a comparative algorithm for the detection of anomalies between the measured and simulated electrical measurements of a PV system; and a statistical algorithm for the identification of outliers, discrepancies, and normal system operation limits. The failure classification stage is otherwise called the decision stage, where developed logic and decision trees are used to perform the classification process. The failure detection stage included a comparing algorithm that discovered disparities between the measured and simulated electrical measurements on the dc side, which were obtained using empirical parametric models for each point. Significant variations between the measured and simulated parameters, indicating a noteworthy performance gap, were classified as failures. The obtained incident global irradiance at the POA, PV system specs, and module temperature were required as inputs for the detection algorithm utilized in this study. Another comparative method utilized in the study is three-sigma limit method. When typical operation limitations set by defined criteria were exceeded, failures were recognized using this approach. The upper and lower control limits in statistical quality control charts were set using three-sigma limits (a statistical calculation that refers to data within three standard deviations of the average), which are commonly used to set the upper and lower control limits in statistical quality control charts. To establish and display the researched operation and boundaries, control charts were used. The typical operation limitations of the PV monitoring system were calculated by dividing the measured and simulated electrical data by the ratio. According to the PV production model, the ratio was used to determine how close the measurements were to their calculated values. The closer the ratio was to 1, the closer the measured parameters were to the modelled values. Since the system performance is affected by sunlight levels, an extra step is performed. To decrease the high bias errors occurring at low irradiance levels, the datasets were filtered at global irradiance levels >50 W/m^2. In the statistical failure detection approach, the local outlier factor (LOF) algorithm was used to find the density-based local outliers. With LOF, each point's local density is compared to its k-nearest neighbors (k-NN), and if the point's density is significantly lower than its neighbors, then the point is in a sparser region than its neighbors, indicating that it is an outlier. Moreover, outlier testing was done using the Bonferroni outlier test algorithm. Based on the linear regression model of the observed and simulated dc power of the array, this program returns the Bonferroni p-value for the most extreme observation. In addition, the seasonal hybrid extreme studentized deviates (S-H-ESD) algorithm was used to find anomalies in a time series dataset that follows an approximately normal distribution. The S-H-ESD technique is an extension of the generalized extreme studentized deviates (ESD) algorithm that uses the time series decomposition and robust statistical measures in conjunction with ESD to detect both global and local anomalies [98]. For the detected failure classification, logic and decision trees were developed. In addition, the decision trees were trained using continuous samples divided in a 70:30% train to test the set ratio utilizing the acquired datasets that included the feature patterns seen during normal and faulty operations. Moreover, they were produced using supervised learning procedures. The accuracy of the proposed FDRs for fault detection and classification was demonstrated in the obtained results for three fault types (inverter failure, bypass diode faults, and partial shading fault), which showed that accuracy rates of 98.7, 95.3, and 96.6% were recorded for inverter failure, bypass fault, and partial shading, respectively. The authors in [96] presented a decision tree-based fault detection and classification for the PV array with an easy and straightforward model training process. Under the creation of a decision tree model for an experimental PV system in both normal and faulty working situations, the authors employed the PV array voltage, current, operating temperature, and irradiance as attributes for the training and test sets. The collected data and pre-processed training set are chosen at random from the experimental data and utilized to create the decision tree model using the WEKA software [99], after which the model is evaluated and validated on unseen real data. Fast training and classification phases, explicit interpretation, and straightforward implementation as an algorithm are all advantages of the proposed decision tree paradigm. Another benefit of the model is that it can detect problems in real time, with detection accuracy ranging from 93.56 to 99.98% and classification accuracy ranging from 85.43 to 99.8%, depending on the model's size. The authors of [97] proposed a defect detection and diagnostic technique for grid connected PV systems (GCPVS) based on the C4.5 decision tree algorithm (which is one of the most popular machine learning algorithms for classification problems [97]), in which a non-parametric model is utilized to forecast the state of the GCPVS through a learning task. Three numerical attributes (ambient temperature, irradiation, and power ratio) which are calculated from the measured and estimated power, as well as two targets (the first of which is either a healthy or a faulty state for detection, and the second of which contains four classes of labels named free fault, string fault, short-circuit fault or line-line fault for diagnosis) are chosen to form the final used data. The dataset was divided into two halves, with 66% utilized for learning and 34% for testing. Then, over the course of 5 days, additional data were collected to measure the robustness, effectiveness, and efficiency of both models. The dataset is required for the learning process in order to construct the decision tree. As a result, an acquisition system is developed to be able to record and store data, such as climatic variation, as well as electrical variables, such as current, voltage, and power at the MPP. Three attributes are chosen, including temperature ambient, irradiation, and the power ratio, which is calculated from the estimated power by the Sandia model and the measured power of GCPVS production. The Sandia model is an empirical relationship that is used to estimate the generated power from a system in a healthy state at MPP using STC data. Since this model has unknown parameters, the flower pollination algorithm (FPA) is used to find the optimal parameter values that correspond to the smallest root mean square error between the estimated Sandia output and the measured power. As a result of the high correlation between the power ratio and the system state, a nominal property called target is constructed as a class label in each instance data in order to accurately forecast these errors. Two major approaches lead to the construction phase. To begin, a splitting criteria is used to select the best split attribute. Thereafter, the tree grows in length as this technique is repeated iteratively in order to categorize all of the instances or to verify one of the stopping criteria. Then, once the tree model has been obtained, a pruning process is carried out to remove the unneeded sub-trees in order to minimize the overfitting phenomena, which can result in a reduction in model complexity due to the reduced tree size. According to the test findings, the models have a great prediction performance in the detection with high accuracy, while the diagnostic model has an accuracy of 99.8%.

Table 3. Summary of decision tree-based methods.

Authors	Reference	Year	Contribution
A. Livera, G. Makrides, J. Sutterlueti, G.E. Georghiou	[67]	2017	Advanced failure detection algorithm and PV performance outlier decision classification
Y. Zhao, L. Yang, B. Lehman, J.F. De Palma, J. Mosesian, R. Lyons	[96]	2012	Detection and classification of PV array faults based on the decision tree algorithm
R. Benkercha, S. Moulahoum	[97]	2018	C4.5 decision tree algorithm-based fault detection and diagnosis for grid-connected PV systems

4.4. Support Vector Machine-Based Methods

The methods reviewed in this subsection are support vector machine-based. Table 4 presents a summary of these methods as well as their contributions.

The authors of [100,101] suggested a system based on support vector machines (SVM) and k-NN tools with the goal of building a fault detection and diagnosis algorithm for PV generators. The algorithms are smart, according to the authors, since they are a hybridization of the SVM approach and the k-NN tool, which are used to improve the classification rate against observations on the classifier itself. The systems' originality is in the construction of a smart classifier based on collected data from the control system, as well as the fault identification and localization of short-circuits in a cell, bypass, and blocking diodes. First, the proposed method utilizes the SVM algorithm, which is a two-class classification technique that seeks hyper-plane separating positive examples from negative ones by ensuring that there is a maximum margin between the nearest positive and negative examples. This ensures that the idea may be applied to new situations, as new examples may not be as similar to those used to determine the hyperplane, but may be on either side of the border. The selection of support vectors, which reflect the discriminate vectors by which the hyper-plane is determined, is a benefit of this method. Only those supporting vectors are utilized to assign a new case, and the examples used in the hyper-plane search are no longer required. Second, k-NN which is a simple and straightforward approach, and does not require learning is utilized to compare new examples of unknown class to old examples in its database. Then, for this new example, k-NN chooses a majority class among its nearest neighbors. In summary, the method uses the activation function of the SVM of Gaussian type and the Euclidean distance between the gravity centers of database observations of the k-NN method. The obtained simulation results, using the proposed smart algorithm in both literatures, exhibit a high classification rate and low error rate. However, the algorithms take a longer processing period due to the mathematical computations. Therefore, future works should focus on improving this aspect of the algorithms. The authors of [48] proposed an algorithm to improve the detection accuracy of line-to-line faults in PV arrays that occur under a wide range of situations, such as low irradiance conditions, high impedance faults, and low mismatch faults. The algorithm is based on pattern recognition (multi-resolution signal decomposition (MSD)) and machine learning techniques (two-stage SVM classifier). It takes advantage of the MSD technique for the extraction of the feature space of line-to-line faults, while the SVM part is essentially for decision making. The system does not require numerous sensors, since it uses measurements of the overall voltage and current of the PV array, thus making it an economical and fast option. It detects line-to-line defects quickly and accurately, and it can be combined with fault location techniques to solve faults quickly. The MSD stage performs digital signal processing (DSP), allowing for the simultaneous time and frequency analysis of a signal, such as the analysis of both stationary and transient components of power quality disturbance. The SVM stage, on the other hand, is carried out to improve accuracy. The two-stage SVM is a binary classifier, which requires training utilizing a minimal amount of historic data from the tested PV system. The authors suggest that the proposed method is not limited to line-line faults only and could be used for the detection of other PV system faults. A method for detecting problems and monitoring the state of PV modules using a two-class data fusion method is introduced in [102]. The approach was created by combining monitoring data from sensor nodes in wireless sensor networks (WSNs) at a monitoring center with a new semi-supervised support vector machine (SVM) classifier, devised and trained using the monitoring center's existing sun irradiance big data. The monitoring center was created in order to access various monitoring data from various PV power stations, and multiple applications were created to use the envisioned system in various platforms. In this paper, a wireless monitoring subnetwork was created to retrieve crucial data from PV modules in power stations, such as the current, voltage, and temperature. The monitoring data received from each sensor was fused by a sink node with sunlight intensity information, and the fusion results were provided to the monitoring center over Internet networks. The data received from the sink node were parsed using the data access interface, and the data from the parsing process was double-checked using the outlier detection technique. The Cloud management module, which was also in charge of data security transmission between the Cloud and the applications, retained the regular data in the private Cloud. The authors built a semi-supervised SVM classifier using historical monitoring sunlight intensity data, which was employed in the outlier identification and solar power forecast algorithms. An outlier identification technique is designed using the prediction model provided by the trained classifier to identify and locate PV module faults by computing the average value of the problematic data. Furthermore, the authors employed a novel application of the PGKA technique to ensure the security of data transmission between the Cloud and apps. The fact that this approach does not require third-party certification to maintain file encryption and encryption keys is an apparent benefit. Despite the fact that this research focuses on crucial PV power station challenges, there is still a long way to go in terms of gathering PV power station data and intelligence properly.

Table 4. Summary of SVM-based methods.

Authors	Reference	Year	Contribution
W. Rezgui, H. Mouss, K. Mouss, M.D. Mouss, M. Benbouzid	[100]	2014	Short-circuit fault diagnosis in PV generators using a smart SVM algorithm
W. Rezgui, N.K. Mouss L.H. Mouss, M.D. Mouss, Y. Amirat, M. Benbouzid	[101]	2014	Short-circuit and impedance faults smart diagnosis in PV generators using the k-NN optimized SVM classifier
T. Hu, M. Zheng, J. Tan, L. Zhu, W. Miao	[102]	2015	PV system intelligent monitoring based on solar irradiance big data and wireless sensor networks

4.5. Neuro-Fuzzy-Based Methods

This subsection presents a review of AI methods of fault detection and diagnosis that are based on the neuro-fuzzy technique. Table 5 presents a summary of the methods as well as their contributions.

The authors of [103] proposed an intelligent system for automatic fault detection in PV fields based on the Takagi-Sugeno-Kahn fuzzy rule-based system (TSK-FRBS) [104]. The method is based on the analysis of recorded voltages and currents collected from a PV plant's inverter. The TSK-FRBS is a power estimator module that estimates the PV field's immediate power production in normal operating conditions (using temperature and irradiance input signals to assess the DC power that the PV plant should produce) and compares it to the real power to check for differences. If there is a large disparity between the two power sources, an alarm signal is issued. In this circumstance, the TSK-FRBS has two advantages. First, it can describe complicated system behaviors without requiring the use of a mathematical model. Second, it is able to deal with noisy and vague data. The schematics of the proposed intelligent system, which is connected to a PV system is shown in Figure 7. It consists of the data acquisition module, the detection module, and the diagnosis module, in a multi-array inverter PV plant. The acquisition module measures the temperature and solar irradiance on the PV plant in real time and extracts the DC current and voltage observed on the respective array from each MPPT. Sensors mounted on the PV field can be used to measure temperature and solar irradiation. They can also be obtained via a remote database linked to a weather station. Then, the acquisition module feeds these measured data to the detection module. The detection module estimates the DC power that each array should output if no fault occurs using these data from the acquisition module. This module compares the estimated and measured powers and generates an alert signal in correspondence with the arrays that provided a lower power than the estimated one; if the difference exceeds a threshold. Finally, the alert signals are sent to the diagnosis module, which can automatically provide information on the type of fault that occurred.

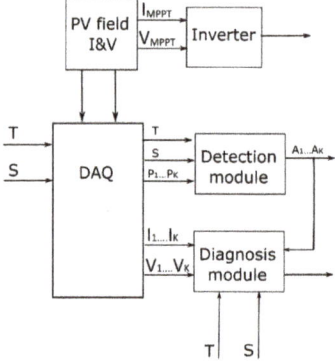

Figure 7. Proposed intelligent system's schematics [103].

The authors of [105] noted that numerous literatures have proposed methods of shadow detection and the reconfiguration of an array. In addition, most of these methods use the voltage, current, and power information to achieve this. The authors saw that monitoring these factors was time consuming and tiresome. Therefore, they presented a novel and effective method of shadow detection for the reconfiguration process, which will help increase the energy production of PV arrays, based on the fuzzy logic and computer vision. The method detects the edge of the object region on the panel from images taken with a camera. Using the background subtraction and object detection method, it then converts the background and foreground image into a grey image format. Object edge detection is performed after the determination of the updated mechanism, which is based on background subtraction and illumination variability. First, dilation and erosion operations are applied to the binary mask in order to determine the object region edge. The image's noise is reduced by these techniques. Following that step, using the Canny edge detection approach, the edges of the objects on the binary image are detected. The final stage in the determination of the subject borders on an image is to use the search and draw contour procedures to determine the related object regions. A relevant pixel region is created for each region whose side information is identified. Finally, the proposed method uses a fuzzy decision-making mechanism to classify the object region as a shadow region utilizing brightness and color distortion values of the object as input parameters. When employed as an input parameter for the reconfiguration operations, the proposed method has a success rate of 98% and increases energy usage performance by roughly 10–15%. The arc defect detection technique described in [106] has a minimal computational requirement, and can function with most of the conventional analog-digital converters (ADC) found in microcontrollers, making it useful in PV applications. When using an inverter rather than a solo device, the algorithm produces better results. This algorithm's integration is also a cost-effective way to detect arc faults and improve the PV system long-term safety. The short- and long-term measurement results are promising, but further long-term experiments are needed to fine-tune the device for phenomena that have yet to be identified. In the proposed method, the detection algorithm uses three indicators, namely frequency analysis, peak detection, and observation of the operating point, as parameters for fault detection. When an arc fault occurs, the indicators display a particular behavior. However, it is not always the same. The signal energy increases slowly at times and quickly at others. When an arc fault occurs in a small plant, the trip is apparent, but not in a large plant. Due to the haziness of the situation, the authors employed fuzzy logic to detect an arc defect. Furthermore, fuzzy logic makes it easier to incorporate the experiences of experts who are unfamiliar with the algorithm. Four sub-detectors, which are followed by the master fuzzy arc fault detector (MFAFD), are created to keep an overview of rules and input variables. Theses sub-detectors are the peak evaluator sub-detector (PESD), which analyzes all of the peaks and delivers a mass output proportional to the probability, indicating that this peak is from an arc fault; window near sub-detector (WNSD), which analyzes the change signal energy over a short timeframe; window wide sub-detector (WWSD), which analyzes the long-term signal energy and can be used when there is no abrupt signal energy growth; and power analyzer sub-detector (PASD), which supervises the power change. Outputs of the four sub-detectors serve as input for the MFAFD. The MFAFD outputs a number between 0 and 1, which represents the mass for the arc fault probability. An arc fault is detected if this probability exceeds a predefined threshold. The authors of [107] presented a method for detecting increases in series resistance using a fuzzy classifier that can distinguish between the increasing series losses and partial shadow situations for resistances greater than 400 W/m^2. As shown in Figure 8, an optional shadow detection algorithm acting before the increased series losses detection system, which could improve the detection accuracy of the system, is also implemented in the diagnostic system. This strategy is especially significant, since the increased series losses and partial shadow circumstances are difficult to discern, as they diminish a PV system's peak output and fill factor. Rather than the controlled laboratory circumstances, the study focuses on estimating the increased series resistance in the field. The method has been tested using experimental measurements. In addition, it has shown good detection rates across a wide range of irradiance levels, as well as in the presence of diverse sizes and patterns of partial shadows. Moreover, the authors showed that a dedicated partial shadow detection algorithm, implemented in the diagnostic system and functioning prior to the higher series losses detection method, improves the overall system's detection accuracy.

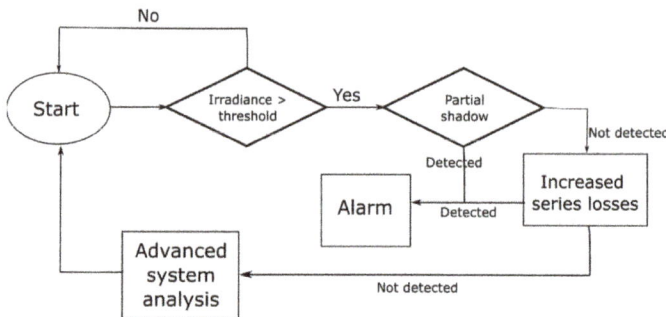

Figure 8. Proposed diagnostic system of partial shadow detection, increased series losses detection, and advanced system analysis and monitoring for the PV module [107].

Under low irradiance conditions, the DC side short-circuit faults in PV arrays consisting of multiple PV panels connected in a series/parallel configuration are nearly undetectable, especially when the MPPT algorithm is in use. In addition, if they go undetected, these faults can significantly reduce the output energy of PV systems, damage the panels, and potentially cause fire hazards. To avoid this, the authors of [108] present a fault detection scheme based on a pattern recognition approach that uses a multi-resolution signal decomposition technique to extract the necessary features, which is then used by a fuzzy inference system to assess whether a defect has occurred. PV array output volts and currents, as well as solar irradiation, are the system's inputs. The multi-resolution signal decomposition technique is used to extract four unique features depending on the above inputs. Following that step, the retrieved characteristics are supplied into the fuzzy inference system, which generates a scalar number based on carefully built membership functions and the associated rule base. Decisions are based on these results. The amount of this output determines whether a line-to-line, line-to-ground or none-of-the-above fault has occurred. In simulation and experiment-based case studies, the performance of the proposed method is demonstrated. Defect identification becomes increasingly difficult when the percentage of fault impedance or mismatch declines. The program also revealed one unintentional operation for a typical case, in which the irradiance had changed drastically within a short period of time, which is unusual in real-world systems. The promising performance of the proposed algorithm is supported by the experimental results. Line-to-line and line-to-ground faults are usually detected using this method. Open-circuits and hot-spot heating, for example, are two further types of PV array faults. As a result, the applicable algorithms for a more exact classification of additional PV concerns could be applied in parallel with or later phases of the detection system. Fuzzy logic-based algorithms are presented in [109,110] to detect the malfunctioning PV modules and partial shadowing circumstances that influence the DC-side of grid-connected PV systems. The authors' algorithm is made up of six layers that work in a sequential order. Input parameters make up the first layer (solar irradiance and module temperature). The theoretical performance analysis of the grid-connected PV system is generated using the LabVIEW virtual instrumentation program in the second layer. The power and voltage ratios are determined in the third layer, and high and low detection limits are set in the fourth layer, which is utilized to apply the 3rd-order polynomial regression model to the power and voltage ratios. The fifth layer contains the input parameters of the examined grid-connected

PV systems, as well as the 3rd-order polynomial detection restrictions. If the measured voltage ratio vs. the measured power ratio is outside of the detection limits, the data will be processed by the sixth layer (which is the last layer), which contains the fuzzy logic categorization system. The suggested method's innovation is proved by the fact that it provides a simple, dependable, and fast fuzzy logic classification system that can be employed with a variety of grid-connected PV systems. The algorithm is also unique in that it is based on fluctuations in the voltage and power of the grid-connected PV system. Few fault diagnosis techniques are capable of being implemented on integrated circuits, according to Sufyan Samara and Emad Natsheh's study [65], and these procedures require expensive and complicated hardware. The authors introduced a unique effective and implementable defect diagnosis approach based on the AI nonlinear autoregressive exogenous NARX neural network and Sugeno fuzzy inference to try to solve this problem. The program employs the Sugeno fuzzy network to isolate and classify errors in PV systems. The NARX network is used to estimate the PV system's maximum output power based on the real-time measured output and surrounding conditions, which is then used by the fuzzy inference algorithm to detect and categorize errors that may develop in the PV system. The algorithm has been demonstrated to work on a low-cost microcontroller. The suggested algorithm will be able to detect a variety of flaws in the PV system, including open- and short-circuit degradation, faulty MPPT, and PS issues. Furthermore, the proposed algorithm can capture non-linear patterns between predictors, such as radiation and temperature, as well as other non-linear correlations of patterns between predictors, to calculate the precise moment of maximum power for the PV system. The actual sensed PV system output power, anticipated PV system output power, and sensed surrounding conditions are all required for fuzzy inference. Using an AI NARX-based neural network, the PV system's output power is projected. The authors concluded that the ability of the proposed method to efficiently diagnose several PV system faults is an important step in achieving a complete system that can diagnose large PV system faults.

Table 5. Summary of neuro-fuzzy-based methods.

Authors	Reference	Year	Contribution
P. Ducange, M. Fazzolari, B. Lazzerini, F. Marcelloni	[103]	2011	Intelligent system for the detection of PV system faults based on TSK-FRBS
K. Karakose, K. Firildak	[105]	2014	Shadow detection system based on fuzzy logic using the obtained PV array images
B. Grichting, J. Goette, and M. Jacomet	[106]	2015	Arc fault detection algorithm in PV systems based on the cascaded fuzzy logic
S. Spataru, D. Sera, T. Kerekes, R. Teodorescu	[107]	2012	Increased series losses detection in the PV array based on fuzzy inference systems
Z. Yi, A.H. Etemadi	[108]	2017	Detection of PV system faults based on the multi-resolution signal decomposition and fuzzy inference system
M. Dhimish, V. Holmes, B. Mehrdadi, M. Dales	[109]	2017	PV system diagnostic technique based on the six layer detection algorithm
M. Dhimish, V. Holmes, B. Mehrdadi, M. Dales, P. Mather	[110]	2017	Fault detection in PV systems based on the theoretical curves modeling and fuzzy classification system
S. Samara, E. Natsheh	[65]	2020	Using the NARX network and linguistic fuzzy rule-based system to diagnose PV panel faults

4.6. Wavelet-Based Methods

In this subsection, the wavelet-based methods are reviewed. A summary of the method contributions presented in the literature is shown in Table 6.

The authors in [111] proposed a system that was designed and tested at various load currents, DC source voltages, and arc lengths to analyze the impact of each parameter on the DC arc, with the goal of providing a thorough understanding of arc behavior that could occur in DC networks. Based on the findings, a detection technique that uses both the time domain and time frequency domain characteristics to distinguish between the DC arc fault and normal conditions has been presented. A simple calculation approach for the arc current variation is utilized to portray the chaotic and dynamic nature of the DC arc physical process. The current variation can indicate different stages of arcing and is used to recognize DC arc faults. Moreover, wavelet decomposition is applied to the arc current signal. The normalized RMS value of wavelet decomposition coefficients demonstrates the ability to detect the arc fault. Finally, a comprehensive DC arc detection algorithm is created using the current change from the time domain analysis and the normalized RMS value from the wavelet analysis. A Texas Instrument's Digital Signal Processors (TI DSP) chip, which has a clock frequency of 150 MHz and has gained popularity in power electronic applications, is used to implement the detection method in the hardware. In the proposed method, the maximum and minimum currents are updated every time new data is inputted at each timer cycle, and the difference in the current is registered at the nth cycle. The value of n is determined by the length of the time window and sampling frequency. Detection accuracy is enhanced if a longer time window is achieved, as it makes the RMS versus time waveform smoother. A long time window, however, leads to a longer time response. When the amount of data points is excessive, the actual implementation must be adjusted to ensure that the calculation load is within the capabilities of the available microprocessor. The wavelet coefficients are calculated by sending a signal through low-pass and high-pass filters, as well as down sampling it by two. The filters are dependent on the used wavelet and are consistent throughout the entire level decomposition. As a result, the detection algorithm can be implemented on a microprocessor in the time domain using convolution and down sampling. The filtering procedure for the first-level DWT to obtain *Coefl*1 is carried out every two timer intervals, for which two new data points are available at this time. The second-level DWT coefficients (*Coefl*2 and *Coefh*2) are calculated whenever two new coefficients from the first-level DWT are received. As the filtering operation was performed for every timer interval instead, this method cuts the calculation load by half. In the main function, the ultimate decision is taken. The main function's flag changes to 1 at the conclusion of each time window. Only two square roots, one division, and two comparisons are performed in the main function, which takes 80 clock cycles in total, according to the benchmarks of the C28x floating point unit fast RTS library. With a clock frequency of 150 MHz, the main function's computation time at the conclusion of each T_{sw} = 25 ms is less than 1 µs, which is negligible. As a result, the detection time is mostly determined by the length of the time window. The results of the studies reveal that the detection algorithm is capable of delivering an alarm in a timely manner following the initiation of an arc fault. Typical operations, such as load changes, can also generate nuisance tripping, which can be reduced with this detection system. In [112], a modified wavelet-based technique termed the wavelet packet transform (WPT) was developed for the detection of diverse disturbances caused by faults in grid-connected solar PV systems. The study continued to evaluate the proposed WPT approach to methods based on the ordinary wavelet transform (WT) under various operating settings. In addition, qualitative and quantitative evaluations suggest that the WPT outperforms WT in terms of detection performance. The WPT uses a set of low-pass and high-pass filters to breakdown the signal retrieved at PCC. It gives both approximate and precise coefficients. Moreover, it extensively decomposes both components in order to determine the signal's frequency agreements. The breakdown procedure carried out in both components is what gives them their value. Wavelet transforms, on the other hand, do only a one-sided decomposition, segmenting only the low-pass frequency components and not the high-pass frequency components. When WT is subjected to noisy or transitory environments, this feature can compromise its performance. In [113], wavelet transformations were used to offer an online fault detection method for power conditioning systems (PCS). Switch open faults and over harmonics are detected using a multi-level decomposition wavelet transform approach. Using the normalized standard deviation of the wavelet coefficient, a quick and accurate diagnostic function is also achievable, allowing the suggested method to detect islanding conditions. Simple calculations (a time correlation generated by sequential multiplication and addition) and exact diagnostic capabilities of fault identification with good simulation outputs to check and evaluate its claims characterize the method's algorithm. The multi-level decomposition wavelet transform provides the method's straightforward calculation characteristic. At each wavelet tree level, the algorithm extracts wavelet coefficients from the measured signal, and errors are discovered and categorized using the wavelet coefficient changes. Using a three-level MLD tree, the fault detection technique was created for PCS fault scenarios. Switch open and over harmonic are the two scenarios considered. The PCS uses a semiconductor switch, such as a field effect transistor (FET) or an IGBT, to convert DC solar voltage to AC grid current. The switch open can be attributed to a switching device failure, whereas the over harmonic can be attributed to a controller or sensor failure. The cases of switch short faults are not taken into account here. The system is protected by the over-current limiting function or melting fuse when the switches are in short fault, and the PCS ceases running. The PCS current has a distorted waveform when the switches are open faulted, and it continues to provide high order harmonics to the grid. An UP and DOWN switch failure in the inverter bridge might be categorized as an open switch problem. The authors of [114] used the discrete wavelet transform (DWT) to analyze the traced I-V curve of a residential PV system and define these coordinated points in the related diagnosis effort. The DWT was utilized to implement the fault diagnosis of residential PV systems as a preprocessing tool. It enables feature extraction through signal decomposition and noise reduction. The reduced short-circuiting current of partially shaded cells is represented by the vertical height or current in a PV string identified by the DWT method, whilst the horizontal or voltage distance from the VOC to the inflex is connected to the number of bypassed modules. The approach is divided into two sections, namely passive and active. In the passive diagnosis section, a residue signal is generated by comparing the measured PV power signal and simulated model in real time to monitor the alarm signal and abnormal condition in the system, using the model base fault diagnosis technique. After the manifest and certainty error signals have been determined, the flash test is used as the active and second portion of the test procedure. During this phase, the step load is separated from the PV and power generation, and the MPPT mode of the inverter is interrupted. The I-V curve of the PV array is tracked and logged by modifying the inverter switching pattern for a deeper inspection and interpretation. The model provided in [115] for the defect diagnosis of PV arrays was improved using improved wavelet neural networks, wavelet neural networks, and back propagation neural networks. The training technique now includes a Gaussian function, which is utilized as an activation function, an additional momentum mechanism, and an adaptive learning rate method. The conclusion is taken from simulation findings that the proposed technique in this study is capable of efficiently diagnosing the PV array problem with good performance accuracy, convergence time, and stability under the identical conditions of the network input and desired output. The proposed fault diagnosis algorithm is summarized in Figure 9. Four PV system fault types are diagnosed using the model, namely short-circuit, open-circuit, abnormal degradation, and partial shading. There are five network output layer variables since the system requires the precise diagnosis of four types of problems and no fault condition. The selection of the number of hidden layer nodes in a neural network is a difficult topic with no theoretical foundation to follow. The number of hidden layer nodes is critical to the network's success. If the number is too little, we may not acquire a network from training, implying that the network's robustness is poor and its anti-noise ability is weak, making it unable to recognize models that have never been seen before. If the hidden layer's node number is too big, the learning time will be too long, and the error will not be minimal. Furthermore, there could be an issue with overfitting. As a result, based on an empirical equation, the trial and error method is commonly employed to identify the appropriate number of concealed nodes.

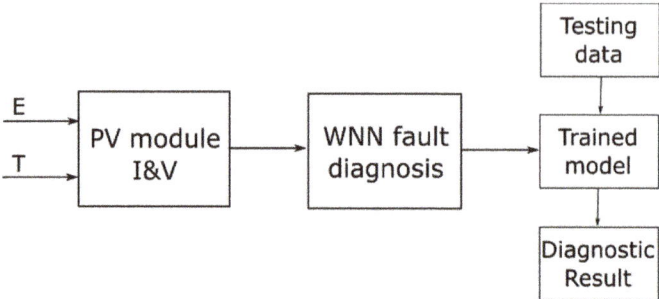

Figure 9. Proposed fault diagnosis method based on the wavelet neural network [115].

In [116], a wavelet optimized exponentially weighted moving average (WOEWMA) monitoring technique based on the principal component analysis (PCA) was created. The presented monitoring approach combines the advantages of exponentially weighted moving averages, MOO, and wavelet representation. MOO is used to tackle the challenge of determining the best strategy for minimizing both the missed detection rate (MDR) and the false alarm rate (FAR). Apart from the decorrelation of auto-correlated observations, the wavelet representation increases the monitoring performance by lowering MDR and FAR, as well as obtaining the exact deterministic features. The proposed method is appropriate for real-time implementation due to its quick calculation time. This approach is robust in a range of faulty conditions, including single and multiple (cascade and simultaneous) occurrences, and can detect faults in both the DC and AC conversion sections. The proposed method has several advantages, including dynamic multiscale representation to extract accurate features and de-correlate auto-correlated measurements; the PCA technique to model PV systems; the EWMA of two parameters in OEWMA are optimized using the multi-objective optimization; and the WOEWMA chart can detect smaller fault shifts, thereby improving the PV model monitoring.

Table 6. Summary of wavelet-based methods.

Authors	Reference	Year	Contribution
X. Yao, L. Herrera, S. Ji, K, Zou, J. Wang	[111]	2014	Detection of DC arc fault based on the characteristic study and time-domain discrete wavelet transform
P.K. Ray, A. Mohanty, B.K. Panigrahi, P.K. Rout	[112]	2018	Fault analysis in solar PV system based on the modified wavelet transform
I.S. Kim	[113]	2016	Online fault detection algorithm in PV system based on the wavelet transform
M. Davarifar, A. Rabhi, A. Hajjaji, E. Kamal, Z. Daneshifar	[114]	2014	Partial shading fault diagnosis based on the discrete wavelet transform
X. Li, P. Yang, J. Ni, J. Zhao	[115]	2014	Improved wavelet neural network algorithm-based method of fault diagnosis in PV array
M. Mansouri, A. Al-khazraji, M. Hajji, M.F. Harkat, H. Nounou, M. Nounou	[116]	2017	Wavelet optimized EWMA-based fault detection method for the PV system application

4.7. Other Methods

This subsection groups all the other reviewed faults that were not classified in any of the six groups. A summary of the contribution of the reviewed methods is presented in Table 7.

The approaches, that use graph-based semi-supervised learning (GBSSL) for fault identification and diagnosis in PV arrays, are available here [13,117]. The authors of [117] presented a sound technique based on GBSSL for recognizing, classifying, locating, and fixing errors in PV arrays, which was improved by expanding the diagnostic space of the GBSSL algorithm and adding more class labels. The model detects and locates a failure, temporarily isolating the system in order for it to continue to operate normally until the problem is resolved. The authors adjusted the way the data were normalized in order to increase the system's ability of finding the unlearned defects, and they were able to detect faults that the algorithm could not detect at first. The functionality of the system to feed the required energy after the defect was eliminated was tested using an interleaved boost converter, and it was discovered that the maximum voltage loss in a standard condition is 1 V, demonstrating the model's high efficiency. The proposed model in [117] was an improvement of a previous model proposed in [13] by the addition of the fault location element and higher fault detection and classification accuracy. The early stage of the algorithm starts with the extraction of a limited number of labeled data with their class labels serving as initial value data and fed into the model. Both the faulted and fault-free data are included in this dataset. To achieve the fault-prone labeled data, faults are purposefully introduced into the system, and the parameters are measured and stored for each data. The data are measured and stored in the fault-free mode in the same way as it is in the normal mode. An important advantage of the proposed GBSSL method is that it only requires a small amount of data for the learning process. To detect PV system faults, the current of all rows of solar cells, as well as the overall voltage of the panel, are constantly measured at any time (this is accomplished by installing 10 current sensors in each row of cells and one voltage sensor at the PV system's output points). As a result, each dataset contains 11 parameters, 10 of which are currents flowing across the cell's rows and one of which is the system's total voltage. In the modeling process, the parameters required to implement the GBSSL algorithm are entered. On the other hand, the authors of [118] proposed an intelligent defect diagnostic approach for PV arrays based on a kernel extreme learning machine (ELM) optimized simulating annealing (SA) algorithm with an improved radial basis function (RBF). Short-circuit faults, aging faults, and shadow faults are among the faults discovered by the proposed method. The results obtained were above 90%, showing that the SA-RBF-ELM fault classification is accurate and stable. According to the acquired simulation findings, the suggested model has three key advantages: (1) The optimal fitness value of the PV array and the model parameters used, as one of the characteristic factors of neural network learning, considerably increase the fault detection accuracy of the four fault kinds described; (2) the RBF-ELM kernel function has strong learning and classification capabilities, making it suitable for detecting and classifying PV array faults; and (3) the SA algorithm can quickly optimize the parameters of the RBF-ELM fault diagnostic model, significantly improving the RBF-ELM model's training accuracy and testing precision. The algorithm proposed is a derivative of the basic ELM algorithm, with kernel function limit learning machine features, which improve its ability to solve the problem of regression prediction leading to higher accuracy and faster calculations. The RBF-ELM model contains a regularization coefficient (C) and radial width (α), which affects the algorithms' performance. The RBF-ELM fault model's training accuracy is employed as the optimization objective function, and the coefficients C and α are the parameters that must be tuned. The simulated annealing approach is then utilized to optimize the parameters of the RBF-ELM fault model, resulting in optimal training and test accuracy for each time. A diagnostic technique for PV systems was developed in [119] using the learning method to take each PV site's condition into account. The technique employs the diagnostic criteria database to analyze the data acquired from the PV system.

The special features of the proposed technique include updating the diagnostic criteria, making it possible to detect normal or abnormal operating conditions of a PV system; the detection of shadow on modules and the pyranometer using the sophisticated verification (SV) method [120]; and the maintenance advice provided by an expert system according to the precise diagnosis. The ratio of acquired data to reference data is calculated to diagnose the system's normality or abnormality. The ratio approaches "1" when actual and average meteorological data are close. For example, when the summer generated power ratio is "1" and the winter generated power ratio is "0.7," a winter shadow or snow on the modules is assumed. The criterion for diagnosis in this situation is "1". The contribution of the proposed method is highlighted in its features, as follows: By updating the diagnostic criteria, it is now feasible to diagnose the normality or abnormality of PV systems, while taking into account the PV system's characteristics, as well as the climate; where a shadow appears on the modules or pyranometer is determined using the SV approach and hourly data analysis; and maintenance recommendations are also given based on the diagnosis outcome. The simulation results of the proposed technique suggest that it offers quick and proper maintenance advice within a short detection period. In [121], a simple short-circuit and open-circuit fault detection approach for PV systems was suggested based on the evaluation of three coefficients. The suggested technique has two steps. First, an offline simulated model for extracting the variation boundaries of the three coefficients for each faulty operation. Second, an online comparison model for comparing real measured coefficients to the simulated coefficients from the offline step. Three coefficients have been established for each fault type in order to detect and diagnose both short-circuit and open- circuit faults, namely the current coefficient, the voltage coefficient, and the power coefficient. The offline step is aimed at extracting the three coefficients' variation boundaries for each type of fault. In order to achieve this, three other operations are conducted. By bringing the detected parameters to a PSIM/MATLAB co-simulation, you may simulate both the healthy and flawed scenarios under a few climatic situations. For each simulated instance, the goal of this stage is to extract a few MPP coordinates. Based on the given equations, determine the three coefficients for each fault situation, then by adding a ±2% offset to the three derived coefficients, you may extract the variation boundaries for each defective type. For the online step, using the various sensors, both the meteorological conditions and MPP values may be detected and monitored during the actual operation of a PV system. The three actual onsite coefficients will be calculated using these measures. Finally, a comparison is made between the real onsite coefficients and the variation boundaries of each faulty case that was previously stored during the offline process. To conclude, the faults detection task will be carried out based on the real onsite monitored power coefficients measured, in a way that if their value exceeds a set threshold, a DC side fault alarm will be triggered. In addition, the faults' type will be determined by comparing the three real onsite coefficients with the variation boundaries of each simulated faulty case. The proposed method is straightforward, efficient, and does not necessitate a large amount of training data. the authors of [71] presented a data-driven anomaly detection and classification system that can accurately detect and categorize a wide range of PV system anomalies. The method consists of two stages. First, the local context-aware detection (LCAD), which is a hierarchical context-aware anomaly detection using supervised learning, and is aimed at identifying possible anomalies in PV strings with current characteristics that are different from the other PV strings under similar environmental conditions. Second, the remote context-aware detection (RCAD), which is a hierarchical context-aware anomaly detection using supervised learning, and is aimed at identifying possible anomalies in PV strings with current characteristics that are different from solar PV farms and benefit from a combination of LCAD and GCAD to detect anomalies at the string level. First, the domain-specific features are designed. To reduce computation complexity and increase classification performance, the multimodal properties are carefully generated and extracted. Then, with the purpose of developing an accurate classification model that is suitable for specific categorization situations, a multimodal model training technique is constructed. The effectiveness, robustness, cost-, and computing efficiency of the suggested strategy are proved by the results of trials conducted over time. The proposed method has the following advantages: A more robust method against irradiance and weather variations that can accurately detect different anomalies without pre-labelled data; 90.2% detection accuracy for the top 100 anomalies that are otherwise nearly undetectable under low irradiance or weather with high cloud cover; the use of SCADA data to classify commonly occurring anomalies at the plant level; and cost- and computation-efficient as it uses readily available data of existing PV systems. Numerous machine learning-based fault detection methods have the following problems, according to the authors in [122]. Fault diagnosis performance is limited due to the insufficient monitored information. Moreover, fault diagnosis models are inefficient to train and update, and labeled fault data samples are difficult to obtain by field experiments. The authors proposed a method with the aim of overcoming these problems and three features were addressed. The first is based on important points and model parameters collected from I-V characteristic curves and environmental factors that are observed. An effective and efficient feature vector of seven dimensions is proposed as input of the model. The second is an emerging kernel based on extreme learning machine (KELM), which features extremely fast learning speed and good generalization performance, utilized to automatically establish the fault diagnosis model. The Nelder-Mead simplex (NMS) optimization method is employed to optimize the KELM parameters, which affect the classification performance. The final aspect is an improved accurate SIMULINK-based PV modeling approach for a laboratory PV array to facilitate the fault simulation and data sample acquisition. There are six steps leading to the establishment of the proposed model, as shown in Figure 10.

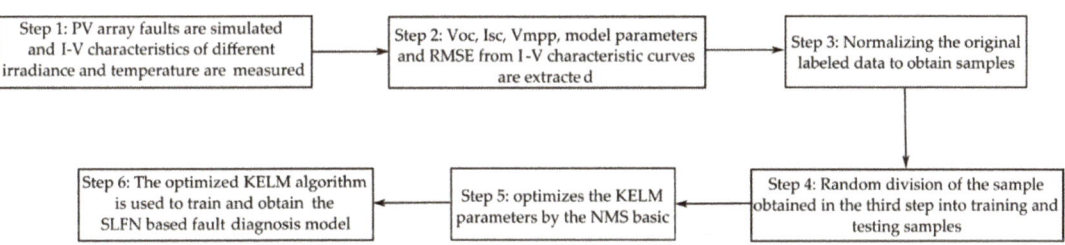

Figure 10. Flowchart on the establishment of the proposed fault diagnosis model [122].

The data samples for each fault condition should cover a wide range of operational irradiance and temperature, in order to make the fault diagnosis model suitable for a variety of operating settings. To begin, certain SIMULINK simulation experiments were used to obtain labeled data samples of normal and problematic situations. Then, on the real laboratory PV array, some field experiments were conducted to achieve some experimentally labeled data samples. Finally, the fault diagnostic model is established using the optimized KELM, which is evaluated and analyzed using both simulated and experimental data samples with known fault kinds. The proposed KELM-based fault detection model is promising in real-time applications due to its exceptionally fast learning speed, simplicity, and high generalization performance. The authors are attempting to apply the fault diagnosis model in digital signal processor (DSP) based embedded real-time systems, in conjunction with an integrated rapid I-V tester that is currently in development. The authors of [123] presented outlier detection rules based on instantaneous PV string current monitoring for failure detection. It is a command to monitor PV functioning and discover faults that may go undetected by overcurrent protection devices (OCPD). Three outlier identification rules were devised and compared by the authors, namely the three-sigma rule, Hampel identifier, and boxplot rule. Weather measurement or model training are not required with the suggested strategy. The Hampel identifier performs well in cases with extremely high contamination levels (33.3% in this investigation), while the boxplot rule performs better under PV faults in cases with relatively high contamination levels (12.5% in the case of this study). The model's reliability improves as the number of PV measurements rises. Despite the fact that the outlier identification methods in this study are based on PV-string level measurements, the authors claim that the proposed approaches should be straightforward to implement with minor modifications on any PV installation level. If the assumption is made that the solar irradiation is identical on the same PV level, the outlier rules can be applied on the PV-module or sub-array level, for example. Aside from the PV string current, the measurement could include PV insulation impedance, output power or energy yield, all of which are commonly used PV metrics. This may provide extra flexibility to fault detection methods. To overcome the limitations of conventional wired monitoring systems, such as physical constraints during data cable laying, high installation and maintenance costs, and reduction in the system lifespan due to the over exposure to extreme weather conditions, a Zigbee-based wireless monitoring system was developed in [124] to replace the conventional systems for online monitoring of parameters, such as temperature, irradiation, PV power output, and grid inverter power output, in grid-connected PV system applications. Moreover, it is equipped with a control function for remote system monitoring and a user-friendly web application, in order for the monitored data to be easily accessible via the Internet. Although the simulation results were satisfactory, the authors pointed out the limitations of the proposed method. These limitations include: (1) The proposed method is location specific. Therefore, before implementing the system in other locations with significantly different weather conditions, the weather factor needs to be taken into consideration. (2) The program used was developed based on the available software and programming language familiar to the authors. In [125], a failure diagnostic algorithm based on an online distributed monitoring system of a PV array of Zigbee wireless sensors network and a genetic algorithm optimization based BP neural network was investigated. The Zigbee wireless network system monitors each module's output current, voltage, and irradiation, as well as the environment's temperature and irradiation. In addition, a simulation PV module is set up, based on which typical problems are simulated and fault training samples are obtained. The fault sample data are then utilized to build and train a generic algorithm optimized BP neural network diagnosis model, which is subsequently used to detect four different PV array operating states (normal, abnormal aging, short-circuit, shadow). Since an open-circuit problem is noticed during the data collecting phase, it is not considered one of the diagnosis model's outputs. According to the simulation data, the proposed defection system has a high level of accuracy. Operators or managers can log in to verify the parameters of each PV module and use the designed mechanism to quickly discover the problematic PV module. In relation to the PV energy conversion system (PVECS), the authors of [126] presented a fault detection system using a fractional-order color relation classifier. The output power degradation is used to monitor the physical circumstances associated with changes in the circuitry of a PV array, such as grounded faults, mismatch faults, bridged faults between two PV panels, and open-circuit faults, using an electrical inspection method. The over-current and ground fault prevention devices can also be used to isolate failures on the AC side. As a result, the grid connection side fault impact can be reduced. Iterative calculations are not required to update the parameters of the inference model in the flexible and inferential model. As a result, it can handle the complexity of an adjustable mechanism in a relatively short design cycle. Embedded system approaches can then be used to implement the proposed detection model. The suggested approach can detect normal conditions, mismatch faults, and four common electrical defects on the DC side, according to the simulation results. For solar radiation of 0.4–1.0 kW/m^2 and temperatures of 25–40 °C, the suggested detection model has an average accuracy of 88.23% in identifying the fault under low/high solar radiation and various temperatures. The authors of [127] presented another interesting fault detection and diagnosis method based on a laterally primed adaptive resonance theory (LAPART) neural network. It is a low-cost way of automatically detecting and diagnosing PV system issues. The LAPART algorithm was taught how to detect fault states using real-world data that were classified as normal system behavior. The algorithm was then given new data and three-fault data points for an initial test. The system was given synthetic data to examine its performance over a statistically significant month-long dataset, and it was able to correctly identify flaws within the dataset. The LAPART algorithm's accuracy is determined by its ability to deliver a high likelihood of detection, while reducing false alarms. The number of true positive values generated by the FDD process is compared to the total number of actual positive values to determine the likelihood of detection. The LAPART architecture combines two fuzzy adaptive resonance theory (ART) algorithms to build a system for predicting outcomes based on the learnt associations. The single fuzzy ART algorithm's fundamental equations include category selection, match criterion, and learning. The goal is to create the optimal template matrix for the provided dataset. The approach employs category selection to discover the existing template matrix that best matches the provided input. In addition, for fast learning applications, the free parameter is frequently set to 10-7. The match criterion then checks to verify if the template matrix and input that is compared fulfill the user-defined vigilance parameter criterion. Depending on the level of intricacy requested, the vigilance free parameter can range from 0 to 1. A high vigilance value of 0.9, for example, yields high complexity but limited generality, whereas a low parameter of 0.5 yields the opposite. Finally, if it passes, the template is changed to reflect what has been learned. The LAPART algorithm is created by linking the two fuzzy ARTs (FARTs), which is seen graphically in Figure 11. The L matrix, which connects the A and B templates, connects the A and B FARTs. Each FART has its own set of vigilance settings, and inputs are delivered to both the A and B sides at the same time during the learning process. The A and B sides work together to generate and update the templates, while also forming links. Testing inputs are only applied to the A side after the training is complete, allowing them to resonate with the already acquired templates. The L matrix's relationships are then used to link with the B side and generate the prediction results.

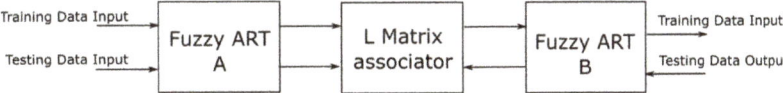

Figure 11. Creation of LAPART algorithm through linking two fuzzy ARTs [127].

The approach provided in [128] is based on monitoring the PV array's output power and is suitable for low irradiance, high impedance, and low mismatch fault circumstances. The irregularity of the time series of the normalized fault-imposed component of PV power is measured using entropy-based complexity as the fault detection criterion. Weather disturbances and partial shade can cause many array faults, which the proposed method can identify. It is applicable to both grid-connected and islanded PV systems and does not require a training set or prior knowledge of the PV array. Moreover, it is an economical strategy as it does not require costly sensors, relying only on the central IED to process the PV voltage and current measurements. The irregularity of the fault imposed power time series is measured by the sample entropy. The study uses the sample entropy-based complexity as the PV array fault detection index (FDI), since the complexity of time series data more effectively captures the behavior of a nonlinear system. The fault-imposed component of the PV array power is zero during the normal operation. As a result, the time series of moving data windows is regular. Therefore, SampEn is equal to zero. The fault-imposed power samples in each moving data window of N points are not identical when the solar irradiance or temperature varies, but they are fairly close together, since weather disturbances are not severe. As a result, FDI will be near-zero in this situation. When a fault develops in the PV array, however, the fault-imposed power samples rapidly shift. As a result, the normal fault-imposed power samples differ dramatically from the post-fault ones. Therefore, non-repetitive patterns can be found in the moving data windows during the initial milliseconds of fault transients, and the estimated SampEn is not zero. The issue is recognized when FDI rises to a non-zero value. It may be concluded that FDI for normal occurrences is approximately zero, whereas FDI for fault events is non-zero. A defined threshold is used to discriminate between non-zero FDI values under fault situations and near-zero FDI values during weather disturbances and partial shadings. As a result, determining the fault detection threshold is simple. It can considerably reduce nuisance tripping. Several experiments that were carried out validate the proposed fault detection algorithm's simplicity, sensitivity, scalability, resilience, and adaptability.

Table 7. Summary of other methods.

Authors	Reference	Year	Contribution
Y. Zhao, R. Ball, J. Mosesian, J. De Palma, B. Lehman	[13]	2015	Graph-based semi-supervised learning algorithm for PV fault detection and classification
H. Momeni, N. Sadoogi, M. Farrokhifar, H.F. Gharibeh	[117]	2020	Graph-based semi-supervised learning method of PV array fault diagnosis and correction
Y. Wu, Z. Chen, L. Wu, P. Lin, S. Cheng, P. Lu	[118]	2017	SA-RBF kernel extreme learning machine-based intelligent fault diagnosis for PV array
Y. Yagi et al.	[119]	2003	Diagnostic technology and expert system for PV system using the learning method
E. Garoudja, K. Kara, A. Chouder, S. Silvestre, S. Kichou	[121]	2016	Short- and open-circuit fault detection in PV system based on the offline and online evaluation of measured coefficient
Y. Zhao, Q. Liu, D. Li, D. Kang, Q. Lu, L. Shang	[71]	2019	Hierarchical context-aware anomaly detection using the unsupervised learning and multimodal anomaly classification method
Z. Chen, L. Wu, S. Cheng, P. Lin, Y. Wu, W. Lin	[122]	2017	Optimized kernel extreme learning machine and I-V characteristics based intelligent fault diagnosis in PV array
Y. Zhao, B. Lehman, R. Ball, J. Mosesian, J.F. De Palma	[123]	2013	Outlier detection based rules for solar PV array fault detection
F. Shariff, N.A. Rahim, W.P. Hew	[124]	2015	Online monitoring of grid-connected PV system based on the Zigbee data acquisition system
H. Lin, Z. Chen, L. Wu, P. Lin, S. Cheng	[125]	2015	Online monitoring and fault diagnosis of PV array based on the genetic algorithm optimized BP neural network
C.L. Kuo, J.L. Chen, S.J. Chen, C.C. Kao, H.T. Yau, C.H. Lin	[126]	2017	PV energy conversion system fault detection using the fractional order color relation classifier
C.B. Jones, J.S. Stein, S. Gonzalez, B.H. King	[127]	2015	Photovoltaic system fault detection and diagnostics using the laterally primed adaptive resonance theory neural network
A. Khoshnami, I. Sadeghkhani	[128]	2018	PV array fault detection based on sample entropy

5. Discussion

This study presented a comprehensive review of PV system fault detection and diagnosis techniques that are based on artificial intelligence and machine learning. Conventional fault detection and diagnosis methods, which equip PV systems with overcurrent protection devices and ground fault detection interrupters, are not sufficient enough to detect certain faults due to low irradiance conditions, nonlinear output characteristics, maximum power point tracker of PV inverters or high fault impedances. This led to the need for more intelligent fault detection and diagnosis methods to replace the conventional methods, in order to improve the PV systems operational efficiency and safety. AI-based methods, which are still currently explored and improved have been found to be the alternative to conventional methods. This paper's contribution outlines the main features of reviewed AI-based methods and the effectiveness of PV fault detection and diagnosis applications. The reviewed methods mostly adopt ML models, such as neural networks, wavelets, fuzzy logics, decision trees, support vector machines, graph-based semi-supervised learning, regression, etc., in order to develop models and algorithms that are trained to learn the relationships between input and output parameters of PV systems. The effectiveness of these methods depends on their ability to detect a fault and pinpoint its location in the shortest possible time; their relative affordability; and ease of use. Of note, there are currently fewer literatures in this area of PV application compared to the other areas, since the topic has only recently been explored, as evident in the oldest paper we could obtain, which dates back to only about 15 years.

Author Contributions: Conceptualization, A.A. and C.F.M.A.; methodology and formal analysis, A.A., C.F.M.A. and M.G.; writing—original draft preparation, A.A.; writing—review and editing, C.F.M.A. and M.G.; supervision, C.F.M.A. and M.G.; funding acquisition, A.A. and C.F.M.A. All authors have read and agreed to the published version of the manuscript.

Funding: This research was partly funded by Coordenação de Aperfeiçoamento de Pessoal de Nível Superior (CAPES) (No. 88887.514132/2020-00).

Institutional Review Board Statement: Not applicable.

Informed Consent Statement: Not applicable.

Data Availability Statement: Not applicable.

Acknowledgments: The authors kindly thank ENERQ-CTPEA Centro de Estudos em Regulação Qualidade de Energia for their support.

Conflicts of Interest: The authors declare no conflict of interest.

References

1. Erdiwansyah; Mamat, R.; Sani, M.S.M.; Sudhakar, K. Renewable energy in Southeast Asia: Policies and recommendations. *Sci. Total Environ.* **2019**, *670*, 1095–1102. [CrossRef]
2. Emodi, N.V.; Chaiechi, T.; Alam Beg, A.R. The impact of climate variability and change on the energy system: A systematic scoping review. *Sci. Total Environ.* **2019**, *676*, 545–563. [CrossRef]
3. Wilberforce, T.; Baroutaji, A.; Soudan, B.; Al-Alami, A.H.; Olabi, A.G. Outlook of carbon capture technology and challenges. *Sci. Total Environ.* **2019**, *657*, 56–72. [CrossRef]
4. Wilberforce, T.; El Hassan, Z.; Durrant, A.; Thompson, J.; Soudan, B.; Olabi, A. Overview of ocean power technology. *Energy* **2019**, *175*, 165–181. [CrossRef]
5. Osmani, K.; Haddad, A.; Lemenand, T.; Castanier, B.; Ramadan, M. A review on maintenance strategies for PV systems. *Sci. Total Environ.* **2020**, *746*, 141753. [CrossRef] [PubMed]
6. Varella, F.K.O.M.; Cavaliero, C.K.N.; Silva, E.P. Regulatory incentives to promote the use of photovoltaic systems in brazil. *Holos* **2012**, *3*, 15–29. [CrossRef]
7. Block, P.A.B.; Salamanca, H.L.L.; Teixeira, M.D.; Dahlke, D.B.; Shiono, O.M.; Donadon, A.R.; Camargo, J.C. Power quality analyses of a large scale photovoltaic system. In Proceedings of the 2014 5th International Renewable Energy Congress (IREC), Hammamet, Tunisia, 25–27 March 2014; pp. 1–6. [CrossRef]
8. Abubakar, A.; Almeida, C.F.M. Analysis of Battery Energy Storage System Sizing in Isolated PV Systems Considering a Novel Methodology and Panel Manufacturers Recommended Methodology. In Proceedings of the 2020 IEEE PES Transmission & Distribution Conference and Exhibition—Latin America (T&D LA), Montevideo, Uruguay, 28 September–2 October 2020; pp. 1–6. [CrossRef]
9. Dulout, J.; Jammes, B.; Alonso, C.; Anvari-Moghaddam, A.; Luna, A.; Guerrero, J.M. Optimal sizing of a lithium battery energy storage system for grid-connected photovoltaic systems. In Proceedings of the 2017 IEEE Second International Conference on DC Microgrids (ICDCM), Nuremburg, Germany, 27–29 June 2017; pp. 582–587. [CrossRef]

10. Abubakar, A.; Gemignani, M.; Almeida, C.F.M. Battery Storage System Sizing Using Synthetic Series Data. In Proceedings of the 2019 IEEE 46th Photovoltaic Specialists Conference (PVSC), Chicago, IL, USA, 16–21 June 2019; pp. 1578–1583. [CrossRef]
11. Feldman, D.; Margolis, R. *Q4 2019/Q1 2020 Solar Industry Update*; National Renewable Energy Laboratory: Denver, CO, USA, 2020.
12. Paul, E.; Bray, D. *Evolution of Solar Operating Practices: Advanced O&M Benefits from Module-Level Monitoring Solution Deployment Brief*; AltaTerra Research: Palo Alto, CA, USA, 2012.
13. Zhao, Y.; Ball, R.; Mosesian, J.; De Palma, J.-F.; Lehman, B. Graph-Based Semi-supervised Learning for Fault Detection and Classification in Solar Photovoltaic Arrays. *IEEE Trans. Power Electron.* **2014**, *30*, 2848–2858. [CrossRef]
14. Maleki, A.; Nazari, M.A.; Pourfayaz, F. Harmony search optimization for optimum sizing of hybrid solar schemes based on battery storage unit. *Energy Rep.* **2020**, *6*, 102–111. [CrossRef]
15. Nkambule, M.S.; Hasan, A.N.; Ali, A.; Hong, J.; Geem, Z.W. Comprehensive Evaluation of Machine Learning MPPT Algorithms for a PV System Under Different Weather Conditions. *J. Electr. Eng. Technol.* **2021**, *16*, 411–427. [CrossRef]
16. Kassim, N.; Sulaiman, S.I.; Othman, Z.; Musirin, I. Harmony search-based optimization of artificial neural network for predicting AC power from a photovoltaic system. In Proceedings of the 2014 IEEE 8th International Power Engineering and Optimization Conference (PEOCO2014), Langkawi, Malaysia, 24–25 March 2014; pp. 504–507. [CrossRef]
17. Abbas, M.; Zhang, D. A smart fault detection approach for PV modules using Adaptive Neuro-Fuzzy Inference framework. *Energy Rep.* **2021**, *7*, 2962–2975. [CrossRef]
18. Lu, X.; Lin, P.; Cheng, S.; Fang, G.; He, X.; Chen, Z.; Wu, L. Fault diagnosis model for photovoltaic array using a dual-channels convolutional neural network with a feature selection structure. *Energy Convers. Manag.* **2021**, *248*, 114777. [CrossRef]
19. Madeti, S.R.; Singh, S.N. Modeling of PV system based on experimental data for fault detection using kNN method. *Sol. Energy* **2018**, *173*, 139–151. [CrossRef]
20. Chen, S.; Li, X.; Meng, Y.; Xie, Z. Wavelet-based protection strategy for series arc faults interfered by multicomponent noise signals in grid-connected photovoltaic systems. *Sol. Energy* **2019**, *183*, 327–336. [CrossRef]
21. Hussain, M.; Dhimish, M.; Titarenko, S.; Mather, P. Artificial neural network based photovoltaic fault detection algorithm integrating two bi-directional input parameters. *Renew. Energy* **2020**, *155*, 1272–1292. [CrossRef]
22. Mellit, A.; Kalogirou, S.A. *A Survey on the Application of Artificial Intelligence Techniques for Photovoltaic Systems*, 3rd ed.; Elsevier Ltd.: Amsterdam, The Netherlands, 2018.
23. Chen, Z.; Yu, H.; Luo, L.; Wu, L.; Zheng, Q.; Wu, Z.; Cheng, S.; Lin, P. Rapid and accurate modeling of PV modules based on extreme learning machine and large datasets of I-V curves. *Appl. Energy* **2021**, *292*, 116929. [CrossRef]
24. Alhejji, A.; Mosaad, M.I. Performance enhancement of grid-connected PV systems using adaptive reference PI controller. *Ain Shams Eng. J.* **2021**, *12*, 541–554. [CrossRef]
25. Jung, S.; Jeoung, J.; Kang, H.; Hong, T. Optimal planning of a rooftop PV system using GIS-based reinforcement learning. *Appl. Energy* **2021**, *298*, 117239. [CrossRef]
26. Vergara, P.P.; Salazar, M.; Giraldo, J.S.; Palensky, P. Optimal dispatch of PV inverters in unbalanced distribution systems using Reinforcement Learning. *Int. J. Electr. Power Energy Syst.* **2022**, *136*, 107628. [CrossRef]
27. Zhou, Y.; Zheng, S.; Zhang, G. Machine learning-based optimal design of a phase change material integrated renewable system with on-site PV, radiative cooling and hybrid ventilations—study of modelling and application in five climatic regions. *Energy* **2020**, *192*, 116608. [CrossRef]
28. Singh, Y.; Pal, N. Reinforcement learning with fuzzified reward approach for MPPT control of PV systems. *Sustain. Energy Technol. Assess.* **2021**, *48*, 101665. [CrossRef]
29. Xie, Z.; Wu, Z. Maximum power point tracking algorithm of PV system based on irradiance estimation and multi-Kernel extreme learning machine. *Sustain. Energy Technol. Assess.* **2021**, *11*, 101090. [CrossRef]
30. Avila, L.; De Paula, M.; Trimboli, M.; Carlucho, I. Deep reinforcement learning approach for MPPT control of partially shaded PV systems in Smart Grids. *Appl. Soft Comput.* **2020**, *97*, 106711. [CrossRef]
31. Behera, M.K.; Saikia, L.C. A new combined extreme learning machine variable steepest gradient ascent MPPT for PV system based on optimized PI-FOI cascade controller under uniform and partial shading conditions. *Sustain. Energy Technol. Assess.* **2020**, *42*, 100859. [CrossRef]
32. Zafar, M.H.; Khan, N.M.; Mirza, A.F.; Mansoor, M.; Akhtar, N.; Qadir, M.U.; Khan, N.A.; Moosavi, S.K.R. A novel meta-heuristic optimization algorithm based MPPT control technique for PV systems under complex partial shading condition. *Sustain. Energy Technol. Assess.* **2021**, *47*, 101367. [CrossRef]
33. Lin, P.; Peng, Z.; Lai, Y.; Cheng, S.; Chen, Z.; Wu, L. Short-term power prediction for photovoltaic power plants using a hybrid improved Kmeans-GRA-Elman model based on multivariate meteorological factors and historical power datasets. *Energy Convers. Manag.* **2018**, *177*, 704–717. [CrossRef]
34. Mishra, M.; Dash, P.B.; Nayak, J.; Naik, B.; Swain, S.K. Deep learning and wavelet transform integrated approach for short-term solar PV power prediction. *J. Int. Meas. Confed.* **2020**, *166*, 108250. [CrossRef]
35. Kow, K.W.; Wong, Y.W.; Rajkumar, R.; Isa, D. An intelligent real-time power management system with active learning prediction engine for PV grid-tied systems. *J. Clean. Prod.* **2018**, *205*, 252–265. [CrossRef]
36. Behera, M.K.; Nayak, N. A comparative study on short-term PV power forecasting using decomposition based optimized extreme learning machine algorithm. *Eng. Sci. Technol. Int. J.* **2020**, *23*, 156–167. [CrossRef]

37. Li, P.; Zhou, K.; Lu, X.; Yang, S. A hybrid deep learning model for short-term PV power forecasting. *Appl. Energy* **2020**, *259*, 114216. [CrossRef]
38. Livera, A.; Florides, M.; Theristis, M.; Makrides, G.; Georghiou, G.E. Failure diagnosis of short- and open-circuit fault conditions in PV systems. In Proceedings of the 2018 IEEE 7th World Conference on Photovoltaic Energy Conversion (WCPEC), Waikoloa, HI, USA, 10–15 June 2018. [CrossRef]
39. Pillai, D.S.; Rajasekar, N. A comprehensive review on protection challenges and fault diagnosis in PV systems. *Renew. Sustain. Energy Rev.* **2018**, *91*, 18–40. [CrossRef]
40. Ghosh, R.; Das, S.; Panizrahi, C.K. Classification of Different Types of Faults in a Photovoltaic System. In Proceedings of the 7th IEEE International Conference on Computation of Power, Energy, Information and Communication, Chennai, India, 28–29 March 2018; pp. 121–128.
41. Nguyen, X.H. Matlab/Simulink Based Modeling to Study Effect of Partial Shadow on Solar Photovoltaic Array. *Environ. Syst. Res.* **2015**, *4*, 20. [CrossRef]
42. Patel, H.; Agarwal, V. MATLAB-Based Modeling to Study the Effects of Partial Shading on PV Array Characteristics. *IEEE Trans. Energy Convers.* **2008**, *23*, 302–310. [CrossRef]
43. Spooner, E.; Wilmot, N. Safety issues, arcing and fusing in PV arrays. In Proceedings of the 3rd International Solar Energy Society Conference—Asia Pacific Region (ISES-AP-08) Incorporating the 46th ANZSES Conference, Sydney, Australia, 25–28 November 2008.
44. Xia, K.; He, Z.; Yuan, Y.; Wang, Y.; Xu, P. An arc fault detection system for the household photovoltaic inverter according to the DC bus currents. In Proceedings of the 2015 18th International Conference on Electrical Machines and Systems (ICEMS), Pattaya City, Thailand, 25–28 October 2015; pp. 1687–1690. [CrossRef]
45. Chen, L.; Li, S.; Wang, X. Quickest Fault Detection in Photovoltaic Systems. *IEEE Trans. Smart Grid* **2018**, *9*, 1835–1847. [CrossRef]
46. Zhao, Y.; Lehman, B.; de Palma, J.-F.; Mosesian, J.; Lyons, R. Challenges to overcurrent protection devices under line-line faults in solar photovoltaic arrays. In Proceedings of the 2011 IEEE Energy Conversion Congress and Exposition, Phoenix, AZ, USA, 17–22 September 2011. [CrossRef]
47. Zhao, Y. Fault Analysis in Solar Photovoltaic Arrays. Ph.D. Thesis, Northeastern University, Boston, MA, USA, 2010.
48. Yi, Z.; Etemadi, A.H. Line-to-Line Fault Detection for Photovoltaic Arrays Based on Multiresolution Signal Decomposition and Two-Stage Support Vector Machine. *IEEE Trans. Ind. Electron.* **2017**, *64*, 8546–8556. [CrossRef]
49. Appiah, A.Y.; Zhang, X.; Ayawli, B.B.K.; Kyeremeh, F. Review and Performance Evaluation of Photovoltaic Array Fault Detection and Diagnosis Techniques. *Int. J. Photoenergy* **2019**, *2019*, 6953530. [CrossRef]
50. Zhao, Y.; De Palma, J.-F.; Mosesian, J.; Lyons, R.; Lehman, B. Line–Line Fault Analysis and Protection Challenges in Solar Photovoltaic Arrays. *IEEE Trans. Ind. Electron.* **2013**, *60*, 3784–3795. [CrossRef]
51. Hua, C.-C.; Ku, P.-K. Implementation of a Stand-Alone Photovoltaic Lighting System with MPPT, Battery Charger and High Brightness LEDs. In Proceedings of the 2005 International Conference on Power Electronics and Drives Systems, Kuala Lumpur, Malaysia, 28 November–1 December 2005; pp. 1601–1605.
52. Wang, Y.; Li, Y.; Ruan, X. High-Accuracy and Fast-Speed MPPT Methods for PV String Under Partially Shaded Conditions. *IEEE Trans. Ind. Electron.* **2016**, *63*, 235–245. [CrossRef]
53. Chan, F.; Calleja, H. Reliability: A New Approach in Design of Inverters for PV Systems. In Proceedings of the 2006 IEEE International Power Electronics Congress, Puebla, Mexico, 16–18 October 2006; pp. 1–6. [CrossRef]
54. Mellit, A.; Tina, G.M.; Kalogirou, S.A. Fault detection and diagnosis methods for photovoltaic systems: A review. *Renew. Sustain. Energy Rev.* **2018**, *91*, 1–17. [CrossRef]
55. Davarifar, M.; Rabhi, A.; El Hajjaji, A. Comprehensive Modulation and Classification of Faults and Analysis Their Effect in DC Side of Photovoltaic System. *Energy Power Eng.* **2013**, *5*, 230–236. [CrossRef]
56. Akram, M.N.; Lotfifard, S. Modeling and Health Monitoring of DC Side of Photovoltaic Array. *IEEE Trans. Sustain. Energy* **2015**, *6*, 1245–1253. [CrossRef]
57. Schimpf, F.; Norum, L.E. Recognition of electric arcing in the DC-wiring of photovoltaic systems. In Proceedings of the Intelec 2009—31st International Telecommunications Energy Conference, Incheon, Korea, 18–22 October 2009; pp. 1–6. [CrossRef]
58. Omer, A. Renewable energy resources for electricity generation in Sudan. *Renew. Sustain. Energy Rev.* **2007**, *11*, 1481–1497. [CrossRef]
59. Russell, S.; Norvig, P. *Artificial Intelligence: A Modern Approach*; Pearson Education: Pearson, London, 2010; Volume 4.
60. Mitchell, T. *Machine Learning*; Springer: New York, NY, USA, 2011; Volume 17.
61. Klass, L. Machine Learning—Definition and Application Examples. Spotlight Metal. 2018. Available online: https://www.spotlightmetal.com/machine-learning--definition-and-application-examples-a-746226/?cmp=go-ta-art-trf-SLM_DSA-20180820&gclid=CjwKCAjwkoz7BRBPEiwAeKw3q30qrjWJ-kiSAkfp6E6Oe_BxzFqk66RL3o2idJPKF1GBXlC94LgOuBoCTwMQAvD_BwE (accessed on 17 September 2020).
62. Simon, H. *Neural Networks and Learning*; Prentice Hall: Hoboken, NJ, USA, 2009; Volume 127.
63. Cormen, T.; Leiserson, C.; Rivest, R.; Stein, C. *Introduction to Algortihms*; MIT Press: Cambridge, MA, USA, 2015; Volume 1.
64. Chang, M.; Hsu, C. PV O&M Optimization by AI practice. In Proceedings of the 36th European Photovoltaic Solar Energy Conference and Exhibition, Marseille, French, 9–13 September 2019. [CrossRef]

65. Samara, S.; Natsheh, E. Intelligent PV Panels Fault Diagnosis Method Based on NARX Network and Linguistic Fuzzy Rule-Based Systems. *Sustainability* **2020**, *12*, 2011. [CrossRef]
66. Chine, W.; Mellit, A.; Lughi, V.; Malek, A.; Sulligoi, G.; Pavan, A.M. A novel fault diagnosis technique for photovoltaic systems based on artificial neural networks. *Renew. Energy* **2016**, *90*, 501–512. [CrossRef]
67. Livera, A.; Makrides, G.; Sutterlueti, J.; Georghiou, G.E. Advanced Failure Detection Algorithms and Performance Outlier Decision Classification for Grid-connected PV Systems. In Proceedings of the 33rd EU PVSEC, Amsterdam, The Netherlands, 25–29 September 2017; pp. 1–7. [CrossRef]
68. Chouder, A.; Silvestre, S. Automatic supervision and fault detection of PV systems based on power losses analysis. *Energy Convers. Manag.* **2010**, *51*, 1929–1937. [CrossRef]
69. Syafaruddin; Karatepe, E.; Hiyama, T. Controlling of artificial neural network for fault diagnosis of photovoltaic array. In Proceedings of the 16th International Conference on Intelligent System Applications to Power Systems, Hersonissos, Greece, 25–28 September 2011; pp. 1–6. [CrossRef]
70. Spataru, S.; Sera, D.; Kerekes, T.; Teodorescu, R. Diagnostic method for photovoltaic systems based on light I–V measurements. *Sol. Energy* **2015**, *119*, 29–44. [CrossRef]
71. Zhao, Y.; Liu, Q.; Li, D.; Kang, D.; Lv, Q.; Shang, L. Hierarchical Anomaly Detection and Multimodal Classification in Large-Scale Photovoltaic Systems. *IEEE Trans. Sustain. Energy* **2018**, *10*, 1351–1361. [CrossRef]
72. Mohamed, A.H.; Nassar, A. New Algorithm for Fault Diagnosis of Photovoltaic Energy Systems. *Int. J. Comput. Appl.* **2015**, *114*, 26–31. [CrossRef]
73. Caruana, R. Multitask Learning. *Mach. Learn.* **1997**, *28*, 41–75. [CrossRef]
74. Dey, A. Machine Learning Algorithms: A Review. *Int. J. Comput. Sci. Inf. Technol.* **2016**, *7*, 1174–1179.
75. Opitz, D.W.; Maclin, R. Popular Ensemble Methods: An Empirical Study. *J. Artif. Intell. Res.* **1999**, *11*, 169–198. [CrossRef]
76. Sharma, V.; Rai, S.; Dev, A. A Comprehensive Study of Artificial Neural Networks. *Int. J. Adv. Res. Comput. Sci. Softw. Eng.* **2012**, *2*, 278–284.
77. Wikipedia. Instance-Based Learning. Available online: https://en.wikipedia.org/wiki/Instance-based_learning (accessed on 31 March 2021).
78. Al-Sahaf, H.; Bi, Y.; Chen, Q.; Lensen, A.; Mei, Y.; Sun, Y.; Tran, B.; Xue, B.; Zhang, M. A survey on evolutionary machine learning. *J. R. Soc. N. Z.* **2019**, *49*, 205–228. [CrossRef]
79. Yar, M.H.; Rahmati, V.; Oskouei, H.R.D. A Survey on Evolutionary Computation: Methods and Their Applications in Engineering. *Mod. Appl. Sci.* **2016**, *10*, 131. [CrossRef]
80. Pavan, A.M.; Mellit, A.; De Pieri, D.; Kalogirou, S. A comparison between BNN and regression polynomial methods for the evaluation of the effect of soiling in large scale photovoltaic plants. *Appl. Energy* **2013**, *108*, 392–401. [CrossRef]
81. Mackay, D.J.C. A Practical Bayesian Framework for Backpropagation Networks. *Neural Comput.* **1992**, *4*, 448–472. [CrossRef]
82. Aziz, F.; Haq, A.U.; Ahmad, S.; Mahmoud, Y.; Jalal, M.; Ali, U. A Novel Convolutional Neural Network-Based Approach for Fault Classification in Photovoltaic Arrays. *IEEE Access* **2020**, *8*, 41889–41904. [CrossRef]
83. Krizhevsky, A.; Sutskever, I.; Hinton, G.E. ImageNet Classification with Deep Convolutional Neural Networks. In Proceedings of the 25th International Conference on Neural Information Processing Systems—Volume 1; Curran Associates Inc.: Red Hook, NY, USA, 2012; pp. 1097–1105.
84. Liu, G.; Yu, W. A fault detection and diagnosis technique for solar system based on Elman neural network. In Proceedings of the 2017 IEEE 2nd Information Technology, Networking, Electronic and Automation Control Conference (ITNEC), Chengdu, China, 15–17 December 2017; pp. 473–480. [CrossRef]
85. Chao, K.-H.; Chiu, C.-L.; Li, C.-J.; Chang, Y.-C. A novel neural network with simple learning algorithm for islanding phenomenon detection of photovoltaic systems. *Expert Syst. Appl.* **2011**, *38*, 12107–12115. [CrossRef]
86. Garoudja, E.; Chouder, A.; Kara, K.; Silvestre, S. An enhanced machine learning based approach for failures detection and diagnosis of PV systems. *Energy Convers. Manag.* **2017**, *151*, 496–513. [CrossRef]
87. Mekki, H.; Mellit, A.; Salhi, H. Artificial neural network-based modelling and fault detection of partial shaded photovoltaic modules. *Simul. Model. Pract. Theory* **2016**, *67*, 1–13. [CrossRef]
88. Coleman, A.; Zalewski, J. Intelligent fault detection and diagnostics in solar plants. In Proceedings of the 6th IEEE International Conference on Intelligent Data Acquisition and Advanced Computing Systems, Prague, Czech Republic, 15–17 September 2011; Volume 2, pp. 948–953. [CrossRef]
89. Nguyen, D.D.; Lehman, B.; Kamarthi, S. Performance evaluation of solar photovoltaic arrays including shadow effects using neural network. In Proceedings of the 2009 IEEE Energy Conversion Congress and Exposition, San Jose, CA, USA, 20–24 September 2009; pp. 3357–3362. [CrossRef]
90. Riley, D.; Johnson, J. Photovoltaic prognostics and heath management using learning algorithms. In Proceedings of the 2012 38th IEEE Photovoltaic Specialists Conference, Austin, TX, USA, 3–8 June 2012; pp. 1535–1539. [CrossRef]
91. Jufri, F.H.; Oh, S.; Jung, J. Development of Photovoltaic abnormal condition detection system using combined regression and Support Vector Machine. *Energy* **2019**, *176*, 457–467. [CrossRef]
92. Spataru, S.; Sera, D.; Kerekes, T.; Teodorescu, R. Photovoltaic array condition monitoring based on online regression of performance model. In Proceedings of the 2013 IEEE 39th Photovoltaic Specialists Conference (PVSC), Tampa, FL, USA, 16–21 June 2013; pp. 815–820. [CrossRef]

93. King, D.; Kratochvil, J.; Boyson, W. *Photovoltaic Array Performance Model*; Department of Energy: Albuquerque, NM, USA, 2004; p. 87105-0752.
94. Shapsough, S.; Dhaouadi, R.; Zualkernan, I. Using Linear Regression and Back Propagation Neural Networks to Predict Performance of Soiled PV Modules. *Procedia Comput. Sci.* **2019**, *155*, 463–470. [CrossRef]
95. Rezgui, W.; Mouss, N.K.; Mouss, L.-H.; Mouss, M.D.; Benbouzid, M. A regression algorithm for the smart prognosis of a reversed polarity fault in a photovoltaic generator. In Proceedings of the 2014 First International Conference on Green Energy ICGE, Sfax, Tunisia, 25–27 March 2014; pp. 134–138. [CrossRef]
96. Zhao, Y.; Yang, L.; Lehman, B.; de Palma, J.-F.; Mosesian, J.; Lyons, R. Decision tree-based fault detection and classification in solar photovoltaic arrays. In Proceedings of the 2012 Twenty-Seventh Annual IEEE Applied Power Electronics Conference and Exposition (APEC), Orlando, FL, USA, 5–9 February 2012; pp. 93–99. [CrossRef]
97. Benkercha, R.; Moulahoum, S. Fault detection and diagnosis based on C4.5 decision tree algorithm for grid connected PV system. *Sol. Energy* **2018**, *173*, 610–634. [CrossRef]
98. Rosner, B. American Society for Quality Percentage Points for a Generalized ESD Many-Outlier Procedure. *Technometrics* **1983**, *25*, 165–172. [CrossRef]
99. Witten, I.; Frank, I.H. *Data Mining—Practical Machine Learning Tools and Techniques with JAVA Implementations*; Morgan Kaufmann: Burlington, MA, USA, 2002; Volume 31.
100. Rezgui, W.; Mouss, L.-H.; Mouss, N.K.; Mouss, M.D.; Benbouzid, M. A smart algorithm for the diagnosis of short-circuit faults in a photovoltaic generator. In Proceedings of the 2014 First International Conference on Green Energy ICGE, Sfax, Tunisia, 25–27 March 2014; pp. 139–143. [CrossRef]
101. Rezgui, W.; Mouss, K.-N.; Mouss, L.-H.; Mouss, M.D.; Amirat, Y.; Benbouzid, M. Optimization of SVM Classifier by k-NN for the Smart Diagnosis of the Short-Circuit and Impedance Faults in a PV Generator. *Int. Rev. Model. Simul.* **2014**, *7*, 863. [CrossRef]
102. Hu, T.; Zheng, M.; Tan, J.; Zhu, L.; Miao, W. Intelligent photovoltaic monitoring based on solar irradiance big data and wireless sensor networks. *Ad Hoc Netw.* **2015**, *35*, 127–136. [CrossRef]
103. Ducange, P.; Fazzolari, M.; Lazzerini, B.; Marcelloni, F. An intelligent system for detecting faults in photovoltaic fields. In Proceedings of the 2011 11th International Conference on Intelligent Systems Design and Applications, Cordoba, Spain, 22–24 November 2011; pp. 1341–1346. [CrossRef]
104. Takagi, T.; Sugeno, M. Fuzzy identification of systems and its applications to modeling and control. *IEEE Trans. Syst. Man Cybern.* **1985**, *SMC-15*, 116–132. [CrossRef]
105. Karakose, M.; Fırıldak, K. A shadow detection approach based on fuzzy logic using images obtained from PV array. In Proceedings of the 2015 6th International Conference on Modeling, Simulation, and Applied Optimization (ICMSAO), Istanbul, Turkey, 27–29 May 2015; pp. 1–5. [CrossRef]
106. Grichting, B.; Goette, J.; Jacomet, M.; Benjamin, G. Cascaded fuzzy logic based arc fault detection in photovoltaic applications. In Proceedings of the 2015 International Conference on Clean Electrical Power (ICCEP), Taormina, Italy, 16–18 June 2015; pp. 178–183. [CrossRef]
107. Spataru, S.; Sera, D.; Kerekes, T.; Teodorescu, R. Detection of increased series losses in PV arrays using Fuzzy Inference Systems. In Proceedings of the 2012 38th IEEE Photovoltaic Specialists Conference, Austin, TX, USA, 3–8 June 2012; pp. 464–469. [CrossRef]
108. Yi, Z.; Etemadi, A.H. Fault Detection for Photovoltaic Systems Based on Multi-Resolution Signal Decomposition and Fuzzy Inference Systems. *IEEE Trans. Smart Grid* **2017**, *8*, 1274–1283. [CrossRef]
109. Dhimish, M.; Holmes, V.; Mehrdadi, B.; Dales, M. Diagnostic method for photovoltaic systems based on six layer detection algorithm. *Electr. Power Syst. Res.* **2017**, *151*, 26–39. [CrossRef]
110. Dhimish, M.; Holmes, V.; Mehrdadi, B.; Dales, M.; Mather, P. Photovoltaic fault detection algorithm based on theoretical curves modelling and fuzzy classification system. *Energy* **2017**, *140*, 276–290. [CrossRef]
111. Yao, X.; Herrera, L.; Ji, S.; Zou, K.; Wang, J. Characteristic Study and Time-Domain Discrete-Wavelet-Transform Based Hybrid Detection of Series DC Arc Faults. *IEEE Trans. Power Electron.* **2014**, *29*, 3103–3115. [CrossRef]
112. Ray, P.K.; Mohanty, A.; Panigrahi, B.K.; Rout, P.K. Modified wavelet transform based fault analysis in a solar photovoltaic system. *Optik* **2018**, *168*, 754–763. [CrossRef]
113. Kim, I.-S. On-line fault detection algorithm of a photovoltaic system using wavelet transform. *Sol. Energy* **2016**, *126*, 137–145. [CrossRef]
114. Davarifar, M.; Rabhi, A.; Hajjaji, A.; Kamal, E.; Daneshifar, Z. Partial shading fault diagnosis in PV system with discrete wavelet transform (DWT). In Proceedings of the 2014 International Conference on Renewable Energy Research and Application (ICRERA), Milwaukee, WI, USA, 19–22 October 2014; pp. 810–814. [CrossRef]
115. Li, X.; Yang, P.; Ni, J.; Zhao, J. Fault diagnostic method for PV array based on improved wavelet neural network algorithm. In Proceedings of the 11th World Congress on Intelligent Control and Automation, Shenyang, China, 29 June–4 July 2014; pp. 1171–1175. [CrossRef]
116. Mansouri, M.; Al-Khazraji, A.; Hajji, M.; Harkat, M.F.; Nounou, H.; Nounou, M. Wavelet optimized EWMA for fault detection and application to photovoltaic systems. *Sol. Energy* **2018**, *167*, 125–136. [CrossRef]
117. Momeni, H.; Sadoogi, N.; Farrokhifar, M.; Gharibeh, H.F. Fault Diagnosis in Photovoltaic Arrays Using GBSSL Method and Proposing a Fault Correction System. *IEEE Trans. Ind. Inform.* **2020**, *16*, 5300–5308. [CrossRef]

118. Wu, Y.; Chen, Z.; Wu, L.; Lin, P.; Cheng, S.; Lu, P. An Intelligent Fault Diagnosis Approach for PV Array Based on SA-RBF Kernel Extreme Learning Machine. *Energy Procedia* **2017**, *105*, 1070–1076. [CrossRef]
119. Yagi, Y.; Kishi, H.; Hagihara, R.; Tanaka, T.; Kozuma, S.; Ishida, T.; Waki, M.; Tanaka, M.; Kiyama, S. Diagnostic technology and an expert system for photovoltaic systems using the learning method. *Sol. Energy Mater. Sol. Cells* **2003**, *75*, 655–663. [CrossRef]
120. Kurokawa, K.; Uchida, D.; Otani, K.; Sugiura, T. Realistic PV performance values obtained by a number of grid-connected systems in Japan. In Proceedings of the NorthSun'99, The 8th International Conference on Solar Energy in High Latitudes, Edmonton, AB, Canada, 11–14 August 1999.
121. Garoudja, E.; Kara, K.; Chouder, A.; Silvestre, S.; Kichou, S. Efficient fault detection and diagnosis procedure for photovoltaic systems. In Proceedings of the 2016 8th International Conference on Modelling, Identification and Control (ICMIC), Algiers, Algeria, 15–17 November 2016; pp. 851–856. [CrossRef]
122. Chen, Z.; Wu, L.; Cheng, S.; Lin, P.; Wu, Y.; Lin, W. Intelligent fault diagnosis of photovoltaic arrays based on optimized kernel extreme learning machine and I-V characteristics. *Appl. Energy* **2017**, *204*, 912–931. [CrossRef]
123. Zhao, Y.; Lehman, B.; Ball, R.; Mosesian, J.; de Palma, J.-F. Outlier detection rules for fault detection in solar photovoltaic arrays. In Proceedings of the 2013 Twenty-Eighth Annual IEEE Applied Power Electronics Conference and Exposition (APEC), Long Beach, CA, USA, 17–21 March 2013; pp. 2913–2920. [CrossRef]
124. Shariff, F.; Rahim, N.A.; Hew, W.P. Zigbee-based data acquisition system for online monitoring of grid-connected photovoltaic system. *Expert Syst. Appl.* **2015**, *42*, 1730–1742. [CrossRef]
125. Lin, H.; Chen, Z.; Wu, L.; Lin, P.; Cheng, S. On-line Monitoring and Fault Diagnosis of PV Array Based on BP Neural Network Optimized by Genetic Algorithm. *Lect. Notes Comput. Sci.* **2015**, *9426*, 102–112. [CrossRef]
126. Kuo, C.-L.; Chen, J.-L.; Chen, S.-J.; Kao, C.-C.; Yau, H.-T.; Lin, C.-H. Photovoltaic Energy Conversion System Fault Detection Using Fractional-Order Color Relation Classifier in Microdistribution Systems. *IEEE Trans. Smart Grid* **2015**, *8*, 1163–1172. [CrossRef]
127. Jones, C.B.; Stein, J.S.; Gonzalez, S.; King, B.H. Photovoltaic system fault detection and diagnostics using Laterally Primed Adaptive Resonance Theory neural network. In Proceedings of the 2015 IEEE 42nd Photovoltaic Specialist Conference (PVSC), New Orleans, LA, USA, 14–19 June 2015; pp. 1–6. [CrossRef]
128. Khoshnami, A.; Sadeghkhani, I. Sample entropy-based fault detection for photovoltaic arrays. *IET Renew. Power Gener.* **2018**, *12*, 1966–1976. [CrossRef]

Communication

Safety Control Architecture for Ventricular Assist Devices

André C. M. Cavalheiro [1,2,†], Diolino J. Santos Filho [2,†], Jônatas C. Dias [2], Aron J. P. Andrade [2,3], José R. Cardoso [2] and Marcos S. G. Tsuzuki [2,*]

1. Departamento de Engenharia Mecatrônica, Centro Universitario da Fundação Santo André, Santo Andre 09060-650, Brazil; andre.cavalheiro@fsa.br
2. Departamento de Engenharia Mecatrônica e de Sistemas Mecânicos, Escola Politécnica da Universidade de São Paulo, Sao Paulo 05508-030, Brazil; diolinos@usp.br (D.J.S.F.); jxdias@usp.br (J.C.D.); aandrade@fajbio.com.br (A.J.P.A.); jose.cardoso@usp.br (J.R.C.)
3. Bioengenharia Instituto Dante Pazzanese de Cardiologia, Sao Paulo 04012-909, Brazil
* Correspondence: mtsuzuki@usp.br
† These authors contributed equally to this work.

Abstract: In patients with severe heart disease, the implantation of a ventricular assist device (VAD) may be necessary, especially in patients with an indication for heart transplantation. For this, the Institute Dante Pazzanese of Cardiology (IDPC) has developed an implantable centrifugal blood pump that will be able to help a diseased human heart to maintain physiological blood flow and pressure. This device will be used as a totally or partially implantable VAD. Therefore, performance assurance and correct specification of the VAD are important factors in achieving a safe interaction between the device and the patient's behavior or condition. Even with reliable devices, some failures may occur if the pumping control does not keep up with changes in the patient's behavior or condition. If the VAD control system has no fault tolerance and no system dynamic adaptation that occurs according to changes in the patient's cardiovascular system, a number of limitations can be observed in the results and effectiveness of these devices, especially in patients with acute comorbidities. This work proposes the application of a mechatronic approach to this class of devices based on advanced control, instrumentation, and automation techniques to define a method to develop a hierarchical supervisory control system capable of dynamically, automatically, and safely VAD control. For this methodology, concepts based on Bayesian networks (BN) were used to diagnose the patient's cardiovascular system conditions, Petri nets (PN) to generate the VAD control algorithm, and safety instrumented systems to ensure the safety of the VAD system.

Keywords: safety instrumented system; ventricular assist device Bayesian network; Petri net

1. Introduction

A ventricular assist device (VAD) has the main function of helping the patient with heart failure to lead a relatively normal life, despite the disease. This blood pump can be used in several cases: during the period the patient is waiting for a heart transplant; during a pre- or postoperative recovery period, or as a destination therapy when the patient has no indication for heart transplantation due to immunological incompatibility, chronic infections, or advanced age [1,2]. VAD projects involve several research areas, such as: mechanical and electromechanical engineering, biomaterials, medicine, and computer technologies for data collection, processing, and decision making. Therefore, sensors are needed to indicate blood pressure, blood flow, body temperature, and heart rate [3]. Two aspects should be taken into account:

- First: the device must perform effectively and accurately; otherwise, if the pump fails during operation and there is no control system capable of interpreting and autonomously handling failures, serious risks to the patient are inevitable [4]. Medical equipment should provide personalized care to reduce such risks [5];

- Second: many VADs keep the blood flow constant regardless of the patient's daily needs, they assist the blood circulation, and do not react adequately to changes in the patient's behavior [6]. If the patient is at rest and needs to perform physical activity, the patient's heart automatically changes its behavior, pumping more blood, but some VADs do not follow the need of the patient's circulatory system by keeping the rotation speed of the VAD fixed.

1.1. Evolution of VADs

With the technological evolution that can be applied in the design of VADs, the use of electric motors to drive blood pumps is highlighted. During the initial phase of the use of these actuators, the rotation control system was focused only on maintaining a reference rotation. Later, control algorithms were developed that could be reconfigured in terms of set-point variation [3,7–9]. Although this evolution was very interesting for the patient to be assisted by the medical team in the postoperative period, the great challenge was to establish the correct time for specialist intervention to reprogram the set point: the patient's return to be reevaluated could be ineffective, if any adverse event occurred that implied an immediate correction of the pump's behavior.

To overcome this limitation, there was the advancement of sensing resources, that is, VADs became automatic devices as it became possible through sensors to monitor the physiological behavior of their patients. This had an impact on the design of control architectures for VADs and on the performance of these devices [10]. In this context, automation has a fundamental characteristic in the development of autonomous devices. Thus, in this work, a methodology is proposed for the development of a VAD rotation control system in a dynamic way, considering: monitoring the rotor speed to act upon the occurrence of failures in the device; reconfiguration of the control system according to the need to supply cardiac output according to the physical activities that the patient is performing at all times and; the possibility of dynamic change in the behavior of the pump, depending on the patient's global status, upon the occurrence of adverse events. For the development of this control solution, not only physical sensors are considered, but also virtual sensors based on specific algorithms to calculate the desired parameters, for example, indicating that the blood flow increased.

1.2. Research Motivation

Therefore, with a VAD with no fault tolerance and no dynamic behavior that adapts to the performance conditions of the cardiovascular system, serious limitations are observed in the results of this application [6]. Thus, this work proposes the application of a mechatronic approach to this class of devices based on advanced control, instrumentation, and automation techniques [11]. These techniques allow in a systematic way to consider in the control architecture the limitations of the current solutions. This methodology has been applied to other medical equipment [12]. In this context, a method is proposed for a VAD supervisory control system that:

- Specifies a logic for the pump speed control, according to the dynamic behavior of the patient. Models based on Bayesian network (BN) [13] should be applied to diagnose the dynamic state of the patient at each moment and to act in controlling the VAD;
- Specifies a safety interlocking logic to prevent failures in the VAD that could generate risks to the patient's life. To do this, the critical states will be diagnosed by means of BN and, from these, implement a real-time diagnostic control system by means of Petri nets (PN).
- Once the diagnostic control system is implemented in parallel, fault handling should be implemented according to the specification of the instrumented functions of security (SIFs) [14,15]. These functions are modeled in PN [16] for generating the control algorithm for fault-handling.

- Check the supervisory model observing the human cardiovascular electronical model, considering the proposed model [17]. Making the supervisory control system validation, the next step is the "in vitro" and "in vivo" validation.

These items are relevant and essential to make the control system of a DAV autonomous and intelligent. However, approaching all these subjects in a single article is a difficult task. Thus, this work approaches the development of a VAD rotation control system in a dynamic way that will solve the local rotation control considering physical sensors. Thus, as a result, an automatic system will be obtained that will allow the speed control to become dynamic. The part of security control with adverse and unknown situations will be the subject of another work.

Thereby, the paper is structured starting with the introduction and research motivation. Section 2 describes some basic concepts: 1. Product Flow Schema (PFS); 2. Hybrid PN (HPN); 3. The PFS/HPN Top-Down approach, and 4. The safety instrumented systems (SIS) and BN/HPN approach. The first approach is used to represent the controller, and the second approach is used to connect the sensors and the treatment modeling to the controller. Section 3 has the proposed VAD supervisory control system design methodology. Section 4 has the results and Section 6 has the conclusions and future works.

2. Basic Concepts

This section is divided into four parts. The first part explains the PFS representation. The PFS is a bipartite graph used to represent a sequence of operations for different situations. The second part explains the HPN used to create formal representations. It also includes the manipulation of continuous variables. The third part explains the top-down PFS/PN approach. The PFS is refined using PN, the dynamics of the logic operations are described using PN. This approach will create the VAD control. The fourth part explains the BN/PN approach where the treatment and sensor information are connected to the controller.

2.1. Product Flow Schema (PFS)

The PFS is a bipartite graph composed of activity elements (action, execution), distribution elements (collect, accumulate, and/or store information or items), and oriented arcs to connect the elements. Figure 1 shows the graphical representation of these elements. It is capable of modeling the specific control functionalities and operations that will be performed by the VAD. Functionality depends on circumstances: normal device behavior, behavior in the face of failure situations, or behavior in the face of adverse events that are associated with the patient's physiology. In this context, the PFS model is descriptive. It is able to represent the sequence of operations for each situation. It also allows representing the causality associated with the occurrence of events. It is represented in terms of oriented arcs that establish a sequential logic for carrying out the operations.

Figure 1. Basic elements of the Production Flow Schema (PFS). The PFS represents the system conceptual description. Arcs connect activity and distributor elements. Activity elements can be refined into a new PFS model or a PN.

2.2. Hybrid Petri Net (HPN)

Murata [18] presents the PN as a promising tool to describe and study systems with concurrency, parallelism, asynchronous, non-deterministic, and/or stochastic. According to Li and Zhao [19], the PN is a suitable mathematical tool to model and analyze discrete event systems that have behaviors such as concurrency, conflict, and casual dependence between events (parallelism). Thus, considering the requirement that VAD needs to perform control functions to adjust pump speed according to changes in cardiac frequency and needs to react against the occurrence of critical faults, a PN is a useful tool to model, analyze, and validate the VAD control. Figure 2 shows the PN elements graphical notation: discrete transaction, discrete place, oriented arc, inhibitor arc, and enabler arc [16]. The continuous transaction and continuous place will be described in the following with the HPN.

According to Li and Zhao [19] a general N PN is a 4-tuple (P, T, F, W). Where P and T are disjoint sets, finite, not empty. P is a set of places, and T is a set of transitions. $F \subseteq (P \times T) \cup (T \times P)$ is called the flow relationship or the set of directed arcs. $W : F \to N^+$ is a mapping that assigns a weight to an arc, that is, it is a weighting function, where $N^+ = \{1, 2, \ldots\}$. The precondition of a node $x \in (P \cup T)$ is defined as $^\circ x = \{y \in P \cup T \mid (y, x) \in F\}$. The postcondition of a node $x \in (P \cup T)$ is defined as $x^\circ = \{y \in P \cup T \mid (x, y) \in F\}$.

The precondition/postcondition of a set is defined as the union of the preconditions and postconditions of its elements. $N = (P, T, F, W)$ is called an ordinary net, and its notation is $N = (P, T, F)$, if $\forall f \in F$, $W(f) = 1$. A token M of $N = (P, T, F)$ is a mapping of $M : P \to N$, where $N = N^1 \cup 0$ and $M(p)$ indicates the number of tokens M in p. Token M is denoted by $\sum_{p \in P} M(p)$.

The HPN model was introduced as an extension of the discrete PN model, allowing the manipulation of real numbers in a continuous form, thus allowing to express explicitly the relation between continuous values and discrete values, keeping good modeling characteristics of discrete systems consented in a consecrated way by PN. The HPN is a bipartite graph that is represented with places and transitions. The components of the HPN are classified into discrete/continuous places and discrete/continuous transitions. A nonnegative real number is represented by a continuous place and the firing rate of the continuous transition is given as a function of the places in the HPN model. Figure 2 shows graphical notations of HPN elements [16]. The HPN has inhibitor and enabler arcs; these arcs enable/disable specific activities. The refinement of a model generated using PFS to a model in HPN is done based on the procedure adopted in Villani et al. [20].

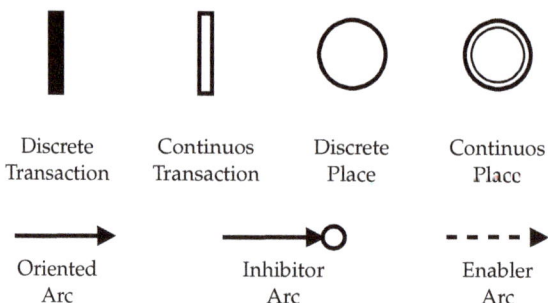

Figure 2. Basic elements of the HPN. The HPN is the lowest level and implements the control algorithm. The continuous transaction and place are responsible of processing continuous values in the HPN. The inhibitor and enabler arcs define a specific flow and consequently enable/disable specific activities. The VAD control is implemented using the HPN.

A system where there are different kinds of variables and that simultaneously presents the evolution of continuous variables and discrete events should be considered a Hybrid System and could be modeled by HPN theory [20]. This behavior, considering the Discrete Event System (DES), unifies with the behavior of the Continuous Variable System (CVS) [21]. Considering the need for the VAD to perform control functions to adjust the pump speed according to changes in heart rate and the need to react to the occurrence of critical failures, a hybrid supervisory control system has been proposed [20].

2.3. The Top-Down PFS/HPN Approach

Details of each activity modeled in the PFS can be refined using PN. The activity element can be refined into another PFS model or a PN using top-down methodology. Figure 3 shows the graphical representation of top-down methodology. Figure 3a shows the initial PFS activity, which will be transformed into the PN representation shown in Figure 3d. It is even possible to create a combined PFS/PN representation as shown in Figure 3c. In turn, after establishing the operation logic using PFS, it is necessary to describe how the operations should be performed dynamically. This description is refined using PN. It is based on the concept of local states associated with structural element markings. Therefore, applying the systematic to generate a PN model from a PFS model, it will be possible to obtain the association between the semantics of PFS and PN. This association will model the execution of each functionality/operation and the local control, according to the state in which the VAD system is at any given time.

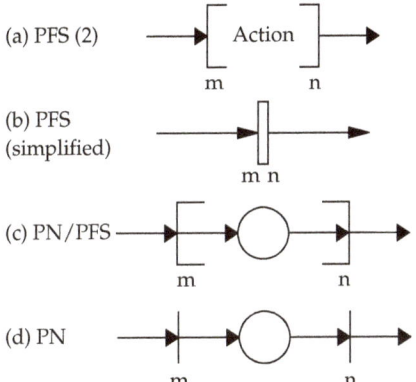

Figure 3. PFS model to PN conversion applying top-down methodology successive refinement. (**a**) PFS graph with just one activity. (**b**) Another possible representation for the same activity element. (**c**) A combined PFS/PN representation. The activity is represented by a discrete place. (**d**) A pure PN representation.

The Product Flow Schema (PFS) tool [20] can be used to systematically model the set of activities that a VAD can perform in a top-down approach. The activity element can be refined into another PFS model or a HPN using top-down methodology. Considering the VAD, some components are the treatment of diagnosed problems, patient diagnosis, control algorithm, validation of the control algorithm, and others. The control algorithm is represented by a new set of activities. The HPN is the lowest level and it represents the control implementation. This approach has been widely used in industrial applications [22]. Therefore, with the mathematical formalism of the HPN, it is possible to create a control algorithm that can be modeled and that complies with the requirements of reachability, liveliness, limitability, conservativeness, and reversibility. This characteristic can be analyzed mathematically in the HPN, qualifying the control to be fault-free, providing adequate security to critical systems such as a VAD control. The HPN generated by PFS model is able to represent the dynamic behavior of VAD system. Since the VAD introduces continuous variables, than HPN is required.

2.4. SIS and BN/PN Approach

To develop the fault diagnosis and treatment model, the concepts of SIS and BN are used. According to Squillante Jr et al. [14] the SIS is a control layer with the objective of mitigating risk or bringing the process to a safe state. The definition of each failure is made from the identification of Safety Instrumented Functions (SIF). In this way, a SIF describes a system failure that must be diagnosed and treated by the SIS. A SIS implements its SIFs by means of sensors and system devices, and performs control by actuators. For each SIF, a parameter called the safety integrity level (SIL) is defined. This parameter is a measure of safety for each component and/or system. BNs provide a formalism for the system to reason about conditions of uncertainty and process failures. In this formalism, propositions are given as numerical parameters that signify the degree of belief according to some evidence or knowledge. Thus, formally, BN $B = (G; Pr)$ are composed of a topological structure G and a set of parameters Pr that represents the probabilistic relationship between its variables [14].

The human body can assume different states depending on the heart disease of each patient. Some states are known and others are unknown, which may vary from patient to patient. Some states can be detected by sensors. Other states, no less important, do not have sensors, which can generate uncertainty in relation to the real state of the variable. However, there are tools that allow the deduction of these states, indirectly, performing an analysis of the global state of the system. Thus, to deal with the uncertainties inherent to this class of systems, the use of BN is proposed. With the BN formalism, it is possible to work with states that are not monitored by sensors and/or unknown states, and with that it is possible to abduct the information and make decisions that would not be possible without the BN analysis.

Thus, in Figure 4 a BN is presented, where two situations are defined: (a) high blood pressure and (b) high blood flow, with their respective probabilities. Depending on the patient's global status, the occurrence of these events can lead to a decrease in VAD rotation. Thus, initially, the probability of 50% of occurring or not the alteration in the rotation of the VAD was considered. If the occurrence of high blood pressure is detected (V), a chance of 20% of this information not influencing the rotation P(V=F) and 80% that the rotation will be decreased to solve this event P(V=V) is defined. Respectively, this analysis can be considered for other combinations of states that the system can assume, following the BN formalism.

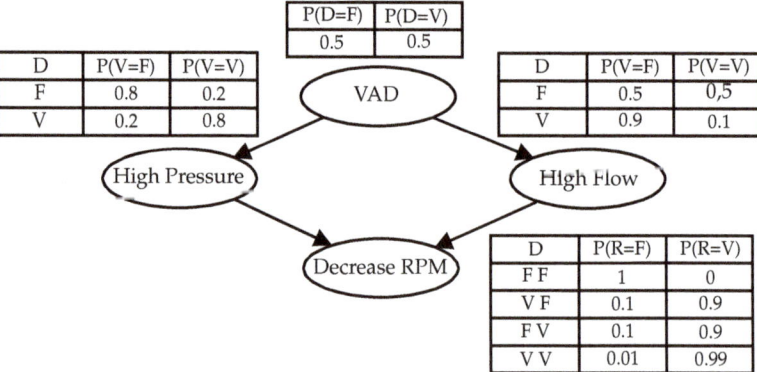

Figure 4. Example of the partial modeling of a safety system using BN. The VAD considers that two events might happen: high arterial pressure and high blood flow. Each event has the same probability of happening. These two events combined can cause a decrease in the VAD motor rotation. This is just an example, and the possible events are in a much larger number.

As BN is an analysis tool, it is necessary to use HPN to generate dynamic control models. Therefore, with the analysis of the incidence of events related to the possible states that the VAD can assume, the refinement of BN into HPN is proposed and, in this way, diagnostic and control models are generated. In Figure 5 the HPN referring to the BN obtained in Figure 4 is represented. In order for the control algorithm to meet the security conditions of a critical system, this activity must be performed, observing the following situations:

- Obtain the causal dependency relationships from the BN for each failure;
- Obtain the logical relationships between external variables (sensors) from the causal relationship;
- Build the HPN from the diagnostic reasoning, obtain an effect versus cause structure to represent the BN diagnostic model in HPN, so that it is possible to carry out the control;
- Consider in the HPN design that it must allow its restart to meet the required restartability property and consider the possibility of the failure being spurious, that is, the possibility of the diagnosis not being carried out considering the calculated uncertainties in the BN;
- Represent in the model the events associated with the transitions and how to represent them, indicating, in particular, the events that require an interaction with the external environment (for example, sensors and actuators).

By following these procedures, it is possible to obtain the PN represented in Figure 5.

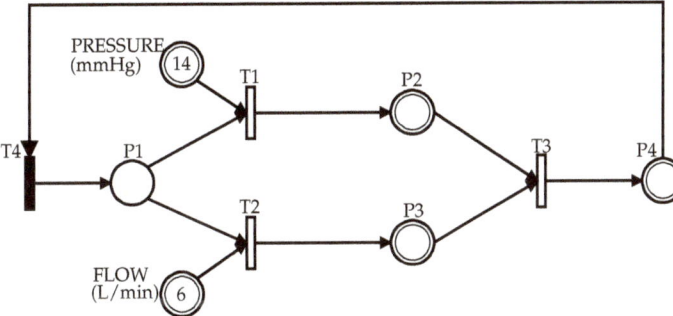

Figure 5. Example of a HPN obtained from BN. In this case, arterial pressure is considered to be high when above 14 mmHg and blood flow is high when above 6 l/min.

3. Proposed VAD Supervisory Control System Design Methodology

To design the VAD supervisory control system systematically according to the techniques presented [23], a set of procedures are proposed below:

- Control System Definition—physicians, engineers, and specialist teams are defined to define the VAD autonomy. This is the team that is responsible for developing the control system. They need to select ideas that can be implemented taking into consideration: available sensors, performance characteristics of the VAD, and technological limitations. The VAD control functions are specified in this phase.
- Patient Diagnosis Modeling—The cause and effect matrix is created considering the definitions made in the last step. Every effect considered is converted to a node of a BN. Then, following the procedures proposed in [24] this fault diagnosis network can be converted into a PN control model.
- Diagnosed Diseases Treatment Modeling—To obtain the HAZOP (hazard and operability) study [15] for VAD, a risk analysis report can be used for the IEC 31010 standard. Thus, the SIL and events (from sensors) and actions (for actuators) for each SIF are obtained. Then, to obtain the safety instrumented systems (SIS), each SIFs are modeled in PN.

- Adopted Solution Analysis—To validate the obtained PN models, first a structural analysis is performed. Then, it is verified by PN deadlock (markings where no transition is enabled). For this, the Visual Object Net simulator [25] can be used.
- Control Algorithm Design—To generate the control algorithm, the validated PN models are translated to a Programmable Controller language following the rules proposed in [24] to generate the control program according to IEC 61131-3 [26] standard.
- Control Algorithm Validation—To make "in vitro" tests, a mathematical model can be used to simulate the human cardiovascular system [17]. Once the control system is validated, the next step is to implement the prototype physically to confirm the functionality of the cardiovascular simulator system. At the Institute Dante Pazzanese of Cardiology (IDPC), there is a programmable mechanical simulator to perform in vitro tests. This equipment allows to simulate a real patient cardiovascular system [27].
- Supervisory System Ready—This step amounts the in vivo validation of the control algorithm. Once the in vitro simulation is validated, the VAD control algorithm is ready for in vivo testing [2].

By defining the design of the supervisory control system method and applying the set of procedures proposed above, it is possible to provide security to the system and enable control of the VAD rotational speed according to changes in the patient's heart rate, which can improve the quality of life and safety of patients, increasing the patient's life and providing time before a heart transplant. The proposed methodology is shown in Figure 6 according to the PFS formalism.

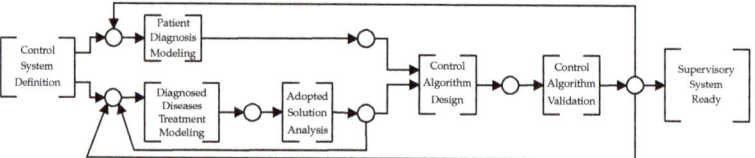

Figure 6. PFS model of the proposed safe control system design method for VAD.

4. Results

By applying the proposed procedure, a method is defined and some characteristics are accomplished: (i) the HAZOP study is applied and considers critical failures to VAD control system, (ii) BN is used to diagnose and decide, (iii) Safety Instrumented Functions (SIF) are used to model HPN and (iv) Supervisory control system modeling is made considering the discrete and continuous variables of VAD.

4.1. Supervisory Design

Tests using a physical prototype are made to confirm the functionality by using a IDPC programmable mechanical simulator to validate the control performance. The local control model is shown in Figure 7. The supervisory modeled in PFS shows four possible heart beating ranges. The sensor information switches the HPN flow to activate the proper range. After switching, the proper control happens.

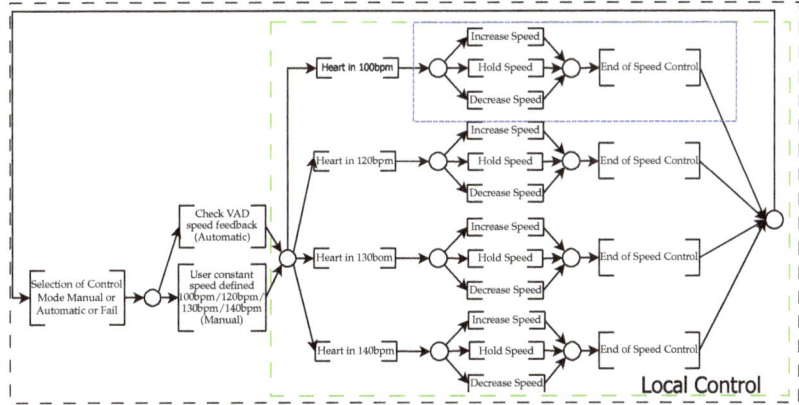

Figure 7. PFS model of the VAD Safe Supervisory control system under study. Four heart beating ranges are defined: 100, 120, 130, and 140 bpm. The module inside the blue box is detailed in Figure 8.

4.2. Control Design

The PFS shown in Figure 7 is refined into a HPN. Figure 8 shows the detailed control of the blue box represented in Figure 7. It has the following dynamics. When the "StartControl" place receives a token, the Speed Control HPN starts and the "LowSpeed" and "HighSpeed" places are monitored. The DAV rotation has a target rotation, considering the specific heart beat. If the speed is not low or high, then the DAV rotation shows the error equals 0, and the speed output remains constant. If the error is different from 0 appears, one of the signals "LowSpeed" or "HighSpeed" has a token. If "LowSpeed" has a token, the speed control token is directed through the enabling arc to the "IncrementSet" place, which has the function of incrementing the controller output to perform the speed adjustment in the driver (local pump control of the DAV) and continues increasing the output until the value is controlled, i.e., the error becomes zero.

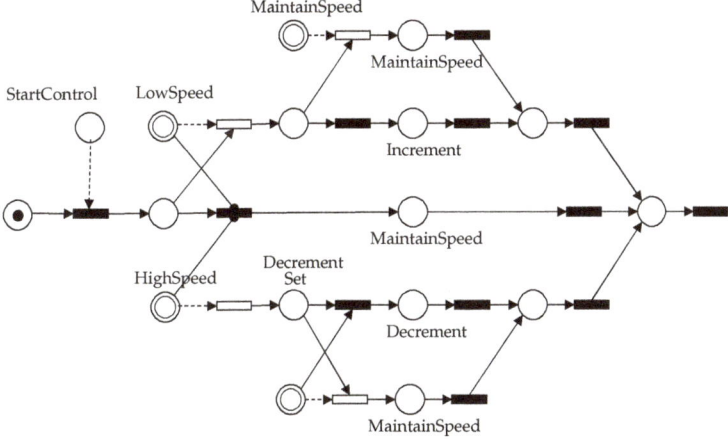

Figure 8. Example of a Speed Control HPN obtained from the blue box shown in the PFS model from Figure 7. The system used to model the HPN has horizontal transitions instead of vertical transitions.

The range of the desired speed value is represented by "LowSpeed" and "HighSpeed" without a token, which enables the token control to the "MaintainSpeed" place, by enabling both inhibiting arcs, keeping the speed value sent to the driver. Another situation expected by the HPN is when the speed increment value reaches the maximum signal value that can be generated by the pump driver, i.e., full rotation or 100% of the activation signal or maximum speed request. This state is represented by the "MaxSpeed" place, which indicates that the signal has reached saturation. In this way, the token is shifted to the "MaintainSpeed" place. This state is implemented to avoid the control's windup problem. The reciprocal situation happens for the state of "HighSpeed", causing the activation of the control of "DecrementSet".

4.3. Numerical Simulation

Numerical results of physical VAD supervisory control system simulation were obtained. Once the modeling and validation of the VAD's control algorithm is performed using HPN and the model is converted to the microcontroller programming language using the procedure of replacing each element of the HPN by elements of the Ladder diagram [24], it enables the implementation of the system and in vitro computational validation, providing an implementation of the device control within a cardiac simulator to validate the computational model. To this end, this algorithm is implemented in the hardware architecture proposed in Figure 9, modeled using the Proteus Design Suite® software. The proposed hardware architecture is composed of the following components:

- Microcontroller: device responsible for executing the security control;
- VAD Motor Driver: device responsible for driving the VAD motor;
- Supervisory: device responsible for the human-machine interface;
- Function Generator: equipment used to simulate the signals coming from the sensors and generate disturbances to carry out analysis of the control system;
- Signal Converter/Filter: device responsible for conditioning the signals generated by the function generator.

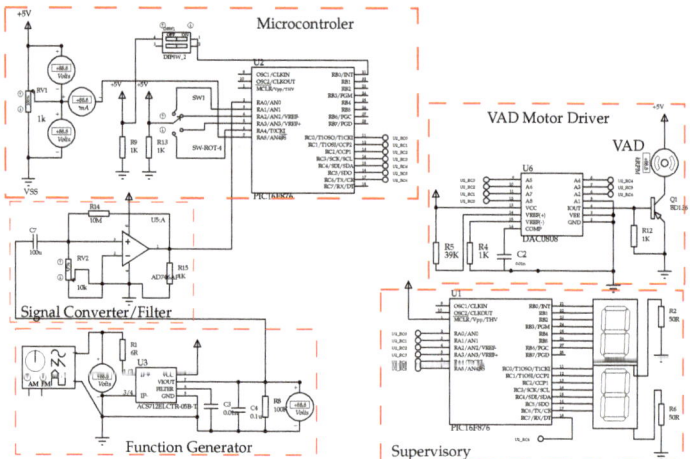

Figure 9. Electronic circuitry developed for the design of the supervisory control. The VAD motor driver controls the blood flow. The VAD control is running at the microcontroller. The function generator creates failures in the system, for example increase the heart beating. The signal converter/filter interfaces the signal to the microcontroller. The supervisory is responsible of the human-machine interface.

In this way, disturbance signals are generated by the Function Generator and sent to the controller through the signal conditioning system. These signals are processed by the VAD control algorithm, which sends, depending on the control model, the appropriate command for the VAD Motor Driver. The VAD Motor Driver is responsible for performing the electrical power interface and activating the VAD motor. In this experiment, the signal range generated by the VAD Motor Driver is from 0 to 5 V DC, where 0 V means 0 RPM and 5 V means the maximum speed, which in the case of this application is 3000 RPM.

To validate the architecture and the control algorithm, a model of the human cardiovascular system [17] was considered, enabling a virtual simulation of the human cardiovascular system (see Figure 10). It is an electrical equivalent model where the signal generated by the pump rotation control is represented in Figure 10 by the positive and negative poles of the VAD. The electronic parameters are correlated with their mechanical parameters as follows: voltage (V) to pressure (mmHg), capacitance (mF) to elasticity (mL/Pa), electrical resistance (kΩ) to resistance to passage of blood (Pa·s/mL), and inductance (µH) to inertia (Pa·s^2/mL). The elements of each artery are represented, including one or two resistors, an inductor, and a capacitor.

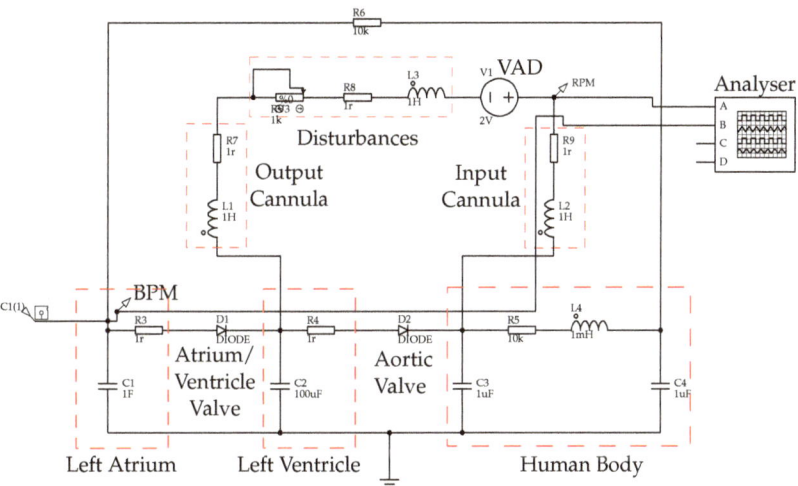

Figure 10. Model adopted to run the cardiovascular system computer simulation. The VAD Motor Pump from Figure 9 is connected at the VAD position.

The model of the human cardiovascular system (see Figure 10) is connected to the supervisory control (see Figure 9). This simulation environment is proposed to analyze the control algorithm. This signal will introduce the voltage relative to the blood flow into the simulation circuit. This increase in voltage together with the characteristics of the human body simulated by the electronic components will produce the reaction curve shown in Figures 11 and 12 as a function of the control model response and the disturbance generated in the simulation.

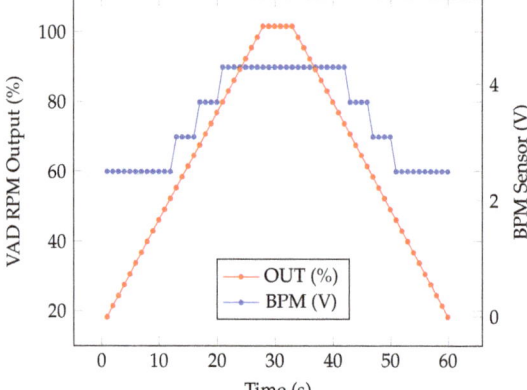

Figure 11. Simulation results considering the proposed architecture of the VAD supervisory and control system. The heart beating increased and decreased (red curve). The speed control set point changed three times up and three times down to adjust the motor peed with the pump rotation (the blue curve).

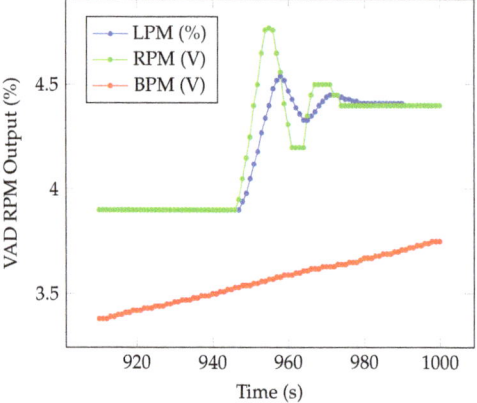

Figure 12. The speed control set point changed, as a consequence of the heart beat increase (red curve, Volts). The control shows the adjustment to new set point. The control algorithm increases the motor rotation (blue curve, %). The increase in the motor rotation increase the blood flow (green curve, Volts). This graph is a detail of a set point change, which happened in Figure 11.

Simulations of heart failure and heart rate oscillations are performed to verify that the control algorithm works correctly, changing the blood pump speed appropriately for each simulated case. The reaction curve of the pump implanted in the human body was used to verify the stability of the control system as a function of the conditions of the cardiovascular system in which the pump will be implanted.

With this, it is possible to validate the states that the system can reach computationally. Consequently, to validate the physical control using the real VAD driver, the designed electronic circuit was prototyped, as shown in Figure 10. Thus, it was possible to test the system in a physical simulator of the cardiovascular system [27].

With the physical test, it is possible to confirm whether the VAD will effectively help a weakened heart to regain its function as a heart pump in a physical way. This physical simulator serves mainly to verify if there is a tendency of the VAD to make a speed correction if there is a load change in the cardiovascular system, that is, if the VAD system controls a rotation rate to maintain the physiological functions independent of the pump rotation working in a synchronized demand with the natural heart. Using a tachograph and the proper instruments, the VAD state and its respective velocity measurements were performed as presented in Figure 11. Therefore, different reaction curves were generated for the different situations required by the control system. Thus, heart rate variation is applied to analyze how the VAD control system must operate in automatic control mode considering the PN model presented in Figure 8. As show in Figure 11, the system responded according to the proposed control model. The red line shows the simulated heart rate starting at zero beats per minute (bpm) and increasing the value to the maximum bpm value that a patient may present (in this case, 200 bpm was considered), then maintaining stability at the upper level and after decreasing the value again until 0 bpm. With this simulation, it is possible to observe how the control should react considering the whole range of flow control considering the bpm patient range. The blue line shows the pump rotation output sent from the supervisory system to the local control, where the change in speed (0–3000 RPM) as a function of heartbeat (bpm) is observed by the control. The green line indicates the blood flow (0–5 LPM), considering the local control of the pump as shown in Figure 12.

These characteristics are confirmed at the time of testing with the physical simulator to validate the model, which will prepare it for the next phase of the proposed method, which is the in vivo test system.

5. VAD Validation

The research project "Electromagnetic Propulsion Systems for Circulatory Support, Implantable Ventricle Assist Device, and Artificial Heart", supported by FAPESP, is currently in the execution phase. This project aims to integrate engineering and medicine to develop solutions for mechanical circulatory support systems. For this, a methodology with the development of 11 subprojects based on applied engineering and medicine was used.

This research work presented is part of the subproject responsible for developing proposals for Safety Supervisory Control Systems for Implantable Blood Pumps. In turn, for the validation of a VAD that has a coupled control system, a test protocol is required to be implemented in human beings [28]. In this context, as an engineering project, a VAD must undergo all verification and validation tests with the help of specialized benches so that in vitro tests are carried out, applying a deductive method based on the formal analysis of appropriate mathematical models to represent the dynamic behavior that is desired [29]. In turn, after this technological validation of the device, it is necessary to carry out the in vivo test steps. Initially, the tests are carried out on animals through procedures established by teams of veterinarians and in accordance with the current code of ethics [30].

Once approved in animals, the last stage of in vivo tests is to carry out the evaluation in human beings. In this context, there is a subproject focused solely on the clinical evaluation study of VAD, which involves the Instituto Dante Pazzanese de Cardiologia [31]. The initial step was to submit a Clinical Assessment Protocol to the National Research Ethics Council and the National Health Surveillance Agency, receiving authorization from both bodies to implant devices in 10 patients. The procedure adopted for analyzing the patient's behavior consists of: (i) performing biochemical studies, tests of the patient's hemodynamic performance; (ii) detailed definition of the surgical procedure; (iii) detailed definition of the postoperative procedure to specify monitoring of physiological signs and administration of compatible medication; (iv) specification of laboratory data to be monitored; (v) planning the indication of heart transplantation and removal of the device in an emergency and; (vi) provision for device removal after satisfactory circulatory assistance tests.

Thus, the procedure for validating a VAD from its technological design to its implantation in patients is explained.

6. Conclusions and Future Works

Currently, the usual VAD control has difficulty to adjust the blood flow according to the patient's condition: there is no supervisory control of the device that adjusts the rotation speed according to the patient's needs, and there is no treatment of VAD failure that helps patient safety. The presented study contributed to a great evolution of the VAD control system, proposing a suppository control layer that identifies and recognizes the adverse event, providing a better quality of life for the patient and increased survival. Thereby, the supervisory system can also assist in diagnosis and interventions to maintain the functions of the VAD. With this study, we can have personalized VAD considering the patient's disease and metabolism. In addition, the concept of SIS is essential to provide a reduction of risks that may interfere with the function of the VAD and the patient's life.

Author Contributions: A.C.M.C. manufactured the sensor; A.C.M.C. and J.C.D. carried out the experiments; A.C.M.C., D.J.S.F. and J.C.D. researched the data processing; All authors participated in the results analysis and discussion; A.C.M.C., A.J.P.A. and M.S.G.T. wrote the manuscript, organized and supervised the research. M.S.G.T. and J.R.C. are responsible for the funding acquisition. All authors have read and agreed to the published version of the manuscript.

Funding: This research was funded by CAPES/PROAP—Grant 817.757/38.860, CAPES PET, FSA, FAPESP (grant 2013/24434-0) and CNPq (MSG Tsuzuki was partially supported by grant 311.195/2019-9).

Institutional Review Board Statement: Not applicable.

Informed Consent Statement: Not applicable.

Data Availability Statement: Not applicable.

Conflicts of Interest: The authors declare no conflict of interest.

References

1. Wada, E.A.E.; Andrade, A.J.P.; Nicolosi, D.E.C.; Bock, E.G.P.; Fonseca, J.W.G.; Leme, J.; Dinkhuysen, J.J.; Biscegli, J.F. Review of the spiral pump performance test, during cardiopulmonary bypass, in 43 patients. In Proceedings of the Technology Meets Surgery International, ABCM, São Paulo, Brazil, 18–19 July 2005.
2. Andrade, A.; Nicolosi, D.; Lucchi, J.; Biscegli, J.; Arruda, A.C.; Ohashi, Y.; Mueller, J.; Tayama, E.; Glueck, J.; Nosé, Y. Auxiliary total artificial heart: A compact electromechanical artificial heart working simultaneously with the natural heart. *Artif. Organs* **1999**, *23*, 876–880. [CrossRef]
3. Ohashi, Y.; de Andrade, A.; Müller, J.; Nosé, Y. Control System Modification of an Electromechanical Pulsatile Total Artifical Heart. *Artif. Organs* **1997**, *21*, 1308–1311. [CrossRef]
4. DeVore, A.D.; Patel, P.A.; Patel, C.B. Medical Management of Patients With a Left Ventricular Assist Device for the Non-Left Ventricular Assist Device Specialist. *JACC Heart Fail.* **2017**, *5*, 621–631. [CrossRef] [PubMed]
5. Chase, J.G.; Tsuzuki, M.S.G.; Benyó, B.; Desaive, T. Editorial: Special Section on Biological Medical Systems. *Annu. Rev. Control* **2019**, *48*, 357–358. [CrossRef]
6. Shafiee, M.; Animah, I. Life extension decision making of safety critical systems: An overview. *J. Loss Prev. Process Ind.* **2017**, *47*, 174–188. [CrossRef]
7. Ghista, D.N.; Patil, K.M.; Gould, P.; Woo, K.B. Computerized left ventricular mechanics and control system analyses models relevant for cardiac diagnosis. *Comput. Biol. Med.* **1973**, *3*, 27–46. [CrossRef]
8. Antaki, J.F.; Boston, J.R.; Simaan, M.A. Control of Heart Assist Devices. In Proceedings of the 42nd IEEE Conference on Decision and Control, Maui, HI, USA, 9–12 December 2003; Volume 4, pp. 4084–4089.
9. Su, S.W.; Wangt, L.; Celler, B.G.; Savkin, A.V.; Guo, Y. Modelling and control for heart rate regulation during treadmill exercise. In Proceedings of the Annual International Conference of the IEEE Engineering in Medicine and Biology, New York, NY, USA, 30 August–3 September 2006; pp. 4299–4304.
10. Leão, T.F. Técnica de Controle Automático da Rotação de Bombas de assistência Ventricular. Ph.D. Thesis, Universidade de São Paulo, São Paulo, Brazil, 2015.
11. Cavalheiro, A.; Fo, D.S.; Andrade, A.; Cardoso, J.R.; Bock, E.; Fonseca, J.; Miyagi, P.E. Design of Supervisory Control System for Ventricular Assist Device. In *IFIP Advances in Information and Communication Technology*; Springer: Berlin/Heidelberg, Germany, 2011; Volume 349 AICT, pp. 375–382.
12. Tsuzuki, M.S.G.; Martins, T.C.; Takimoto, R.T.; Tanabi, N.; Sato, A.K.; Scaff, W.; Johansen, C.F.D.; Campos, C.A.T.; Kalynytschenko, E.; Silva, H.F.; et al. Mechanical Ventilator VENT19. *Polytechnica* **2021**, *4*, 33–46. [CrossRef]
13. Cooper, G.F.; Herskovits, E. A Bayesian method for the induction of probabilistic networks from data. *Mach. Learn.* **1992**, *9*, 309–347. [CrossRef]
14. Squillante Jr, R.; Santos Filho, D.J.; Garcia Melo, J.I.; Junqueira, F.; Miyagi, P.E. Safety instrumented system designed based on Bayesian network and Petri net. In Proceedings of the 8th International Conference on Mathematical problems in Engineering, Aerospace and Sciences (ICNPAA), São José dos Campos, Brazil, 30 June–3 July 2010.

15. IEC. *Functional Safety of Electrical/Electronic/Programmable Electronic Safety-Related Systems (IEC 61508)*, 1st ed.; International Electrotechnical Commission: Geneva, Switzerland, 1998.
16. Matsuno, H.; Tanaka, Y.; Aoshima, H.; Doi, A.; Matsui, M.; Miyano, S. Biopathways representation and simulation on hybrid functional Petri Net. *Silico Biol.* **2003**, *3*, 389–404.
17. Abdolrazaghi, M.; Navidbakhsh, M.; Hassani, K. Mathematical Modelling and Electrical Analog Equivalent of the Human Cardiovascular System. *Cardiovasc. Eng.* **2010**, *10*, 45–51. [CrossRef] [PubMed]
18. Murata, T. Petri Nets: Properties, Analysis and Applications. *Proc. IEEE* **1989**, *77*, 541–580. [CrossRef]
19. Li, Z.; Zhao, M. On controllability of dependent siphons for deadlock prevention in generalized Petri nets. *IEEE Trans. Syst. Man, Cybern. Part A Syst. Hum.* **2008**, *38*, 369–384.
20. Villani, E.; Miyagi, P.E.; Valette, R. Landing system verification based on petri nets and a hybrid approach. *IEEE Trans. Aerosp. Electron. Syst.* **2006**, *42*, 1420–1436. [CrossRef]
21. Ho, Y.C.; Society, I.C.S. *Discrete Event Dynamic Systems: Analyzing Complexity and Performance in the Modern World*; Institute of Electrical and Electronics Engineers: Piscataway, NJ, USA, 1992; p. 291.
22. Barari, A.; Tsuzuki, M.S.G.; Cohen, Y.; Macchi, M. Intelligent manufacturing systems towards industry 4.0 era. *J. Intell. Manuf.* **2021**, *32*, 1793–1796. [CrossRef]
23. Cavalheiro, A.C.; Fo, D.J.S.; Andrade, A.; Cardoso, J.R.; Horikawa, O.; Bock, E.; Fonseca, J. Specification of supervisory control systems for ventricular assist devices. *Artif. Organs* **2011**, *35*, 465–470. [CrossRef] [PubMed]
24. Cavalheiro, A.C.M. Sistema de Controle para Diagnóstico e Tratamento de Falhas em Dispositivos de Assistência Ventricular. Ph.D. Thesis, Universidade de São Paulo, São Paulo, Brazil, 2013.
25. Petri Net Based Engineer Tool; Version 2.7a, Copyright Dr. Rainer Drath. Visual Object Net. 2007. Available online: https://www.r-drath.de/Drath/Home/Visual_Object_Net++.html (accessed on 15 October 2021).
26. IEC. *Programmable Controllers Part 3, Programming Languages (IEC1131-3)*, 1st ed.; International Electrotechnical Commission: Geneva, Switzerland, 1993.
27. Felipini, C.L.; de Andrade, A.J.; Lucchi, J.C.; da Fonseca, J.W.; Nicolosi, D. An electro-fluid-dynamic simulator for the cardiovascular system. *Artif. Organs* **2008**, *32*, 349–354. [CrossRef] [PubMed]
28. Bock, E.; Andrade, A.J.; Dinkhuysen, J.; Arruda, C.; Fonseca, J.; Leme, J.; Utiyama, B.; Leao, T.; Uebelhart, B.; Antunes, P.; et al. Introductory tests to in vivo evaluation: Magnetic coupling influence in motor controller. *ASAIO J.* **2011**, *57*, 462–465. [CrossRef] [PubMed]
29. Silva, B.U.D.; Jatene, A.D.; Leme, J.; Fonseca, J.W.; Silva, C.; Uebelhart, B.; Suzuki, C.K.; Andrade, A.J. In vitro assessment of the apico aortic blood pump: Anatomical positioning, hydrodynamic performance, hemolysis studies, and analysis in a hybrid cardiovascular simulator. *Artif. Organs* **2013**, *37*, 950–953. [CrossRef] [PubMed]
30. Silva, C.D.; Silva, B.U.D.; Leme, J.; Uebelhart, B.; Dinkhuysen, J.; Biscegli, J.F.; Andrade, A.J.; Zavaglia, C. In vivo evaluation of centrifugal blood pump for cardiopulmonary bypass-spiral pump. *Artif. Organs* **2013**, *37*, 954–957. [CrossRef] [PubMed]
31. Dinkhuysen, J.J.; de Andrade, A.J.P.; Leme, J.; Silva, C.; Medina, C.S.; Pereira, C.C.; Biscegli, J.F. Clinical evaluation of the Spiral Pump® after improvements to the original project in patients submitted to cardiac surgeries with cardiopulmonary bypass. *Braz. J. Cardiovasc. Surg.* **2014**, *29*, 330–337.

Article

Water Content Monitoring in Water-in-Oil Emulsions Using a Piezoceramic Sensor

Carlos A. B. Reyna [1], Ediguer E. Franco [2], Alberto L. Durán [1], Luiz O. V. Pereira [3], Marcos S. G. Tsuzuki [1] and Flávio Buiochi [1,*]

[1] Department of Mechatronics and Mechanical Systems Engineering, Escola Politécnica da Universidade de São Paulo, São Paulo 05508-030, Brazil; carlosburbano@usp.br (C.A.B.R.); duran@usp.br (A.L.D.); mtsuzuki@usp.br (M.S.G.T.)
[2] Facultad de Ingeniería, Universidad Autónoma de Occidente, Cali 760030, Colombia; eefranco@uao.edu.co
[3] Research, Development and Innovation Petrobras S.A., Ilha do Fundão, Rio de Janeiro 21941-915, Brazil; luizoctavio@petrobras.com.br
* Correspondence: fbuiochi@usp.br

Abstract: This work deals with a transmission-reception ultrasonic technique for the real-time estimation of the water content in water-in-crude oil emulsions. The working principle is the measurement of the propagation velocity, using two in-house manufactured transducers designed for water coupling, with a central frequency of about 3 MHz. Water-in-crude oil emulsions with a water volume concentration from 0% to 40% were generated by mechanical emulsification. Tests were carried out at three temperatures. The results showed that the propagation velocity is a sensitive parameter that is able to determine the water content, allowing for differentiating the concentrations of up to 40% of water. The main motivation is the development of techniques for non-invasive and real-time monitoring of the water content of emulsions in petrochemical processes.

Keywords: water-in-crude oil emulsion; water content; ultrasound; propagation velocity

Citation: Reyna, C.A.B.; Franco, E.E.; Durán, A.L.; Pereira, L.O.V.; Tsuzuki, M.S.G.; Buiochi, F. Water Content Monitoring in Water-in-Oil Emulsions Using a Piezoceramic Sensor. *Machines* **2021**, *9*, 335. https://doi.org/10.3390/machines9120335

Academic Editor: Dan Zhang

Received: 17 October 2021
Accepted: 2 December 2021
Published: 6 December 2021

Publisher's Note: MDPI stays neutral with regard to jurisdictional claims in published maps and institutional affiliations.

Copyright: © 2021 by the authors. Licensee MDPI, Basel, Switzerland. This article is an open access article distributed under the terms and conditions of the Creative Commons Attribution (CC BY) license (https://creativecommons.org/licenses/by/4.0/).

1. Introduction

Petroleum is a mixture of organic compounds formed by the anaerobic decomposition of organic sediments in natural geological cavities or wells. The extraction techniques significantly modify its composition, adding an aqueous phase not miscible with the organic mixture and producing a complex emulsion [1]. Water is a contaminant in all oil derivatives, decreasing the quality of the resulting fuels [2], also implying higher pollution, which is a hindrance to the processes of transport, storage, and distillation. In addition, the water carries a wide variety of mineral solutes (sea salt, magnesium, and silicon, mainly) that damage pipes, valves, and pumps. The water is withdrawn from the raw oil, inducing coalescence, a process that consists of merging two or more droplets during contact to form a larger droplet [3,4]. This process may be enhanced by chemical substances. The final water content is a critical parameter in the subsequent petrochemical processing.

Characterization techniques for emulsions include electron and light microscopy, light and neutron scattering, electrical conductivity, and nuclear magnetic resonance [5,6]. Most of these techniques are only suitable for diluted and non-opaque emulsions, conditions not met by water-in-crude oil emulsions [6]. Other common methods for determining water content in crude oil, such as centrifugation [7], Karl Fisher's distillation, and grinding methods [8], require extracting a sample from the pipeline for further processing in a laboratory. These laboratory tests are time consuming and delay the processing and transportation of crude oil.

The real-time monitoring of water concentration before and after the coalescence process is of particular interest to the petrochemical industry [1]. Ultrasound can be useful for characterizing emulsions because it is robust, relatively inexpensive, easy to operate, allows characterizing opaque liquids and it provides in-line and real-time monitoring of emulsion stability evolution using a multi-backscattering sensor [9] and sand production monitoring using a wideband vibration sensor [10]. Ultrasound-based techniques have been used for characterizing liquids, such as edible oils, honey, polymer resins and motor oils [11–13], as well as for process monitoring, such as polymer and concrete curing [14,15]. As regards wave generation and reception methods, the literature reports the use of compact thick-film piezoelectric transducers [16], laser techniques for both the generation and reception of ultrasonic pulses [17] and conventional ultrasonic transducers [18].

In the case of multi-phase fluids, ultrasonic techniques have been used to characterize and to monitor some physical properties of emulsions, suspensions and slurries [19–21]. Other interesting works deal with the monitoring of a multiphase oil-water-gas flow directly in a pipe [18] and the detection of oily contaminants in water courses [22]. These works show the interest in the subject and the variety of possible approaches.

Water-in-crude oil emulsions showed a water concentration in a volume of 40% for droplet size distribution ranging from 0.4 µm to 40 µm [23] and a water concentration in a volume of up to 30% for droplet size distribution from 1 µm to 10 µm [1,5]. This characteristic, in addition to a high variation in their chemical composition, could lead to a dispersive medium with high attenuation. Under these conditions, the best approach must be the monitoring of an average acoustic parameter at a suitable operating frequency. Ultrasonic spectroscopy allows the determination of the distribution of droplet sizes and concentration by measuring the propagation velocity and attenuation spectra [24,25]. This is a well-established technique, useful in monomodal droplet size distribution, and restricted to diluted emulsions (volume fraction of the dispersed phase is less than 1%) [26]. In a recent work [27], the authors used acoustic models and measured attenuation spectra to estimate the droplet size distribution in water-in-sunflower oil emulsions. They reported droplet size distributions of 0.4–5, 0.4–8, 0.4–15, 0.4–12 and 0.4–100 µm for water volume fractions of 0.1, 0.2, 0.3, 0.4 and 0.5, respectively. The droplet size distribution results of the ultrasonic spectroscopy for emulsions of moderate concentrations up to 20% were very close to the experimental data obtained by using laser diffraction.

The measurement of the propagation velocity of ultrasonic waves has been used to infer the physical properties of water-in-crude oil emulsions. In 2021, a novel multi-backscattering sensor with a simple signal processing methodology, which allows the measurement of the propagation velocity, was proposed to monitor water-in-crude-oil emulsions [28]. The ultrasonic multiple-backscattering sensor consists of a 3.5-MHz transducer and a set of thin cylindrical scatterers located in the near field. The results from this experimental arrangement showed an almost linear behavior of the propagation velocity over a volumetric water concentration from 0% to 50%. This interesting result was corroborated in the present work.

This paper presents an ultrasonic technique to estimate water concentration in water-in-crude oil emulsions. The working principle is the determination of the time of flight of ultrasonic waves between two custom-made transducers. The sensing device developed was initially tested with static samples to establish the measurement methodology. Other measurements were carried out with the sample being stirred by a laboratory mixer. This was done to maximize the droplet interaction, accelerating the coalescence process. The main motivation is the development of compact, inexpensive, and chemically and mechanically resistant devices, which could be attached to pipes or valves in the oil process lines for on-line and real-time monitoring.

2. Theoretical Background

Although emulsions are classified as continuous materials, they have local effects that generate a complex acoustic behavior. If the mean diameter of the dispersed phase droplets is smaller than the wavelength, the local acoustic phenomena converge to a wavefront traveling through the mixture with constant velocity. In this case, a simple acoustic propagation model relates the propagation velocity and the concentration, establishing that the total propagation time of an ultrasonic wave through a heterogeneous mixture is the sum of the times in each phase [29] (see Figure 1):

$$t_e = t_w + t_o = \frac{X_w}{c_w} + \frac{X_o}{c_o} = \frac{X_e}{c_e}, \quad (1)$$

where t is the propagation time, X is the wave path length, c is the propagation velocity and subscripts e, w and o refer to emulsion, water and oil, respectively. The relationship between propagation velocity c in the emulsion and water volume fraction ϕ is:

$$c_e = \frac{1}{\frac{\phi}{c_w} + \frac{1-\phi}{c_o}}. \quad (2)$$

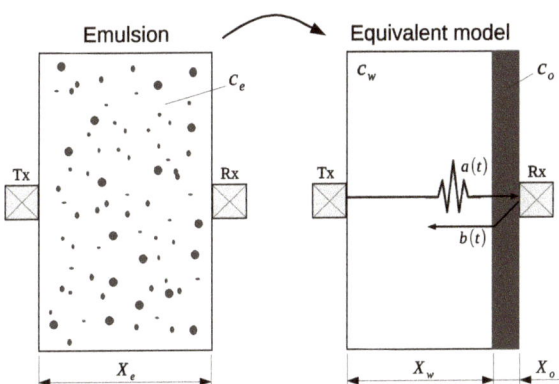

Figure 1. Measurement scheme, showing the mixture model for the propagation velocity in the emulsion (Urick's model) and the arrangement of the ultrasonic transducers and the transmitted $a(t)$ and reflected $b(t)$ signals.

Figure 1 also shows the ultrasonic transducer arrangement, composed of an emitter/receiver (Tx) and a receiver (Rx) and reflector. Transducer Tx is used in pulse-echo mode while transducer Rx operates as a receiver. The excitation of Tx generates wave $a(t)$ that propagates through the sample and reaches Rx. The part of the wave reflected from the Rx face generates signal $b(t)$, which returns to Tx. This configuration allows the correlation of $a(t)$ and $b(t)$ with a shortest path (X_e) between them and minimizes the insertion loss reflections by using other materials (steel or aluminum). As distance X_e is known, time delay t_e between signals $a(t)$ and $b(t)$ allows the determination of propagation velocity c_e in the emulsion.

3. Materials and Methods

3.1. Ultrasonic Probe

The measurement probe, shown in Figure 2a, was manufactured using two square piezoelectric ceramics (10 mm sides and 0.36 mm thickness) of Pz37 (Ferroperm Piezoceramics A/S, Kvistgard, Denmark). The resulting transducers were tuned close to 3 MHz. Each probe transducer was manufactured in an ABS housing made in a 3D printer. The backing layer was made of epoxy resin (Araldite GY 279BR and Aradur HY 9BR, Huntsman, Brazil, in a ratio 10:1 by weight) mixed with alumina powder (1 μm, Buehler, IL, USA) in a concentration of 30% by weight to increase the attenuation in the backing layer, avoiding the overlapping signals in reception and increasing the axial resolution. The matching layer was made with the same epoxy resin used in the backing layer, but without alumina powder, such that the matching was not complete. This way, the receptor can receive part of the energy but still reflect part of it to the emitter. There is a compromise here, because the pair of transducers can be used interchangeably (emitter or receptor). The matching layer thickness was approximately a quarter wavelength ($\lambda/4 \cong 1.5$ mm). The transducer elements were aligned with a metallic spacer to ensure good parallelism. The distance between the transducers was 30 mm, for acoustic echoes not to overlap when inserted into water. The propagation velocity in water is a well known property [30–34]. Then, the ultrasonic sensor was calibrated using distilled water as a reference substance. The calibration provides a more accurate value of the distance between transducers Rx and Tx, which is required in the signal processing algorithm. Both transducers were tested in emission and in reception mode with slight differences in the measured delay.

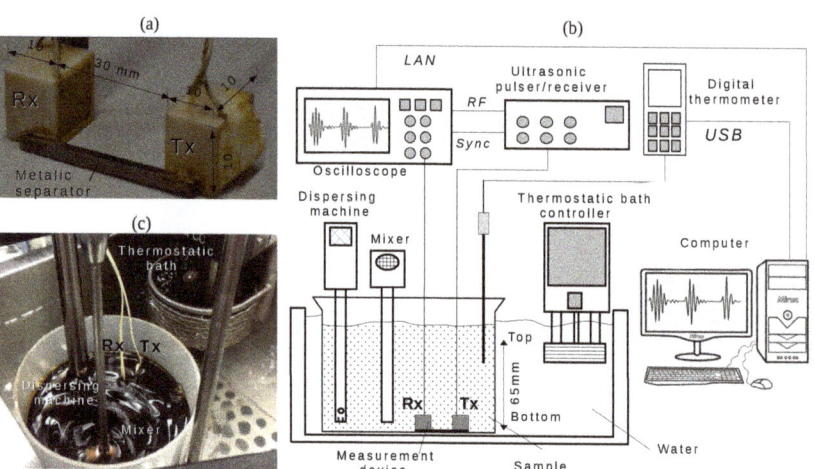

Figure 2. Experimental setup: (**a**) Image of the in-house manufactured ultrasonic sensor (**b**) schematic representation and (**c**) image of the thermostatic bath with the emulsion under test.

3.2. Experimental Setup

Figure 2b shows a schematic representation of the experimental setup. The emitter transducer (Tx), working in pulse-echo mode, was connected to an ultrasonic pulser/receiver (Olympus Panametrics model 5077-PR, Waltham, MA, USA), and the receiver transducer (Rx) was directly connected to a channel of the oscilloscope. The digital oscilloscope (Agilent Technologies, model 5042, Santa Clara, CA, USA) was used to digitize the ultrasonic signals. The sample temperature was measured using a digital thermometer (DeltaOHM, model HD2107.2, Caselle di Selvazzano (PD), Italy). Both the oscilloscope and the digital thermometer were connected to a desktop computer allowing the simultaneous acquisition and storage of the ultrasonic signals and temperature via LAN network.

The temperature of the experiment was controlled by a thermostatic bath (Huber, CC-106A) with an accuracy of 0.1 °C. The emulsion was stored in an 800-mL beaker partially immersed in the thermostatic bath. The ultrasonic sensor and the thermometer were inserted into the beaker. For homogenizing the emulsion, a dispersing machine (IKA Labortechnick, model T25, Staufen, Germany) was used at 8600 rpm. In addition, a 200-rpm mixer (Fisatom, model 711, São Paulo—SP, Brazil) was used to agitate the sample when measurements with moving fluid were desired. The mixer and emulsifier blades were completely inserted in the fluid to reduce the effect of air bubbles. Figure 2c shows an image of the experiment in the thermostatic bath. The sample can be seen spinning by the action of the mixer.

Figure 3 shows the waveforms of the ultrasonic pulses received by both the Rx and Tx transducers in pure water. The signal received by Tx has a different waveform and has a lower amplitude when compared to the signal received by Rx. The lower amplitude can be attributed to the energy loss inside transducer Rx, which is a good receiver. Reverberation inside the transducer layers results in a set of signals that are added to produce the distorted waveform reflected by Rx and received by Tx.

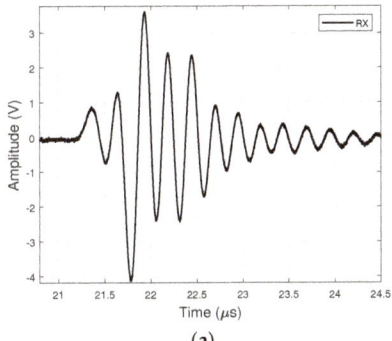
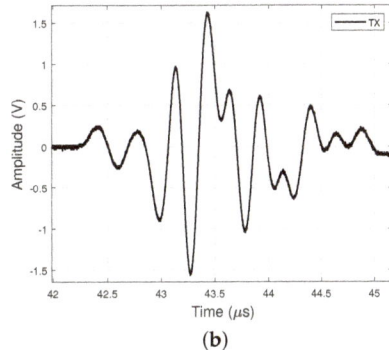

Figure 3. Signals obtained from transducers with water as the propagation medium: (**a**) Rx and (**b**) Tx.

Salt-water mixtures of different concentrations were used to test the measurement setup. The tests were carried out at 25 °C and the results were compared with others reported in the literature [30]. In the experiment, successive concentrations were obtained by adding 5 g of salt to a solution previously prepared. It started with 300 g of water and ended by achieving a salt concentration of 18.9% by weight. The solutions for each concentration were mixed for 150 s to completely dissolve the salt. At rest, ultrasonic signals were acquired to determine the propagation velocity of each solution. This procedure was repeated three times, and the averages and standard deviations of the propagation velocities were obtained.

3.3. Experimental Procedure

In petrochemical processes, measurements are required under flow conditions. The flow can affect the path of the acoustic waves, distorting the recorded signals. Experiments carried out at rest and with stirring allow comparison of the stability of the measurements obtained in these conditions. In this work, the flow condition was simulated by a stirring impeller (mixer) with three blades of radius 30 mm placed inside a beaker of diameter of 150 mm. The angular velocity of the mixer was 200 rpm, a relatively low value to reduce the probability of cavitation in the sample, which could lead to the evaporation of the most volatile compounds in the crude oil.

The first experimental procedure used to test the emulsions started by pouring 300 mL of crude oil (30.5° API, well LL83, Mangaratiba, Petrobras, Brazil) into a beaker partially immersed in the thermal bath. The amount of water required for the desired concentration was added to the beaker. The dispersing machine (8600 rpm) was turned on for 80 s. Then, the mixture was left to stand for 40 s to minimize the effect of the heat and most air bubbles generated by the dispersing machine. The ultrasonic sensor was placed on the bottom of the beaker. Finally, 310 acquisitions, considering ultrasonic signals and temperature data, were taken during approximately 160 min. The measurement process was carried out for three water volume fractions ($\phi = 0.12, 0.21, 0.29$) at 25 °C. All these measurements were repeated with the mixer turned on at 200 rpm.

For the concentration of $\phi = 0.29$, the experiments were repeated with the emulsion samples at rest and in motion with the probe positioned close to the top surface of the emulsion (65 mm from the bottom of the beaker, Figure 2b). For the signals to be acquired, the mixer was turned off for static measurements in the sample and turned on for moving measurements in the same sample.

To test a greater amount of concentration, a second experimental procedure was carried out. The measurements of water-in-crude oil emulsion were performed in the range of water volume fraction from 0 to 0.4 and three different temperatures (20, 25 and 30 °C). Each measurement was made after adding water to a previously prepared emulsion and turning on the dispersing machine for 80 s, obtaining a new concentration. The emulsification process of the samples, followed by a rest period, was the same as the first experimental procedure described above, but in this case, only 30 ultrasonic signals were acquired during 110 s. The measurements were repeated with the mixer turned on at 200 rpm. For each temperature, two samples were used for static acquisitions. After that, for each of the same three temperatures, two samples were used for dynamic acquisitions (mixer turned on at 200 rpm).

4. Results and Discussion

In Figure 4, the propagation velocities measured for salt-water mixtures were plotted as a function of the salt concentration and were compared with the theoretical velocity curve [35]. The measurements obtained by the ultrasonic sensor are in good agreement with the values in the literature [30] and the standard deviation is small. The linear behavior of the propagation velocity as a function of salt content was confirmed. However, there is a difference in the slope, although the maximum deviation between the theoretical and experimental cases is 0.6% at the highest concentration (19% by weight). These results show the measurement technique has good agreement in the case of solutions.

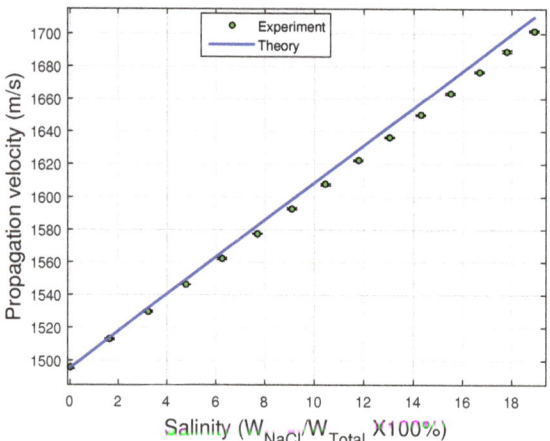

Figure 4. Propagation velocity for saline solutions at 25 °C as a function of salt concentration by weight compared to the theoretical model obtained from [35].

Figure 5a shows the propagation velocity in the emulsion as a function of the acquisition time. The acquisition time was 160 min, beginning when the rest time ends (40 s). The results for three concentrations ($\phi = 0.12, 0.21, 0.29$) and for both the static and moving cases are shown. In all these cases, a reduction in propagation velocity is observed in the first 10–20 min. From the minimum value observed at the beginning of the curve, the velocity rises; at the end of the acquisition time, there is an increase of 0.3% and 0.7% for the static and moving cases, respectively. This increase in velocity is relatively small and can be explained by physical changes in the emulsion, mainly coalescence. The behavior is similar in all three concentrations. However, it is evident that a steady state value was not reached at the end of the acquisition time.

Figure 5b shows the propagation velocity as a function of the acquisition time for the transducer placed at the top (close to the free surface of the sample) and at the bottom of the beaker, and for static and moving cases. Measurements at the top and bottom of static and moving samples were performed using a water volume fraction of 0.29. The results show that, in the static case, the propagation velocity curves are almost identical for both sensor positions, whereas in the case of the emulsion under stirring, the curves were different. When the sensor is located at the bottom of the container, the propagation velocity is slightly higher throughout the experiment. Furthermore, after the minimum value presented at the beginning of the curve, the propagation velocity value increases at a higher rate than in the other three cases and presents less dispersion than in the static case. When the sensor is located at the top, the curve is similar to that observed in the static case. The increase in velocity after the minimum value is linear and presents less dispersion.

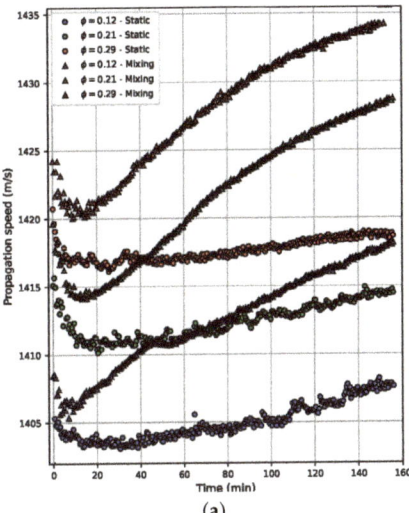
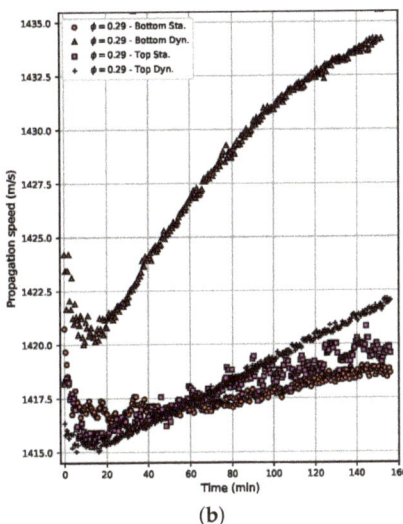

Figure 5. Propagation velocity as a function of acquisition time at 25.0 °C: (**a**) comparison of static and in motion cases and (**b**) measurements at top (near free surface) and bottom of the sample for $\phi = 0.29$.

The propagation velocity increased at a higher rate in the moving emulsion with the sensor located at the bottom of the container, when compared to the moving emulsion with the sensor located at the top. This happened due to the increase in coalescence. The movement of the emulsion increases the probability of the water droplets to collide and merge, generating larger water droplets that move towards the bottom of the container under the action of gravity [36].

Figure 6 shows the propagation velocity in the emulsions as a function of water volume fraction ($\phi = 0.12, 0.21$ and 0.29) at three time instants ($t = 2, 80, 160$ min) for both static and moving cases. The theoretical propagation velocity according to the Urick mixture model is shown as a solid line. The results for the static case are closer to the theoretical behavior than the moving case for all instants of time. With the mixer turned on (in the motion case), an almost constant velocity difference (offset value) that depends on time is observed. At the first time instant (2 min), the velocity values have a maximum difference of just 0.3% with respect to the theoretical values. Thus, to ensure that the emulsion concentration does not change due to coalescence, the characterization could be carried out two minutes after the emulsification process and the rest time.

Figure 7a shows the average of the 30 measurements taken during the first 110 s (after emulsification and resting steps) at three temperatures for the static case. The measurement results were compared with their respective theoretical curves obtained with Urick's model (solid lines). The results are in agreement with the theoretical values, showing the same almost linear trend of the model and allowing a clear differentiation of measurements at the three temperatures. The differences in the mean values of the propagation velocity between static and in-motion emulsions (see Figure 7b) at the same concentration are less than 0.16%. These results show that the difference in the propagation velocity of static and moving emulsions is small when the characterization is made immediately after the emulsification and resting (40 s) processes ends. The propagation velocities measured at 20 °C and 25 °C (red square and blue triangle data) in the static emulsions are similar to the experimental results of propagation velocity in water-in-crude-oil emulsions shown in [28], in which the sensor uses a single transducer, a set of steel bars as a reflector and its volume is ten times greater in comparison to the prototype proposed herein. The same base crude oil used to produce the samples here were used in [28], but there the experiments were carried out at 22 °C with water content from $\phi = 0$ to $\phi = 0.37$, with an ultrasonic multiple-backscattering sensor.

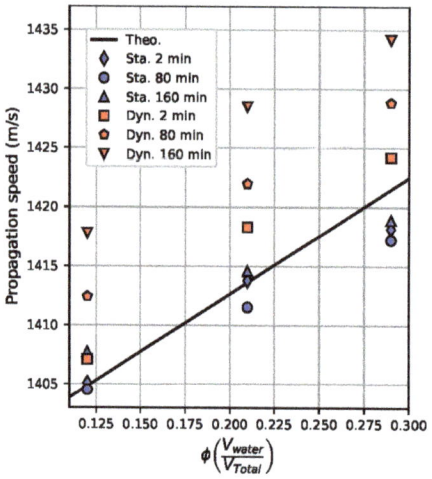

Figure 6. Urick model comparison at $T = 25$ °C. Considering the propagation velocity, the maximum absolute deviation from the mean is 8.1 $\frac{m}{s}$ ($dev_{max} = 0.6\%$) at $\phi = 0.29$.

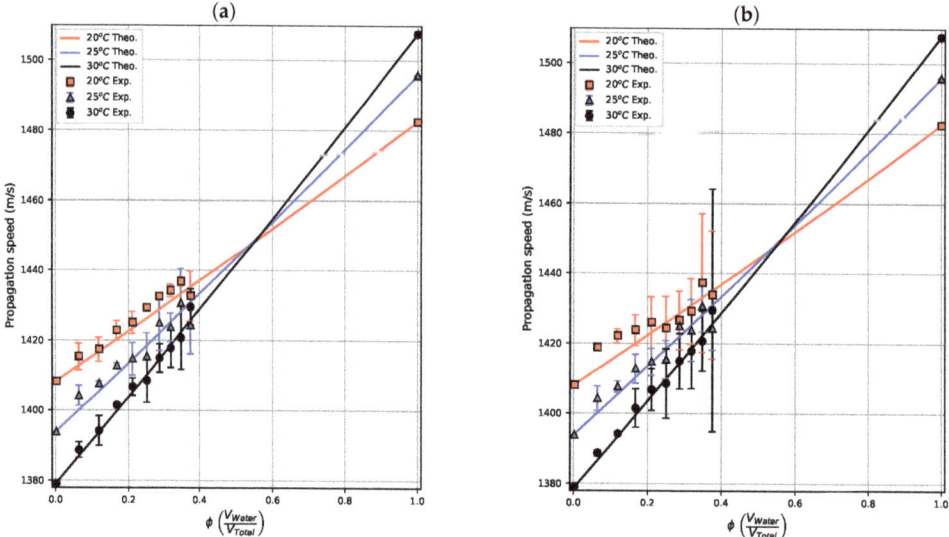

Figure 7. (**a**) Mixer turn off ($dev_{max} = 1.3\%$) and (**b**) Mixer turn on ($dev_{max} = 4.8\%$).

The results from the emulsion in motion showed greater standard deviations at the highest water concentrations. This behavior shows greater instability of the physical properties of the emulsion at these concentrations. Tests carried out with the mixer turned on and with a blade tip speed greater than 1 m/s showed widely dispersed results (not shown here) for all concentrations.

In the case of moving samples, the propagation velocity variations in the emulsions for $\phi > 0.25$ may be a consequence of the formation of water droplets with sizes close to wavelength, increasing the acoustic diffraction. This effect can also increase when temperature or flow velocity increases because the organic gases dissolved in the oil can form bubbles that can affect the wave propagation.

In Figure 5a, the tested emulsions were prepared from mixtures of pure oil and water. In Figure 7a,b, the tested samples were prepared by adding water to the emulsion previously used, obtaining new concentrations. In Figure 5a, the propagation velocity values for $\phi = 0.12$, 0.21 and 0.29 obtained during the first 5 min at 25 °C are similar to those obtained for the same concentrations at 25 °C in Figure 7 (blue triangles). Although the method of emulsion preparation was different from the results presented in Figures 5a and 7, close values were obtained, suggesting that the droplet behavior is similar regardless of the method of emulsion preparation.

5. Conclusions

A feasibility study of the real-time estimation by ultrasound of the water content in oil in static and moving emulsions, taking into account the temperature effect, was presented. The methodology employed made it possible to measure the emulsion properties under the proposed conditions. The presented technique is a variation of acoustic characterization of materials, widely used in liquids and solids. It allows the measurement of water content emulsified with crude oil, showing results close to those described by the Urick model. Water concentration measurements in stirring emulsions show a greater standard deviation compared to emulsions at rest. However, the mean velocity values are similar for both static and moving emulsions. Although this technique is not useful to obtain the correct instantaneous values of water concentration before 3 min after emulsification, an average of 30 measurements or more, obtained over that initial stage, could allow the measurement. The measuring accuracy of samples at different temperatures can be significantly increased in static emulsions. The depth positioning of the probe is relevant in the case of movement.

The determination of the water concentration in the emulsions within small time intervals (minutes) in relation to other methods (hours) indicates that the acoustic technique proposed herein could be useful for monitoring water in crude oil emulsions in real-time, before and after the induced coalescence processes that occur in the petrochemical industry.

Author Contributions: C.A.B.R. manufactured the sensor; C.A.B.R. and A.L.D. carried out the experiments; C.A.B.R., E.E.F. and F.B. researched the data processing; All the authors participated in the results analysis and discussion; C.A.B.R., E.E.F. and F.B. wrote the manuscript, organized and supervised the research. M.S.G.T. and L.O.V.P. are responsible for the funding acquisition. All authors have read and agreed to the published version of the manuscript.

Funding: This research was funded by Petrobras/ANP (grant 5850.0108871.18.9), CAPES/PROAP— Grant 817.757/38.860 and CNPq (MSG Tsuzuki was partially supported by grant 311.195/2019-9). EE Franco had financial support from the Universidad Autónoma de Occidente (Colombia) by grant 19INTER-306.

Conflicts of Interest: The authors declare no conflict of interest.

References

1. Shah, A.; Fishwick, R.; Wood, J.; Leeke, G.; Rigby, S.; Greaves, M. A review of novel techniques for heavy oil and bitumen extraction and upgrading. *Energy Environ. Sci.* **2010**, *3*, 700–714. [CrossRef]
2. Thomas, J.E. *Fundamentos de Engenharia de Petróleo*; Interciência: Rio de Janeiro, Brazil 2001.
3. Rio, E.; Biance, A.L. Thermodynamic and mechanical timescales involved in foam film rupture and liquid foam coalescence. *ChemPhysChem* **2014**, *15*, 3692–3707. [CrossRef]

4. Critello, D.C.; Pullano, S.A.; Gallo, G.; Matula, T.J.; Fiorillo, A.S. Low frequency ultrasound as a potentially viable foaming option for pathological veins. *Colloids Surf. A Physicochem. Eng. Asp.* **2020**, *599*, 124919. [CrossRef]
5. Morgan, V.G.; Sad, C.; Constantino, A.F.; Azeredo, R.B.; Lacerda, V.; Castro, E.V.; Barbosa, L.L. Droplet size distribution in water-crude oil emulsions by low-field NMR. *J. Braz. Chem. Soc.* **2019**, *30*, 1587–1598. [CrossRef]
6. Raigan, K. Ultrasonic Techniques for Characterizations of Oils and Their Emulsions and Monitoring Oil Layer Depth of Spill Emulsions and Monitoring Oil Layer Depth of Spill. Ph.D. Thesis, The University of Western Ontario, London, ON, Canada, 2020.
7. Jadoon, S.; Malik, A.; Amin, A.A. Separation of Sediment Contents and Water from Crude Oil of Khurmala and Guwayer Oil Fields in Kurdistan Region by using Centrifuge Method. *Int. J. Adv. Eng. Res. Sci.* **2017**, *4*, 2919–2922. [CrossRef]
8. Ivanova, P.G.; Aneva, Z.V. Assessment and assurance of quality in water measurement by coulometric Karl Fischer titration of petroleum products. *Accredit. Qual. Assur.* **2006**, *10*, 543–549. [CrossRef]
9. Perez, N.; Blasina, F.; Buiochi, F.; Duran, A.; Adamowski, J. Evaluation of a multiple scattering sensor for water-in-oil emulsion monitoring. In Proceedings of the Meetings on Acoustics ICU. Acoustical Society of America, San Diego, CA, USA, 2–6 December 2019; Volume 38, p. 055007.
10. Wang, K.; Liu, Z.; Liu, G.; Yi, L.; Yang, K.; Peng, S.; Chen, M. Vibration sensor approaches for the monitoring of sand production in Bohai bay. *Shock Vib.* **2015**, *2015*, 591780.
11. Runrun, H.; Runyang, M.; Chenghui, W.; Jing, H. Ultrasonic shear-wave reflectometry applied to monitor the dynamic viscosity of reheated edible oil. *J. Food Process. Eng.* **2020**, *43*, e13402. [CrossRef]
12. Franco, E.E.; Adamowski, J.C.; Higuti, R.T.; Buiochi, F. Viscosity measurement of Newtonian liquids using the complex reflection coefficient. *IEEE Trans. Ultrason. Ferroelectr. Freq. Control* **2008**, *55*, 2247–2253. [CrossRef]
13. Kazys, R.; Voleisis, A.; Sliteris, R. Investigation of the Acoustic Properties of Viscosity Standards. *Arch. Acoust.* **2016**, *41*, 55–58. [CrossRef]
14. Alig, I.; Lellinger, D.; Sulimma, J.; Tadjbakhsch, S. Ultrasonic shear wave reflection method for measurements of the viscoelastic properties of polymer films. *Rev. Sci. Instrum.* **1997**, *68*, 1536–1542. [CrossRef]
15. Wang, X.; Subramaniam, K.V.; Lin, F. Ultrasonic measurement of viscoelastic shear modulus development in hydrating cement paste. *Ultrasonics* **2010**, *50*, 726–738. [CrossRef] [PubMed]
16. Meng, G.; Jaworski, A.J.; White, N.M. Composition measurements of crude oil and process water emulsions using thick-film ultrasonic transducers. *Chem. Eng. Process. Process. Intensif.* **2006**, *45*, 383–391. [CrossRef]
17. Lu, Z.Q.; Yang, X.; Zhao, K.; Wei, J.X.; Jin, W.J.; Jiang, C.; Zhao, L.J. Non-contact Measurement of the Water Content in Crude Oil with All-Optical Detection. *Energy Fuels* **2015**, *29*, 2919–2922. [CrossRef]
18. Chillara, V.K.; Sturtevant, B.T.; Pantea, C.; Sinha, D.N. Ultrasonic sensing for noninvasive characterization of oil-water-gas flow in a pipe. *AIP Conf. Proc.* **2017**, *1806*, 090014.
19. Higuti, R.T.; Bacaneli, E.; Furukawa, C.M.; Adamowski, J.C. Ultrasonic characterization of emulsions: Milk and water in oil. In Proceedings of the 1999 IEEE Ultrasonics Symposium, Caesars Tahoe, NV, USA, 17–20 October 1999; International Symposium (Cat. No. 99CH37027); IEEE: Piscataway, NJ, USA, 1999; Volume 1, pp. 779–782.
20. McClements, D.J. Ultrasonic characterizations of emulsions and suspensions. *Adv. Colloid Interface Sci.* **1997**, *37*, 33–72. [CrossRef]
21. Stolojanu, V.; Prakash, A. Cgaracterization of slurry systems by ultrasonic techniques. *Chem. Eng. J.* **2001**, *84*, 215–222. [CrossRef]
22. Franco, E.E.; Adamowski, J.C.; Buiochi, F. Ultrasonic sensor for the presence of oily contaminants in water. *Dyna* **2012**, *79*, 4–9.
23. Maia Filho, D.C.; Ramalho, J.B.; Spinelli, L.S.; Lucas, E.F. Aging of water-in-crude oil emulsions: Effect on water content, droplet size distribution, dynamic viscosity and stability. *Colloids Surf. A Physicochem. Eng. Asp.* **2012**, *396*, 208–212. [CrossRef]
24. Richter, A.; Voigt, T.; Ripperger, S. Ultrasonic attenuation spectroscopy of emulsions with droplet sizes greater than 10 microm. *J. Colloid Interface Sci.* **2007**, *315*, 482–492. [CrossRef] [PubMed]
25. Su, M.; Cai, X.; Xue, M.; Dong, L.; Xu, F. Particle sizing in dense two-phase droplet systems by ultrasonic attenuation and velocity spectra. *Sci. China Ser. E Technol. Sci.* **2009**, *52*, 1502–1510. [CrossRef]
26. Dukhin, A.S.; Goetz, P.J. (Eds.) *Characterization of Liquids, Nano- and Microparticulates, and Porous Bodies Using Ultrasound*, 2nd ed.; Studies in Interface Science 24; Elsevier: Amsterdam, The Netherlands, 2010.
27. Silva, C.A.; Saraiva, S.V.; Bonetti, D.; Higuti, R.T.; Cunha, R.L.; Pereira, L.O.; Silva, F.V.; Fileti, A.M. Application of acoustic models for polydisperse emulsion characterization using ultrasonic spectroscopy in the long wavelength regime. *Colloids Surf. A Physicochem. Eng. Asp.* **2020**, *602*, 125062. [CrossRef]
28. Durán, A.L.; Franco, E.E.; Reyna, C.A.; Pérez, N.; Tsuzuki, M.S.; Buiochi, F. Water Content Monitoring in Water-in-Crude-Oil Emulsions Using an Ultrasonic Multiple-Backscattering Sensor. *Sensors* **2021**, *21*, 5088. [CrossRef] [PubMed]
29. Urick, R.J. A Sound Velocity Method for Determining the Compressibility of Finely Divided Substances. *J. Appl. Phys.* **1947**, *18*, 983–987. [CrossRef]
30. Hubbard, J.C.; Loomis, A. CXXII. The velocity of sound in liquids at high frequencies by the sonic interferometer. *Lond. Edinb. Dublin Philos. Mag. J. Sci.* **1928**, *5*, 1177–1190. [CrossRef]
31. Adamowski, J.C.; Buiochi, F.; Tsuzuki, M.S.G.; Pérez, N.; Camerini, C.S.; Patusco, C. Ultrasonic measurement of micrometric wall-thickness loss due to corrosion inside pipes. In Proceedings of the IEEE International Ultrasonics Symposium (IUS), Prague, Czech Republic, 21–25 July 2013; pp. 1881–1884.

32. Takimoto, R.Y.; Matuda, M.; Lavras, T.; Adamoswki, J.; Pires, G.; Peres, N.; Ueda, E.; Tsuzuki, M.S.G. An echo analysis method for the ultrasonic measurement of micrometric wall-thickness loss inside pipes. In Proceedings of the 2018 13th IEEE International Conference on Industry Applications, São Paulo, Brazil, 12–14 November 2018; pp. 999–1003.
33. Duran, A.L.; Sato, A.K.; Silva, A.M., Jr.; Franco, E.E.; Buiochi, F.; Martins, T.C.; Adamowski, J.C.; Tsuzuki, M.S.G. GPU Accelerated Acoustic Field Determination for a Continuously Excited Circular Ultrasonic Transducer. In Proceedings of the 21st IFAC World Congress, Berlin, Germany, 11–17 July 2020; pp. 10480–10484.
34. Takimoto, R.Y.; Matuda, M.Y.; Oliveira, T.F.; Adamowski, J.C.; Sato, A.K.; Martins, T.C.; Tsuzuki, M.S.G. Comparison of optical and ultrasonic methods for quantification of underwater gas leaks. In Proceedings of the 21st IFAC World Congress, Berlin, Germany, 11–17 July 2020; Volume 53, pp. 16721–16726.
35. Kleis, S.; Sanchez, L. Dependence of speed of sound on salinity and temperature in concentrated NaCl solutions. *Sol. Energy* **1990**, *45*, 201–206. [CrossRef]
36. Alshaafi, E. Ultrasonic Techniques for Characterization of Oil-Water Emulsion and Monitoring of Interface in Separation Vessels and Monitoring of Interface in Separation Vessels. Master's Thesis, The University of Western Ontario, London, ON, Canada, 2017.

Article

Data Analytics for Noise Reduction in Optical Metrology of Reflective Planar Surfaces

Cody Berry [1], Marcos S. G. Tsuzuki [2] and Ahmad Barari [1,*]

[1] Advanced Digital Design, Manufacturing, and Metrology Laboratories (AD2MLabs), Department of Mechanical and Manufacturing Engineering, Ontario Tech University, Oshawa, ON 2000, Canada; cody.berry@uoit.ca

[2] Laboratory of Computational Geometry, Department of Mechatronics and Mechanical Systems Engineering, Escola Politécnica da Universidade de São Paulo, São Paulo 05508-030, Brazil; mtsuzuki@usp.br

* Correspondence: ahmad.barari@uoit.ca

Abstract: On-line data collection from the manufactured parts is an essential element in Industry 4.0 to monitor the production's health, which required strong data analytics. The optical metrology-based inspection of highly reflective parts in a production line, such as parts with metallic surfaces, is a difficult challenge. As many on-line inspection paradigms require the use of optical sensors, this reflectivity can lead to large amounts of noise, rendering the scan inaccurate. This paper discusses a method for noise reduction and removal in datapoints resulting from scanning the reflective planar surfaces. Utilizing a global statistic-based iterative approach, noise is gradually removed from the dataset at increasing percentages. The change in the standard deviation of point-plane distances is examined, and an optimal amount of noisy data is removed to reduce uncertainty in representing the workpiece. The developed algorithm provides a fast and efficient method for noise reduction in optical coordinate metrology and scanning.

Keywords: coordinate metrology; optical scanning; noise reduction; digital manufacturing; integrated inspection system; data analytics; uncertainty

1. Introduction

The fourth industrial revolution demands for intelligence in manufacturing when dynamic data collection and data analytics are needed to support learning the production condition, prognostics, and production health monitoring. Intelligence in these complex processes is generated based on accurate knowledge about the process. Digital metrology of the geometric and dimensional characteristics of the workpiece can be a very useful feature in this paradigm to assist the creation of knowledge about the process and product. Typically, inspection is a human-driven process that is conducted by using cyber-physical systems including Coordinate Metrology Machines (CMM), optical and tactile scanners, and vision systems. Ideally, the human element could be removed entirely, and cyber intelligence could be used to determine whether a manufactured product is up to standards or not. The removal of human subjectivity from the inspection process could lead to better finished parts overall. Therefore, it is important that computers can be taught how to inspect a workpiece, as well as make important decisions about its quality, without the need for human intervention.

To allow learning about the part that is being inspected, multiple cyber tools are used. Whether the inspection is through laser scanning, photogrammetry, structured light scanning, etc., a 3D coordinate representation of the workpiece is generated. Today's coordinate metrology sensors can collect thousands of 3D data points in a portion of a second from a finished or semi-finished surface in a production line. However, the collected data includes a combination of the real geometric information of the measured object, inspection errors, and the noises resulting from the physical nature of the sensing process. In order to "extract" the desired geometric information of the workpiece from this amalgamation of data, strong data analytics methodologies are needed.

Often the optical metrology data has no underlying information to provide prior knowledge to the data analytics processes, as to the exact orientation of the part, the existence of noise within the scan, or what a defect looks like. It is up to the programmer implementing the system to teach the computer how to do these things. In this paper, a method to remove noise from a laser scan of planar data is introduced.

This task is especially important in workpieces with highly reflective surfaces. As the laser scanning process emits a line of laser light that must be detected by a receiving camera, it is possible for this light to be scattered or for other sources of light to be detected, which causes noise in the resulting scan. This noise will show up in the data set as points that are deviated from the actual surface being scanned. Detecting these points can range in difficulty, as some points may exist far above or below a surface and are then easy to detect. However, other noisy data points may exist much closer to the real data, which can make them near impossible to detect.

2. Literature Review

As Industry 4.0 further becomes the norm for the manufacturing sector, employing intelligent inspection systems is required. Automated inspection has been an important topic for many industries for the past decades, to allow a highly consistent, unaided inspection process while maintaining the desired levels of uncertainties and precision [1–4]. Controlling the inspection uncertainty is always a challenging task in automated inspection. The robust design of inspection equipment by modeling the deformations, displacements, vibration, and other sources of the imperfection of the components [5], or by creating mechanisms with the capabilities for self-calibration [6] are among the major approaches in reducing the inspection uncertainty by improving the physical inspection components. However, controlling the inspection uncertainties by only focusing on the hardware and the physical equipment is always limited and can become very expensive. Today's metrology equipment are complex cyber-physical systems, and as it is demonstrated in [7] the cyber components contribute to controlling the inspection uncertainty no less than the hardware components. It has been discussed in previous research how highly valuable information about the manufacturing process can be extracted from the produced parts directly [8,9]. While there is a long history of developments in the inspection and metrology of manufacturing and assemblies, there is a lot of work to be done in reducing the uncertainties in digital metrology [10]. The new paradigm of digital metrology for inspection of geometric features and dimensional accuracies is described as a cyber-physical system with three major cyber components. These three cyber components are described as Point Measurement Planning (PMP), Substitute Geometry Estimation (SGE), and Deviation Zone Evaluation (DZE) [11,12]. In several previous research works the effects of sampling strategy including the number and procedure of data collection on the inspection uncertainty are investigated [4,13,14] and several methodologies for selection of the best set of data points in the inspection process are developed. The main approach in these contributions has been a closed-loop of DZE and PMP. The DZE-PMP loop allows using the gradually learned knowledge about the inspected entity to dynamically decide for revising the data set. Among the developed methodologies, the neighborhood search and the data reduction methodologies for virtual sampling from large datasets have shown very promising results with great potentials for further development and implementation [4]. In addition, interesting results have been achieved considering the upstream manufacturing process data for PMP to allow modeling the actual manufacturing errors for error compensation or any downstream post-processing operation. The approach is referred to as Computer-Aided Manufacturing (CAM)—based inspection [15], instead of the typical Computer-Aided Design (CAD)—based inspection.

Estimating the Minimum Deviation Zone (MDZ) based on a set of discrete points for non-primitive geometries is a very challenging task. The problem is even more complex when constrains such as the tolerance envelopes are imposed, for freeform surfaces, and for multi-feature cases. The Total Least Square (TLS) fitting criteria is becoming more popular in coordinate metrology since it has a statistical nature and it is computationally less expensive to solve. Various successful methodologies for TLS and weighted TLS (WTLS) are developed which can be used for error compensation, repair, or post-processing in the manufacturing systems [2,8,10,16]. Various works have been conducted to develop reliable and quick algorithms for TLS of complex geometries, freeform curves, and sculptured surfaces. As an example, Ref. [17] presents a strong and fast approach for TLS fitting of Non-Uniform Rational B-Spline (NURBS) surfaces using a method referred to as Dynamic Principal Component Analysis (DPCA).

Dynamic completion of DZE by closed-loop of DZE with PMP and SGE has been the subject of several recent research works. In these contributions, the distribution of geometric deviations gradually evaluated by DZE are used for dynamic refinement of the sampling data points and estimation of fitted substitute geometry [18–21]. Intelligence is needed to address the requirement of the three main cyber components of an integrated inspection system, developing the point measurement strategy based on an estimation of the manufacturing errors [13,16], or using search-guided approaches to find the best representatives of the manufactured surface [18] are among the main approaches to assist PMP. The former approach relies on significant knowledge of the manufacturing process and demands for employment of digital twins or the detailed simulation of the manufacturing process. As a result, the solutions can be computationally very expensive and logically neglect the effect of non-systematic manufacturing errors. The latter approach requires a loop of PM-SGE-DZE tasks using learning mechanisms, statistical tools, and artificial intelligence. The efficiency of this approach highly depends on the convergence of the iteration process which in difficult cases it may result in a very time-consuming process.

This paper presents an approach of using a PMP-SGE-DZE iterative loop toward solving a challenging problem in the scanning of highly reflective surfaces. These surfaces can refract and distort optical scanning techniques, leading to noise [22,23]. While switching to other methods of examination is possible, these methods can be less efficient, be difficult to automate, or be inappropriate for the geometry being examined [24,25]. Noise can lead to inaccurate results in an inspection. While these changes can be seemingly small, they can lead to perfectly acceptable parts being rejected and scrapped. It has been shown in multiple research work how significantly the results of the inspection may vary due to these noises particularly by affecting SGE and DZE evaluations [20,21,26]. As the goal is to have near-perfect inspection without human intervention, it is important that this noise can be accurately removed from the scan data.

There are multiple different noise reduction techniques that have been developed for a variety of situations [27,28]. Whether through segmentation of the dataset, non-iterative approaches, or intelligent search algorithms, there are many advantages or disadvantages to the methods. This leads to the need for multiple different algorithms to be developed to suit individual situations. Weyrich et al. [29] looked at the nearby groupings, or neighborhoods, of points in order to determine whether or not individual points were noise. By using three different criteria, the probability of a point being noise could be determined, and by setting a threshold, the severity of noise reduction could be changed. This is an example of a very localized method, but as it required in-depth analysis of every point within a point cloud, it could take a long time to complete the analysis. Zhou et al. [30] introduced a non-iterative method that separated the data set into small and large threshold regions and treated them using separate algorithms. Their method was very successful in noise reduction of 3d surfaces and being non-iterative, it ran very quickly. Ning et al. [31] looked at density analysis for outlier detection in points clouds. By examining the density of points in small areas of high-density point clouds, a reasonable estimation of noise in each area could be conducted. This method was quick and highly effective at removing outliers but could possibly struggle in areas of high-density noise. This is because these areas of noise may have a similar density to the overall point cloud, rendering them similar in the eyes of the algorithm. Wang and Feng [32] looked specifically at reflective surfaces and utilized the scan orientation of multiple scans to best determine where noise exists in scans of parts with higher complexity. This method had a very high success rate for removing noise, but the requirement of extra scans increases scan time for large parts significantly. Rosman et al. [33] broke down the data set into smaller, similar, overlapping patches. By examining these patches concurrently, noise could be removed from all patches. This method of denoising was focused more on surface reconstruction than analysis and could possibly smooth real errors within the scan, an undesirable result while searching for errors along the surface. Schall et al. [34] examined point clusters in a scan. By using a kernel density estimation technique, the likelihood of a point existing on the real surface was determined. Similar to the technique developed by Weyrich et al. [29], this likelihood was used to classify a point as noise or real data. Like a few of these options, our developed method looks at the point cloud globally. Additionally, the developed method is iterative, but the run time is small due to the relatively low computational complexity.

3. Methodology

The developed methodology is explained in this section. Although the methodology is implemented for laser scanning using a robotic arm, it can be used for any other coordinate metrology setup. In the current setup, an ABB robotic arm is programmed to iterate through several different motion paths. These paths are designed so that the arm rotates the camera 5 degrees with each pass while maintaining the same vertical distance from the camera center to the workpiece's surface. As most cameras will provide feedback on the optimal distance of the camera to the workpiece via a bar or color coding, this is used to set the initial distance. In these experiments, an LMI Gocator 2410 with an x-resolution of 0.01 mm is used. The setup is tuned with the camera parallel to the surface in question, so that the 0-degree position will provide the optimal results for the scan, with minimal amounts of noise. The scanning initially begins at the −25 degree points and iterates 5 degrees positively until the −25 degree mark, this process is shown graphically in Figure 1. The movement path is determined so that the entire workpiece will be captured regardless of the scan angle. All parameters regarding motion speed and path planning, other than the start and end points and the height of the scanner are determined automatically by the robotic control system.

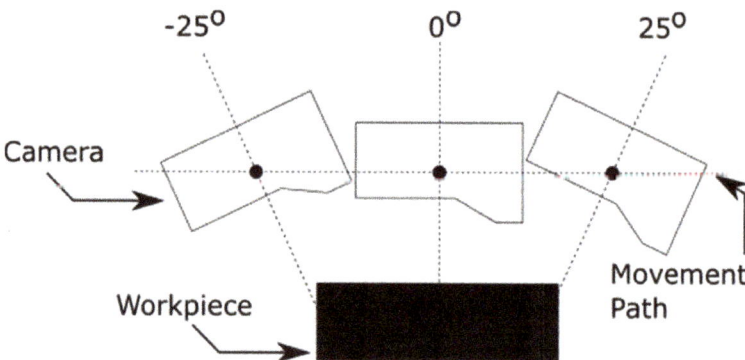

Figure 1. Camera setup and different angle extremes.

The camera scanning parameters will need to be adjusted for each material scanned, including if the workpiece height changes. These values must be determined for each part used due to variations in material properties and will not be consistent between different workpieces. With an automated algorithm on the robotic arm used in this paper, only the initial position had to be set, then all other positions were calculated automatically based on the programmed scanning pattern. This allowed for all motions to be consistent.

Another important aspect of the experiment is the lighting conditions. The scans were all conducted in a controlled setting, with minimal effects of outside lighting present. As this scanning process is an optical one, abrupt changes in lighting conditions can cause a lot of noise to be captured. Another method used to reduce the impact of lighting conditions involved redoing all scans for a workpiece after rotating it 90 degrees. This allowed for the determination of the effects of lighting conditions, for if lighting conditions were an issue, the errors seen would not rotate with the workpiece.

In order to ensure different situations are represented, these tests will need to be rerun for both under and over-exposed conditions. In the underexposed tests, the amount of data captured will be much smaller than in a regular scan. The camera filters out areas of its view that are not a laser and uses the intensity of light of certain wavelengths in order to determine where the laser is. As such, by allowing a smaller amount of light into the viewfinder, more of the laser will not be processed. This leads to datasets without a lot of useful data. This can be beneficial for the reduction of noise, but also leads to situations with a small amount of actual surface information, which could mean the scan misses imperfections on the surface. In the overexposed condition, the opposite occurs, and more light is allowed into the viewfinder. This can lead to noisier point clouds as lower intensity areas of the laser that would typically be filtered out by the software would now be processed as the real surface, while not necessarily being on the real surface.

Once the scans were completed, the background data was removed. This consisted of the plate that the workpiece was laying on. This surface was matte black, and so the data captured for the surface was very consistent with very little noise. A large distance was also maintained between the inspected workpiece surface and the support surface. These factors allowed for the background data to be removed by simply fitting a plane to the data not associated with the actual workpiece surface, then removing it. The collected points (Ps), which will be the input data set, were then exported to the XYZ filetype, which consists of rows of X, Y, and Z coordinates.

Once data collection has been completed, the datasets are imported to the developed software environment to be analyzed. To begin, a plane is fit to the dataset using Total Least Squares (TLS) fitting using the Principal Component Analysis (PCA) method. This is a commonly used algorithm for planar fitting and returns a normal vector and point that defines the fit plane [35]. The distance between each point of the dataset and the fit plane is then calculated using the point-plane distance calculation shown in Equation (1).

$$d = \frac{|Ax+By+Cz+D|}{\sqrt{A^2+B^2+C^2}} \qquad (1)$$

where A, B, and C are components of the fit plane's normal vector, x, y, and z are the coordinates of a point, and D is equal to the following.

$$D = -Ax_0 - By_0 - Cz_0 \qquad (2)$$

where x_0, y_0, and z_0 are the coordinates of a point on the plane. With the point-plane distances calculated, a statistical analysis is conducted to determine how many points are beyond the 6σ range. Points beyond this range will be far from the planar surface, so there is a high likelihood that they are noise. The percentage of data in this range is calculated against the entire data set, and this value is divided by a Minimization Factor (MF). This value can control the amount of data being removed at each step. Another check made to ensure too much data is not removed was ensuring that the reduction step did not exceed a Maximum Reduction (MR) step, which is a percentage of the overall data set. The filtered percentage, which tracks how much data is going to be removed, is then set to 0 to initialize the data removal loop. This process is seen in Figure 2.

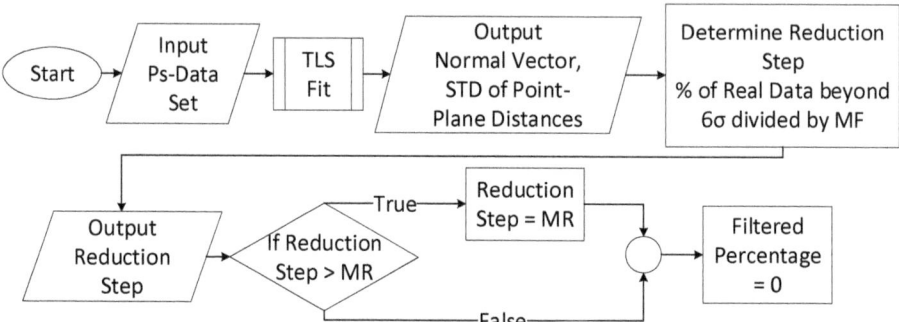

Figure 2. Data initialization process with point dataset (Ps).

Once the reduction step has been determined, the filtering process begins. In each iteration, the amount of data removed (filtered percentage) increases by the reduction step. This amount of data is removed from the points furthest away from the fit plane. This new data set (PsT), is then fit again with a plane, using TLS. The change in the standard deviation of the point-plane distances is recorded, as well as the rate of change of the same value, these are the standard deviation (STD) history and delta graphs. This process repeats for a set number of iterations (nIT), with each iteration removing more data, as defined by the reduction step. Once the number of iterations has become larger than nIT, a check occurs to determine whether the algorithm can stop with data reduction. This check involves examining the STD delta graph, as it shows the rate of change for the STD. Once the STD delta graph has reached a steady state, determined by the rate of change being less than a minimum rate of change (mROC), the algorithm is stopped. This indicates that the points being removed from the point cloud are likely no longer having a large effect on the STD of the entire point cloud, and thus are unlikely to be outliers. Once this is true, the STD of the point-plane distances of the data set has stabilized. This data reduction algorithm is laid out in Figure 3.

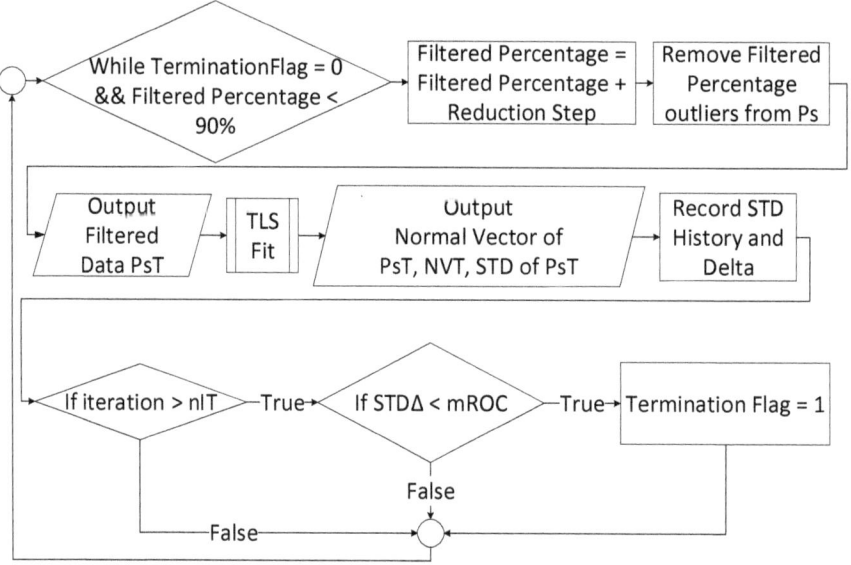

Figure 3. Main loop for the data reduction algorithm, with filtered output from step 1 (PsT).

After the data reduction loop has been completed, the amount of data to be removed for the final data set is calculated. This method takes advantage of the general shape of the STD delta graph. An example is shown in Figure 4.

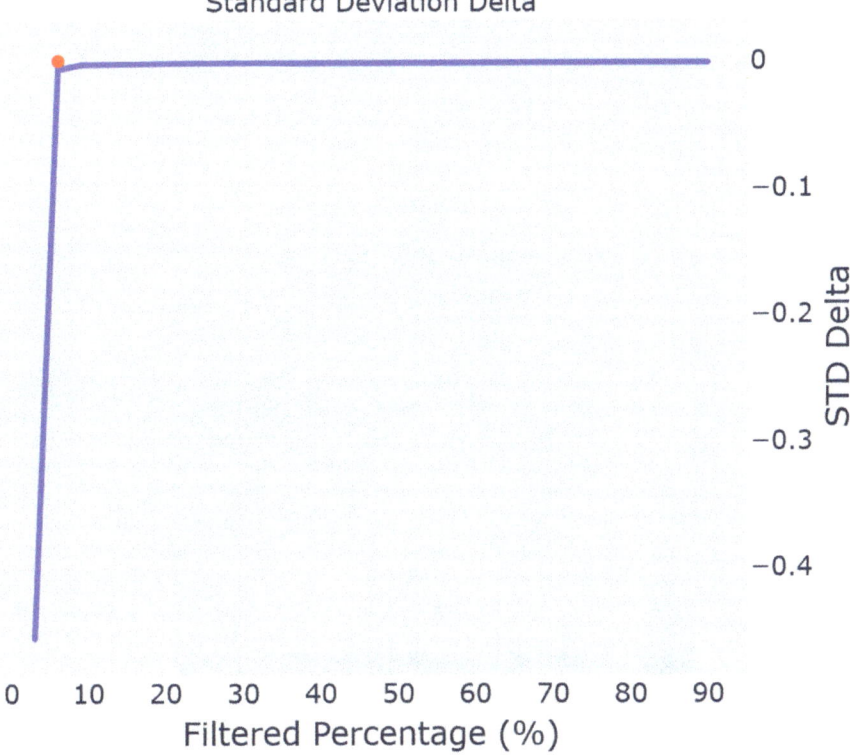

Figure 4. Ideal STD delta graph after data reduction process, with a red dot to represent the intercept of the linear sections.

In the STD delta graph, there is a steep linear section as the STD of the data set decreases, and another linear section with a very flat slope after most outliers have been removed where the STD of the data set is not changing by a significant amount. Once this steady state has been reached, the optimal number of points to be filtered must be determined. In order to accomplish this, the shape of the ideal result is exploited. As the point where the two linear sections meet is the point where the large change in the STD value occurs, the graph is treated as though it is a triangle. The two end points of the STD delta graph are connected to form a line, and the Euclidean distance of each point of the STD delta graph to the line is determined. The furthest point from this line, which would be the vertex opposite the side in the triangle, is selected. The chosen point is the filtered percentage at which the noise removal stopped removing extreme outliers. As the STD delta value begins to remain constant, the points being removed lie closer and closer to the plane. If the filtered percentage is chosen beyond this leveling-off point, actual surface data will likely be removed. The distance is calculated by treating each filtered percentage amount as a point and calculating the Euclidean distance between the intersection point and each filtered percentage amount. The value with the shortest distance is then chosen. The full dataset is then filtered using the specified percentage of removed points, and finally, the filtered data set is returned. This process is fully outlined in Figure 5.

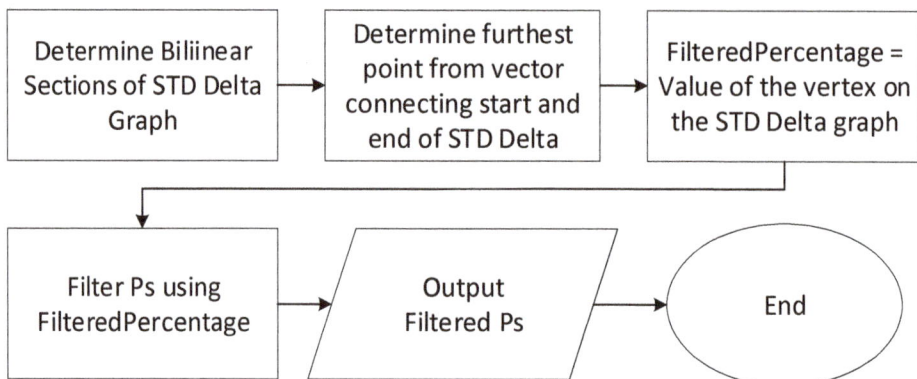

Figure 5. Final noise reduction pass to determine the accurate data.

4. Implementation and Results

Based on the developed methodology, software was developed to verify its ability to remove noise. Multiple workpieces were analyzed in this study to get a sample of different reflectivity. The variables defined in the methodology were set as follows; minimization factor was 10, maximum reduction step was 0.3, the number of iterations was 10, and the minimum rate of change of 0.005%. These variables were determined through experimentation for the samples chosen. Three samples were used:

- a piece of brushed aluminum with flatness of 0.12 mm (Figure 6a),
- a steel gauge block, workshop grade, with flatness of 0.5 µm (Figure 6b), and
- a selective lased sintering (SLS) manufactured part with flatness of 0.08 mm (Figure 6c).

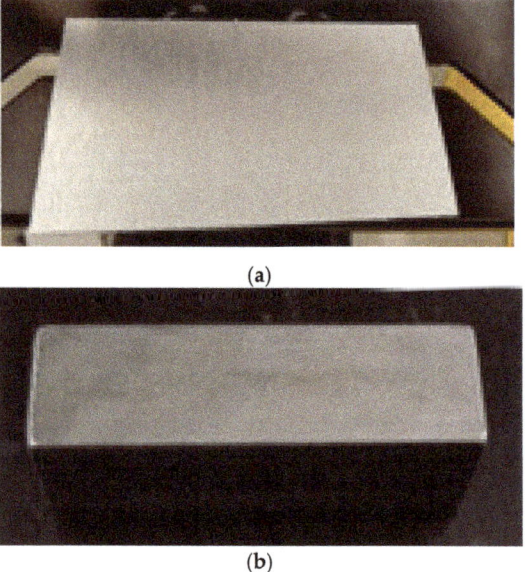

Figure 6. *Cont.*

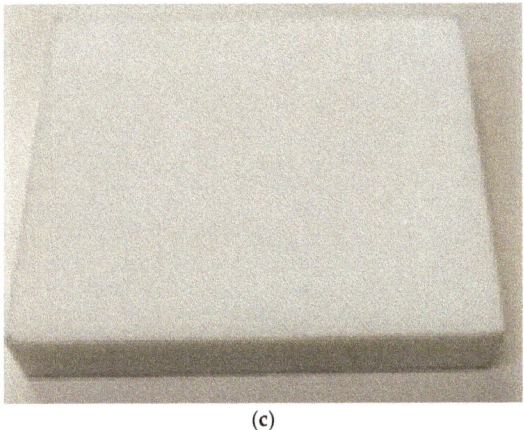

(c)

Figure 6. Images of three test samples: (**a**) gauge block; (**b**) brushed aluminum; (**c**) SLS printed part.

All objects were selected to have minimal form error to reduce the effects these errors may have on noise generation. These three samples provided unique challenges for scanning. The flatness of the parts is according to the manufacturers' specifications and verified by a tactile inspection coordinate metrology probe, where applicable. The gauge block had a reflective surface, yet also had a thin film of oil on its surface which reduced the amount of reflection. The SLS part, which is comprised of small, melted granules of plastic, absorbed a lot of the laser light due to its color and had granules on the surfaces which scattered light depending on how they were hit. Finally, the brushed aluminum surface structure had differently oriented grains that reflected light in different manners based on how the light was interacting with them. This was different from the SLS part as the aluminum contained long sections of reflectivity, so it was possible for bands of noise to appear on the part. All these possible sources of error can contribute to the uncertainty of measured values. When determining uncertainty, one of the areas that must be considered is the manufacturer specifications for the device being used to measure the object [36]. When in a non-standard situation, these specifications may not be accurate. Processing of the data through the use of knowledge of the measurement and conditions can help to reduce the uncertainty of the measurement [37]. This method may be a useful tool in the reduction of uncertainty of optical scanning of non-ideal material. The results of a few of these tests will be discussed in detail, then the results for the entire test will be shown.

The first sample to be examined is the brushed aluminum at an angle of +15°. There is a band of noise along the center of the piece, highlighted in red in Figure 7. This is an example of the best-case scenario for the algorithm. The STD delta graph is of the optimal shape, so there is a clear point where the effect of point removal becomes minimal. The deviation zone estimated by the data was reduced by 90.8% through this method with 6% of the point cloud filtered. The STD of deviations was 0.515 mm which was reduced by only 6% filtration to 0.047 mm. The total deviation zone was estimated as 0.113 mm which is fairly close to the known flatness of the part.

The next sample is the gauge block at +10° shown in Figure 8. The noise in this scan is more distributed throughout the entire surface of the part, both above and below. Again, this situation is ideal. The STD delta graph follows with the shape that is sought after, and a large majority of the noise is filtered from the scan. For this sample, the flatness was reduced by 97.4%. The STD of deviations was 0.512 mm which was reduced by filtration to 0.013 mm. The total deviation zone was estimated as 0.024 mm. Obviously, the gauge block is more precise than the level of accuracy in a laser scanner and the uncertainties in the mechanical and optical features of the measurement device do not allow accurate measurement of a highly precise gauge block. However, the results show that the algorithm was successful in filtering the dominating outlier noises from the data.

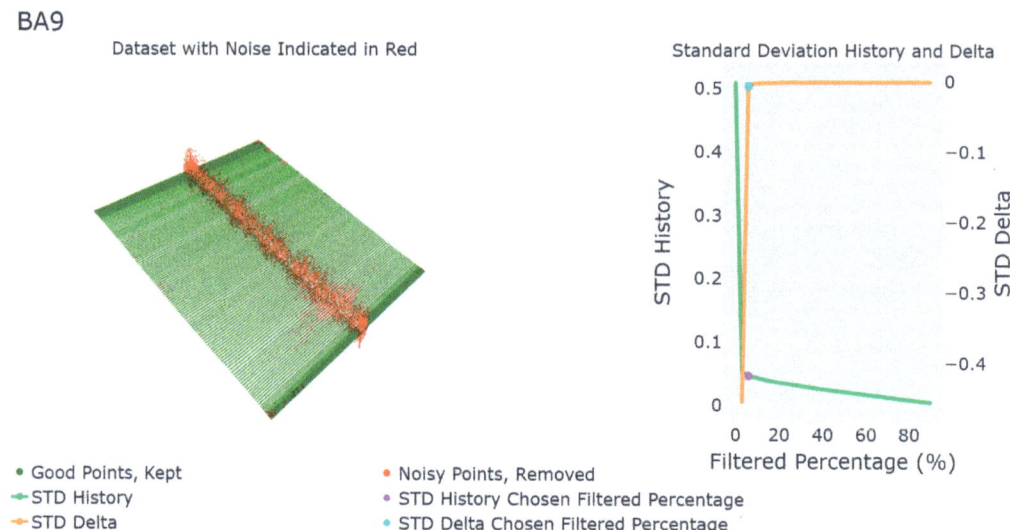

Figure 7. Results for brushed aluminum at +15° with STD history and delta graphs combined.

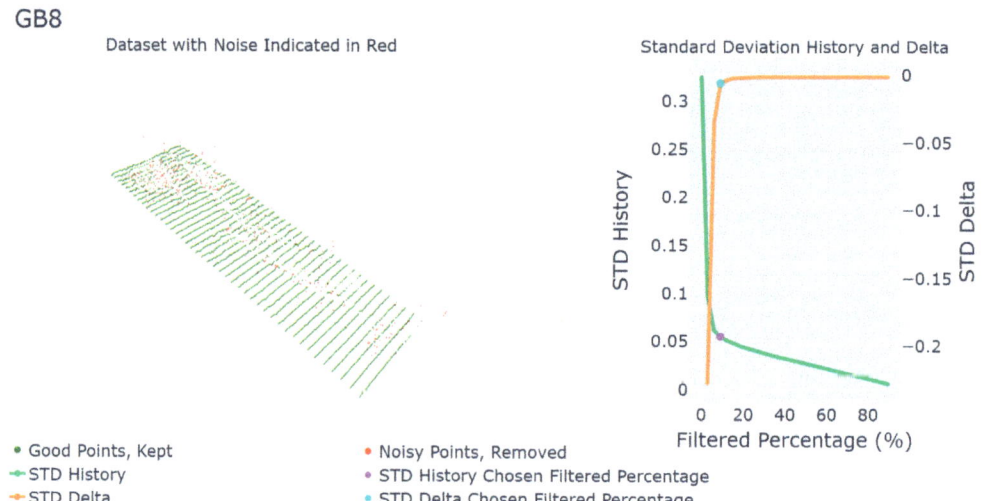

Figure 8. Results for gauge block at +10°.

The next sample is the SLS piece, after a 90° rotation, at a scan angle of 0° (Figure 9). The scan for this part looks cupped due to the capture of the edges of the surface. There is noise shown under the scan. However, the density of the noise is very low. For this scan, there is no initial quick drop in the standard deviation of point-plane distances, which is a less than ideal result. This is due to the relatively small amount of noise having a small effect on the flatness measured when compared to the rest of the data set. There was a reduction of only 38.2% for this sample. The STD of deviation was 0.101 mm which was reduced by filtration to 0.062 mm. The total deviation zone was estimated as 0.153 mm.

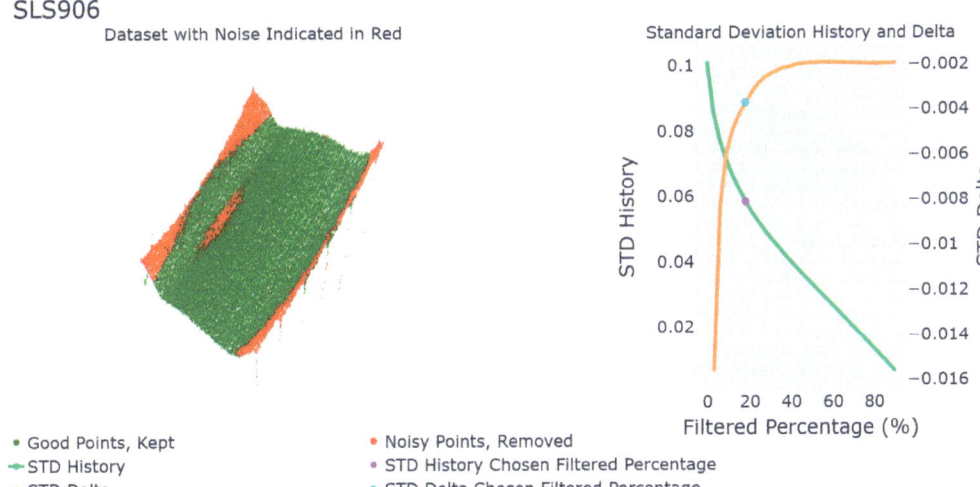

Figure 9. Results for the SLS part at 0° after 90° rotation.

After reviewing fairly straightforward cases, a few rare problematic cases are presented in the following. As can be seen, these problematic cases require overly aggressive filtrations with some results that may cause misleading. The next sample is the brushed aluminum piece, with overexposed lighting conditions, after being rotated 90°, at a scan angle of +25° (Figure 10). This scan had very little noise, however, the removal of the background data was not entirely completed to determine the effect of additional planar data on the algorithm. This resulted in a "stepped" planar surface with two separate heights. In a situation like this, the fit plane would be angled to capture both high-density planar areas. As the surface of the workpiece dominated the scan, a portion of it survived the filtering process, however, a large number of correct scan points was removed due to the initial plane fitting. The STD delta graph shows the wildly varying changes in the data set as points were filtered. Once the secondary background plane was eliminated from the data set, there was no further reduction in the flatness measurement, as the rest of the scan was already ideal. The STD of deviation was 4.51 mm which was reduced by filtration to 0.057 mm. The total deviation zone was estimated as 0.108 mm. While this behavior may seem desirable, it is possible that it could lead to misleading results for an operator or other decision-making process. Fortunately, it is easy to tell from the STD delta graph that something went wrong. This could be used as a diagnostic to ensure scans are occurring correctly.

The next example of the problematic cases is brushed aluminum at +25° (Figure 11). At first glance, the STD delta graph looks to have two bilinear sections, but with a large region of slowing change. The other issue is that the change is quite small to begin with, by only thousandths of a millimeter as indicated by the STD delta scale. This is because the scan had very little noise. As this process is a statistical analysis of the point-plane distances for a data set, if that data set is already very well formed, with only a couple of noisy points standing out from the main data set, the algorithm will overcompensate and begin to remove useful data from the set. This can be seen in the top left picture of Figure 11, where pieces of the plane have been removed without there being any noise in those areas. Like the previous issue, this can be detected by examining the change in STD over time. If there is a very small change, or if the STD history graph is nearly linear, it is likely that the data set is already very clean. The STD of deviation was 0.058 mm which was reduced by filtration to 0.039 mm. The total deviation zone was estimated as 0.104 mm.

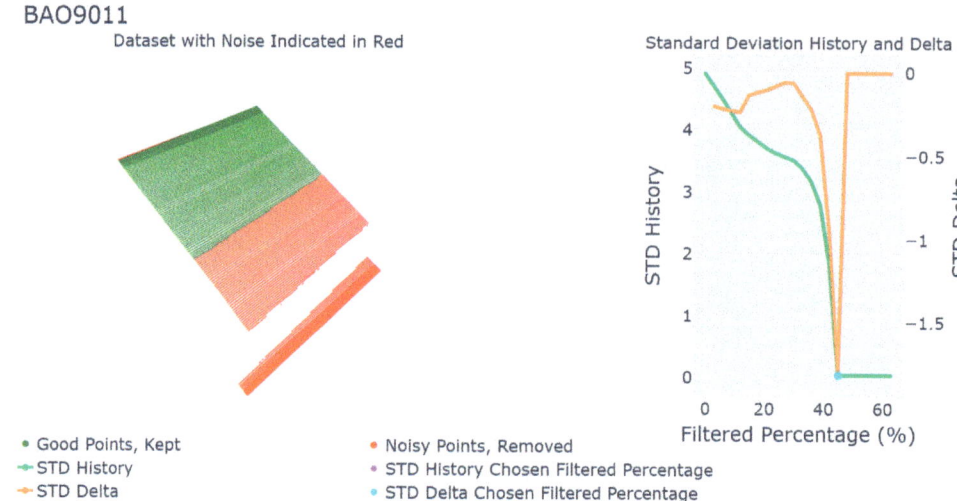

Figure 10. Results for the brushed aluminum part at +25° after 90° rotation, overexposed.

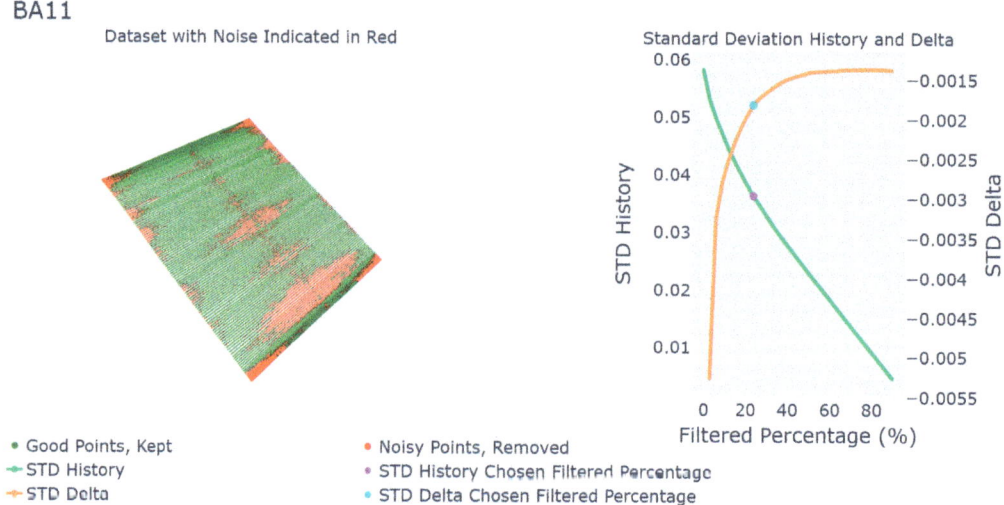

Figure 11. Results for the brushed aluminum part at +25°.

Finally, the SLS part after a 90° turn, at −10° while overexposed (Figure 12). This data set was quite noisy and contained a "ghost" layer of data below the actual surface. This is possibly due to the light absorption properties of the material. As the laser light hits the surface, it is diffused throughout the surface, causing a "glow". The scanner's receiver could still pick this refracted light up, resulting in a surface below the actual surface. This added layer of data caused a shift in the STD delta graph. There are not two linear sections; instead, the line has steps in it after the initial steep increase. Due to the algorithm only using a section from either end of the STD delta graph to determine the intersection point, shown in red in the ideal STD delta graph from Figure 4, the steps are not considered when determining the correct percentage of data to remove. This allowed most real data to remain while removing most of the noise. The STD of deviation was 0.823 mm which was reduced by filtration to 0.525 mm. The total deviation zone was estimated as 0.734 mm.

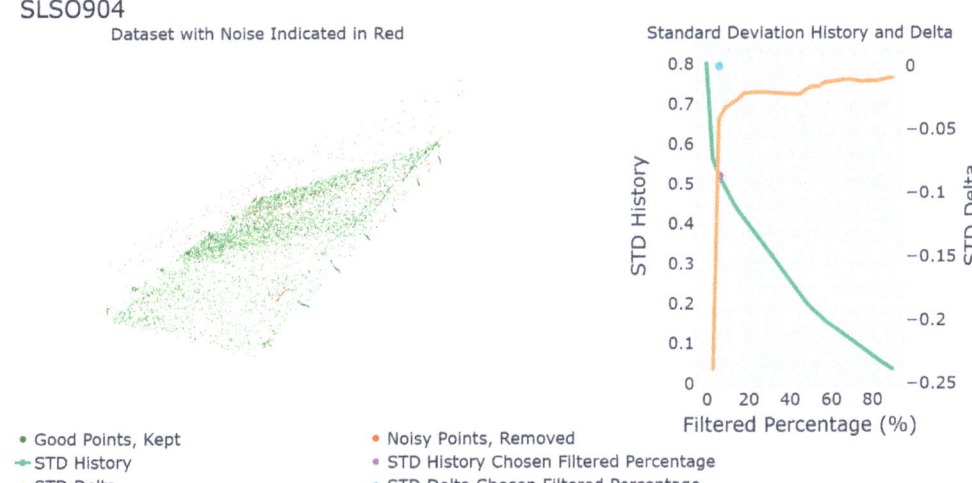

Figure 12. Results for the SLS part at −10° after 90° rotation, overexposed.

Table 1 shows the overall results for each part to four significant digits, due to the number of points used and the accuracy of the used scanner. Regardless of the sample or scan angle, there was a decrease in the detected flatness, showing removal of noise. In the naming of the cases, BA stands for brushed aluminum, GB stands for gauge block, and SLS stands for the selective laser sintered part. The number 90 in the name indicates that this part is rotated 90 degrees for the second set of measurements. The overexposed and underexposed items are defined with the letters "O" and "U" respectively, at the end of each name in this table. Generally, the overexposed condition benefitted the most from noise reduction, likely due to the added light allowed into the receiver causing more noise to be recorded. Conversely, the underexposed condition did not see as great a benefit as less light is allowed into the receiver.

Table 1. Results for all samples.

	Average Deviation Zone Reduction	Maximum Decrease	Minimum Decrease
BA	54.73%	90.88%	33.63%
BA90	51.19%	98.71%	28.93%
BA90O	68.35%	99.18%	33.50%
BA90U	43.84%	74.81%	26.05%
BAO	70.24%	95.58%	37.79%
BAU	40.29%	53.69%	27.72%
GB	62.60%	88.24%	19.12%
GB90	69.11%	88.22%	31.67%
GB90O	68.55%	89.01%	36.01%
GB90U	40.42%	89.52%	24.34%
GBO	66.69%	94.31%	21.55%
GBU	39.79%	95.41%	24.49%
SLS	40.26%	60.67%	14.88%
SLS90	38.45%	54.00%	15.70%
SLS90O	59.49%	81.55%	35.23%
SLS90U	34.52%	45.58%	24.72%
SLSO	50.53%	65.27%	31.59%
SLSU	41.30%	50.06%	27.77%

5. Behavior Analysis

In order to determine how effective the developed methodology is, it is important to study how it is successful to determine the behavior of data. In order to do this, different quantitative parameters for effectiveness were determined. These were determined based on the ideal shape of the STD delta graph. As the graph ideally was comprised of two linear sections, with the intersection of these sections determining the amount of data to remove, parameters evaluating the closeness of this model were chosen. These six parameters are the slopes of the two fit lines, the distance of the chosen filtered percentage and the fit lines, and finally the standard deviation of the Euclidean distance of the sample points to the fit lines. In the ideal situation, the slope parameters will be maximum for the initial line, and 0 for the end line, the distances will both be 0, and the STD values will also both be 0. The following Figures 13 and 14 show a RadViz for the test results, to determine the similarities in the parameters for each material and angle tested. RadViz is a multivariate visualization algorithm that allows for different variables to sit along the outside of a circle, where inside the circle different datapoints, for this purpose test cases, are placed. These data points are pulled to each of the outer variables as though attached by a spring and using the value of the variable as a spring constant, the datapoint is placed where the force would be equal to zero [38]. This method can be useful to determine if your data clusters well, or if there is excessive variation within your data set.

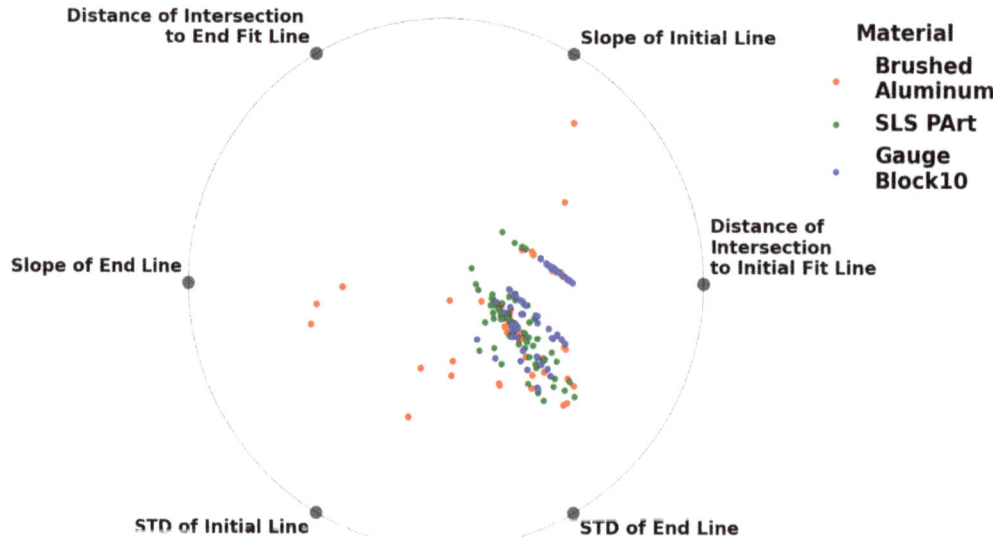

Figure 13. RadViz of parameters for each test, separated by material.

In these visualizations, clear clustering can be seen. In this visualization, clustering is indicative of the correlation between different parameters. If for multiple, or many tests, the results cluster to one section of the circle, the items closest have the greatest effect on the result of the test. For all but the outliers, there was a very clear pull to both the distance of intersection to the initial line fit and to the STD of the end line. This means these two parameters likely have the greatest effect on the result of this filtering method and can be used to judge if a particular set of parameters for a particular work piece is optimal. For both the SLS Part and gauge block, the results were all very closely clustered together. These parts did not have the extreme reflectivity of the brushed aluminum piece, and so the results are very closely correlated. For the brushed aluminum sample, the test results outside of the cluster area all correspond to extreme angles of a scan. This is likely due to the extreme angle of the scan and the high reflectivity of the material causing only a small amount of light to enter the receiver. In this case, filtering is not an effective strategy as the data collected will be inherently wrong. To fix these cases, different scanning parameters would need to be chosen entirely. However, in most cases, the tight clustering shows that the filtering scheme is effective, even at extreme angles, for two of the three surfaces tested, and for non-extreme angles for the brushed aluminum sample.

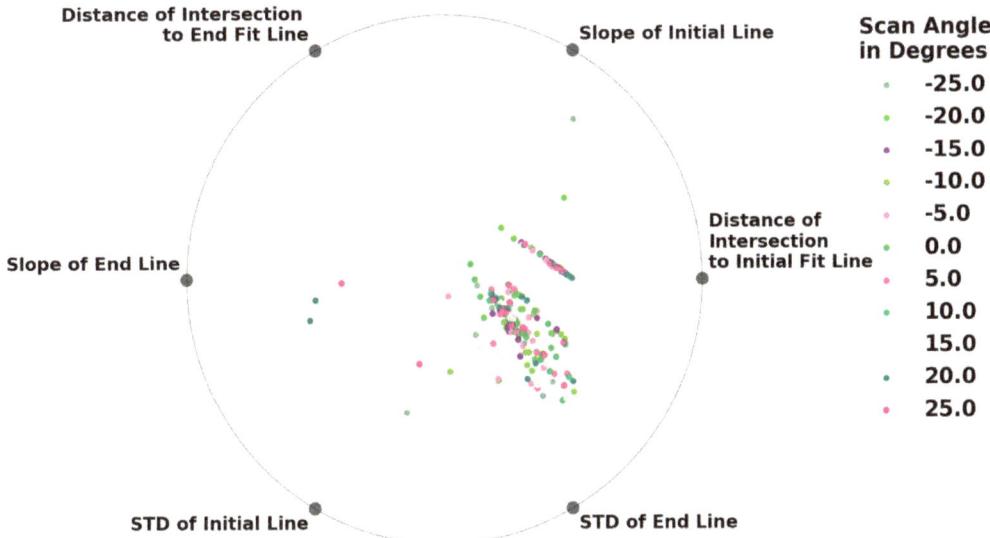

Figure 14. RadViz of parameters for each test, separated by angle.

6. Conclusions

In this paper, a noise reduction algorithm for planar systems was introduced. This system examined datasets globally in order to remove noisy data points from 3D point cloud datasets using a statistics-based approach. The methodology and algorithm were introduced and explained, and three sample parts were measured using a detailed procedure by robotic laser scanner 33 times under various conditions, for a total of 198 scans. These datasets were then processed using the algorithm in order to determine if the noise reduction was effective. Overall, the results are highly satisfactory and in the majority of cases, only with automatic filtration of a small amount of data, the estimated deviation zones are significantly improved. Examples of these cases are provided in the paper. In datasets where there was a large amount of noise, the algorithm was very effective at minimizing the effect of noise on the results. However, when the data set was already noise-free, the algorithm tended to slightly overcompensate and remove actual data. There are rare problematic datasets that are severely affected by the environmental scanning condition. As a result, a group of false data were introduced in these cases which challenged the algorithm in its filtration process. A few worst cases of these kinds are also presented in the paper. An example can be the cases where the background data was left in the scan, the fit plane was at an angle with the actual surface and so some real data was removed during the filtration process. Regardless of the error, there was evidence of these issues existing in both the STD delta and STD history graphs. It has been demonstrated that the algorithm was fairly successful to filter the false data even in these problematic cases. However, since the false data already form some patterns in these cases, the recommendation is the remove the false data by employing a pattern recognition method prior to using the developed noise filtration algorithm. In general, the developed algorithm and methodology are evaluated to be very efficient in removing the noisiness in optical metrology data. The methodology can also be employed in various scales of data and for various industrial applications. It is computationally very efficient and can be easily used for on-line noise removal during the inspection process.

Author Contributions: Conceptualization, M.S.G.T. and A.B.; Data curation, C.B. and A.B.; Formal analysis, C.B. and A.B.; Funding acquisition, A.B.; Investigation, C.B. and A.B.; Methodology, M.S.G.T. and A.B.; Project administration, C.B.; Resources, A.B.; Software, C.B.; Supervision, A.B. All authors have read and agreed to the published version of the manuscript.

Funding: There is no specific funding for this research.

Institutional Review Board Statement: Not applicable.

Informed Consent Statement: Not applicable.

Data Availability Statement: Not applicable.

Acknowledgments: The research support provided by the Natural Science and Engineering Research Council of Canada (NSERC) is greatly appreciated.

Conflicts of Interest: The authors declare no conflict of interest.

References

1. Zhao, H.; Kruth, J.P.; Van Gestel, N.; Boeckmans, B.; Bleys, P. Automated dimensional inspection planning using the combination of laser scanner and tactile probe. *Meas. J. Int. Meas. Confed.* **2012**, *45*, 1057–1066. [CrossRef]
2. Berry, C.; Barari, A. Closed-Loop Coordinate Metrology for Hybrid Manufacturing System. *IFAC-PapersOnLine* **2018**, *51*, 752–757. [CrossRef]
3. Wells, L.J.; Megahed, F.M.; Niziolek, C.B.; Camelio, J.A.; Woodall, W.H. Statistical process monitoring approach for high-density point clouds. *J. Intell. Manuf.* **2013**, *24*, 1267–1279. [CrossRef]
4. Lalehpour, A.; Berry, C.; Barari, A. Adaptive data reduction with neighbourhood search approach in coordinate measurement of planar surfaces. *J. Manuf. Syst.* **2017**, *45*, 28–47. [CrossRef]
5. Askari, H.; Esmailzadeh, E.; Barari, A. A unified approach for nonlinear vibration analysis of curved structures using non-uniform rational B-spline representation. *J. Sound Vib.* **2015**, *353*, 292–307. [CrossRef]
6. Mahboubkhah, M.; Barari, A. Design and development of a novel 4-DOF parallel kinematic coordinate measuring machine (CMM). *Int. J. Comput. Integr. Manuf.* **2019**, *32*, 750–760. [CrossRef]
7. Barari, A. Automotive body inspection uncertainty associated with computational processes. *Int. J. Veh. Des.* **2011**, *57*, 230–241. [CrossRef]
8. ElMaraghy, H.A.; Barari, A.; Knopf, G.K. Integrated inspection and machining for maximum conformance to design tolerances. *CIRP-Annal.* **2004**, *53*, 411–416. [CrossRef]
9. Lalehpour, A.; Barari, A. Developing skin model in coordinate metrology using a finite element method. *Measurement* **2017**, *109*, 149–159. [CrossRef]
10. Barari, A.; ElMaraghy, H.A.; Knopf, G.K. Integrated Inspection and Machining Approach to Machining Error Compensation: Advantages and Limitations. In Proceedings of the Flexible Automation and Intelligent Manufacturing (FAIM2004), Toronto, ON, Canada, 12–14 July 2004; pp. 563–572.
11. Berry, C.; Barari, A. Cyber-physical system utilizing work-piece memory in digital manufacturing. *IFAC-PapersOnLine* **2019**, *52*, 201–206. [CrossRef]
12. Gohari, H.; Berry, C.; Barari, A. A digital twin for integrated inspection system in digital manufacturing. *IFAC-PapersOnLine* **2019**, *52*, 182–187. [CrossRef]
13. Martins, T.C.; Tsuzuki, M.S.G.; Takimoto, R.; Barari, A.; Gallo, G.B.; Garcia, M.; Tiba, H. Algorithmic iterative sampling in coordinate metrology plan for coordinate metrology using dynamic uncertainty analysis. In Proceedings of the 2014 12th IEEE International Conference on Industrial Informatics (INDIN), Porto Alegre, Brazil, 27–30 July 2014; pp. 316–319.
14. Barari, A.; Mordo, S. Effect of sampling strategy on uncertainty and precision of flatness inspection studied by dynamic minimum deviation zone evaluation. *Int. J. Metrol. Qual. Eng.* **2013**, *4*, 3–8. [CrossRef]
15. Barari, A. CAM-Based Inspection of Machined Surfaces. In Proceedings of the 5th International Conference on Advances in Production Engineering—APE 2010, Warsaw, Poland, 11–13 June 2010.
16. Barari, A.; Elmaraghy, H.A.; Orban, P. NURBS representation of estimated surfaces resulting from machining errors. *Int. J. Comput. Integr. Manuf.* **2009**, *22*, 395–410. [CrossRef]
17. Gohari, H.; Barari, A. A quick deviation zone fitting in coordinate metrology of NURBS surfaces using principle component analysis. *Meas. J. Int. Meas. Confed.* **2016**, *92*, 352–364. [CrossRef]

18. Barari, A.; ElMaraghy, H.A.; Knopf, G.K. Evaluation of Geometric Deviations in Sculptured Surfaces Using Probability Density Estimation. In *Models for Computer Aided Tolerancing in Design and Manufacturing*; Springer: Dordrecht, The Netherlands, 2007; pp. 135–146.
19. Jamiolahmadi, S.; Barari, A. Study of detailed deviation zone considering coordinate metrology uncertainty. *Measurement* **2018**, *126*, 433–457. [CrossRef]
20. Jamiolahmadi, S.; Barari, A. Convergence of a finite difference approach for detailed deviation zone estimation in coordinate metrology. *Acta Imeko* **2015**, *4*, 20–25.
21. Jamiolahmadi, S.; Barari, A. Estimation of Surface roughness of Additive Manufacturing Parts Using Finite Difference Method. In Proceedings of the ASPE 2014 Spring Topical Meeting: Dimensional Accuracy and Surface Finish in Additive Manufacturing, Berkeley, CA, USA, 13–16 April 2014; American Society of Precision Engineering (ASPE): Albany, NY, USA, 2014.
22. Wang, Y.; Feng, H.Y. Modeling outlier formation in scanning reflective surfaces using a laser stripe scanner. *Meas. J. Int. Meas. Confed.* **2014**, *57*, 108–121. [CrossRef]
23. Wang, Y.; Feng, H.Y. Effects of scanning orientation on outlier formation in 3D laser scanning of reflective surfaces. *Opt. Lasers Eng.* **2016**, *81*, 35–45. [CrossRef]
24. Mian, S.H.; Al-Ahmari, A. Comparative analysis of different digitization systems and selection of best alternative. *J. Intell. Manuf.* **2019**, *30*, 2039–2067. [CrossRef]
25. Mohib, A.; Azab, A.; Elmaraghy, H. Feature-based hybrid inspection planning: A mathematical programming approach. *Int. J. Comput. Integr. Manuf.* **2009**, *22*, 13–29. [CrossRef]
26. Zahmati, J.; Amirabadi, H.; Mehrad, V. A hybrid measurement sampling method for accurate inspection of geometric errors on freeform surfaces. *Meas. J. Int. Meas. Confed.* **2018**, *122*, 155–167. [CrossRef]
27. Kobbelt, L.; Botsch, M. A survey of point-based techniques in computer graphics. *Comput. Graph.* **2004**, *28*, 801–814. [CrossRef]
28. Han, X.F.; Jin, J.S.; Wang, M.J.; Jiang, W.; Gao, L.; Xiao, L. A review of algorithms for filtering the 3D point cloud. *Signal Process. Image Commun.* **2017**, *57*, 103–112. [CrossRef]
29. Weyrich, T.; Pauly, M.; Keiser, R.; Heinzle, S.; Scandella, S.; Gross, M.H. Post-processing of Scanned 3D Surface Data. In Proceedings of the IEEE eurographics symposium on point-based graphics, Grenoble, France, 8 August 2004; pp. 85–94. [CrossRef]
30. Zhou, S.; Liu, X.; Wang, C.; Yang, B. Non-iterative denoising algorithm based on a dual threshold for a 3D point cloud. *Opt. Lasers Eng.* **2020**, *126*, 105921. [CrossRef]
31. Ning, X.; Li, F.; Tian, G.; Wang, Y. An efficient outlier removal method for scattered point cloud data. *PLoS ONE* **2018**, *13*, 1–22. [CrossRef] [PubMed]
32. Wang, Y.; Feng, H.Y. A rotating scan scheme for automatic outlier removal in laser scanning of reflective surfaces. *Int. J. Adv. Manuf. Technol.* **2015**, *81*, 705–716. [CrossRef]
33. Rosman, G.; Dubrovina, A.; Kimmel, R. Patch-collaborative spectral point-cloud denoising. *Comput. Graph. Forum* **2013**, *32*, 1–12. [CrossRef]
34. Schall, O.; Belyaev, A.; Seidel, H.-P. Robust filtering of noisy scattered point data. In Proceedings of the Proceedings Eurographics/IEEE VGTC Symposium Point-Based Graphics, Stony Brook, NY, USA, 21–22 June 2005; pp. 71–144.
35. Srinivasan, V.; Shakarji, C.M.; Morse, E.P. On the Enduring Appeal of Least-Squares Fitting in Computational Coordinate Metrology. *J. Comput. Inf. Sci. Eng.* **2012**, *12*, 011008. [CrossRef]
36. Kacker, R.; Sommer, K.D.; Kessel, R. Evolution of modern approaches to express uncertainty in measurement. *Metrologia* **2007**, *44*, 513–529. [CrossRef]
37. Kessel, W. Measurement uncertainty according to ISO/BIPM-GUM. *Thermochim. Acta* **2002**, *382*, 1–16. [CrossRef]
38. Hoffman, P.; Grinstein, G.; Marx, K.; Grosse, I.; Stanley, E. DNA visual and analytic data mining. *Proc. IEEE Vis. Conf.* **1997**, 437–441. [CrossRef]

Article

Unmanned Aerial Vehicles Motion Control with Fuzzy Tuning of Cascaded-PID Gains [†]

Fabio A. A. Andrade [1,2,*], Ihannah P. Guedes [3], Guilherme F. Carvalho [3], Alessandro R. L. Zachi [3], Diego B. Haddad [3], Luciana F. Almeida [3], Aurélio G. de Melo [4] and Milena F. Pinto [3]

- [1] Department of Microsystems, Faculty of Technology, Natural Sciences and Maritime Sciences, University of South-Eastern Norway (USN), 3184 Borre, Norway
- [2] NORCE Norwegian Research Centre, 5838 Bergen, Norway
- [3] Federal Center of Technological Education of Rio de Janeiro (CEFET/RJ), Rio de Janeiro 20271-110, Brazil; ihannah.guedes@aluno.cefet-rj.br (I.P.G.); guilherme.carvalho@aluno.cefet-rj.br (G.F.C.); alessandro.zachi@cefet-rj.br (A.R.L.Z.); diego.haddad@cefet-rj.br (D.B.H.); luciana.almeida@cefet-rj.br (L.F.A.); milenafaria@ieee.org (M.F.P.)
- [4] Department of Electrical Engineering, Federal University of Juiz de Fora, Juiz de Fora 36036-900, Brazil; aurelio.melo@engenharia.ufjf.br
- * Correspondence: fabio.a.andrade@usn.no
- [†] This paper is an extended version of the conference paper published in: Carvalho, G.; Guedes, I.; Pinto, M.; Zachi, A.; Almeida, L.; Andrade, F.; Melo, A.G. Hybrid PID-Fuzzy controller for autonomous UAV stabilization. In Proceedings of the 2021 14th IEEE International Conference on Industry Applications (INDUSCON), São Paulo, Brazil, 16–18 August 2021; IEEE: Piscataway, NJ, USA, 2021; pp. 1296–1302.

Abstract: One of the main challenges of maneuvering an Unmanned Aerial Vehicle (UAV) to keep a stabilized flight is dealing with its fast and highly coupled nonlinear dynamics. There are several solutions in the literature, but most of them require fine-tuning of the parameters. In order to avoid the exhaustive tuning procedures, this work employs a Fuzzy Logic strategy for online tuning of the PID gains of the UAV motion controller. A Cascaded-PID scheme is proposed, in which velocity commands are calculated and sent to the flight control unit from a given target desired position (waypoint). Therefore, the flight control unit is responsible for the lower control loop. The main advantage of the proposed method is that it can be applied to any UAV without the need of its formal mathematical model. Robot Operating System (ROS) is used to integrate the proposed system and the flight control unit. The solution was evaluated through flight tests and simulations, which were conducted using Unreal Engine 4 with the Microsoft AirSim plugin. In the simulations, the proposed method is compared with the traditional Ziegler-Nichols tuning method, another Fuzzy Logic approach, and the ArduPilot built-in PID controller. The simulation results show that the proposed method, compared to the ArduPilot controller, drives the UAV to reach the desired setpoint faster. When compared to Ziegler-Nichols and another different Fuzzy Logic approach, the proposed method demonstrates to provide a faster accommodation and yield smaller errors amplitudes.

Keywords: control strategy; UAV; fuzzy; PID controller; ROS

Citation: Andrade, F.A.A.; Guedes, I.P.; Carvalho, G.F.; Zachi, A.R.L.; Haddad, D.B.; Almeida, L.F.; de Melo, A.G.; Pinto, M.F. Unmanned Aerial Vehicles Motion Control with Fuzzy Tuning of Cascaded-PID Gains. *Machines* 2022, 10, 12. https://doi.org/10.3390/machines10010012

Academic Editors: Marcos de Sales Guerra Tsuzuki, Marcosiris Amorim de Oliveira Pessoa and Alexandre Acássio

Received: 18 November 2021
Accepted: 20 December 2021
Published: 23 December 2021

Publisher's Note: MDPI stays neutral with regard to jurisdictional claims in published maps and institutional affiliations.

Copyright: © 2021 by the authors. Licensee MDPI, Basel, Switzerland. This article is an open access article distributed under the terms and conditions of the Creative Commons Attribution (CC BY) license (https://creativecommons.org/licenses/by/4.0/).

1. Introduction

Recently, the use of autonomous vehicles and robotics technologies has increased significantly. Such systems are now being used to perform a great number of tasks in an optimized manner. Most traditional solutions demanded human resources, which may provide gaps and cause unsafe working places or human workers' depletion due to repetitive tasks. Human safety issues are taken into account in some autonomous unmanned vehicle-related tasks in [1–4].

The field of Unmanned Aerial Vehicles (UAVs) is gaining a growing interest over the past years due to the possibility of enabling new services that help modernize transportation tasks [5], inspection [6], supply chain support [7], search and rescue activities [8], change detection in water scenes [9], air quality assessment (e.g., by measurements of gaseous elemental mercury) [10], early wildfire detection [11], delivery goods tasks [12,13], information warfare [14], topographic surveys in active mines [15], plant genotyping [16], documentation and inspection of historical buildings [17], among others. The reason for using them in several applications can be explained by the UAV's ability to perform complex activities with low-cost flight operation and maneuvering flexibility [18,19].

Autonomous or semi-autonomous UAVs have been used to substitute human workers in different tasks in order to reduce maintenance costs and intervention times, especially in inspections [20,21]. There are different ways of performing such inspections with a UAV. In these all different flying situations, the UAV can fly very close to the object with a slow speed, producing valuable and reliable information about the inspected area. According to [1], the use of UAV can help to reduce the the mission's complexity for data gathering due to its high versatility, and the possibility of attaching new technologies into it. In this work, this and some other aspects are also taken in place to use as work motivation.

The UAV should also be capable of performing these missions stably. Therefore, there is also great importance in controlling the UAV's movement itself. For example, the authors of [22] have used a self-tuned PID control method to deal with external disturbances in a quadrotor UAV. In [23], the authors have proposed a hybrid PID control strategy to overcome sensor noise and strong wind disturbances. Several works in the literature have been proposed to make UAVs more robust to disturbances, parametric uncertainties, among other problems. For example, the control method proposed in [24] is a fault-tolerant strategy that takes into account system uncertainties and actuator failures. In [25], the authors have proposed a flight control for a quadrotor UAV for hovering with a slung load attached to it. The mathematical model was simplified to several controllable linear subsystems via reasonable assumptions. A robust H_∞ controller was designed by utilizing the estimated states of a state observer. Other works have proposed robust control techniques for compensating for the effects of external disturbances and uncertainties in the UAV model parameters [26–29].

The physical instability of the UAV's platform causes motion in the acquired videos, which imposes harmful impacts on the accuracy of camera-based measurements [30]. These issues, among others, motivate the adoption of flight stabilization techniques, which allow the adaptation to operational changes based on the knowledge of dynamical properties [31]. They commonly use a navigation system to feed a classical PID controller, which has a simple structure, good stability, and less dependence on the exact system model. Although the PID controller has a simple structure to be implemented, the process of adjusting its parameters requires attention from the designer, particularly when nonlinearities are present. This is an issue that has been receiving growing attention. Such dynamical characteristics force the PID design and tuning to become even more complex, demanding an additional control approach. Computational Intelligence techniques can be used to optimize the PID gains, as seen in [32], where the PID gains were tuned by using Particle Swarm Optimization technique, and in [33], where Genetic Algorithm was used. The Fuzzy Logic Theory has also emerged as a solution for dealing with systems that are not easy to be modeled because of their nonlinearities and undetermined states. In this sense, many researchers have applied fuzzy controllers to obtain improved performance, and robustness properties compared to those that use pure classical control algorithms [34–38].

Note that the fuzzy-based control is considered as a control scheme that can improve the system's robustness and adaptability. This approach can be used to dynamically adjust the controller parameters in accordance with the output [39]. The authors in [40], proposed the use of a fuzzy PID scheme to control the attitude of a UAV. They used the fuzzy to adjust the controller parameters by inference rules. A similar scheme was proposed in [41]. The results showed that the UAV obtained better dynamics and stable performance.

In this work, a hybrid approach composed of a Cascaded-PID and a Fuzzy Logic controller is implemented. Due to the Cascaded-PID module, the proposed approach offers the system adaptive capabilities engendered by the Fuzzy Logic part and a robustness property against parametric uncertainties.

The focus of the devised approach is to propose a controller that has the robustness of a PID, but also that could be applied to many different scenarios. PIDs are widely used in the context of machine control and stabilization. However, the values of the Proportional, Integrative, and Derivative gains rely directly on the plant model. It can be very challenging to choose values that will fit the best way in all operational situations that this kind of robot can use. In order to amplify the range of use for those UAVs, the insertion of a fuzzy-based algorithm is implemented. In this case, the fuzzy would not control the movement speed and position itself, as usually is seen in state of the art, but would provide the adapted PID gains to the system and develop an optimized new PID controller to it.

Therefore, the main contributions of this work are:

- To propose a novel control strategy for UAVs, combining a Cascaded-PID with a fuzzy logic controller;
- To provide a method for fine tuning the fuzzy range of the PID gains values;
- To present a solution that can be embedded on UAV's companion computers using ROS;
- To provide a testing solution using a state-of-the-art high-fidelity simulation engine such as Unreal Engine with AirSim.

This paper is structured as follows. Section 2 described the advanced method, whereas Section 3 describes some promising results and Section 4 the discussion. At last, Section 5 concludes the paper by furnishing the final conclusions.

2. Materials and Methods

The proposed scheme combines a Cascaded-PID with a Fuzzy algorithm that is responsible for calculating the PID gains. In the so-called fuzzy controller, the control strategy is described through linguistic rules that imprecisely connect various situations with the actions taken. Different from the traditional PID controller, a formal mathematical model of the plant is not necessary. Approximately knowing the UAV's behavior when exposed to different inputs is enough for defining the fuzzy rules, which is a feasible task to UAV specialists. Therefore, these linguistic rules that define the control strategy represent the linguistic model of the plant. Note that Fuzzy and the PID can provide an effective solution to the system's non-linearity. As a result, the system can accurately converge to the desired position in fewer iterations.

In the beginning, suitable PID values may be defined for the PID controller. As time goes by, such gains are updated dynamically by the Fuzzy algorithm, whose rules are only dependent on the position error and its derivative.

Figure 1 depicts the designed iterative learning control algorithm, along with the real-time management of Fuzzy gain computation. The UAV Desired Position is the commanded waypoint that the UAV should go to. The desired position can be changed at any time during the process, allowing the system to follow a moving goal or a trajectory, for example.

It is important to note that one independent Fuzzy Logic controller must be set for each control axis (x, y and z). In this work, the methods, figures and tables will be only relative to the x-axis to avoid unnecessary repetition.

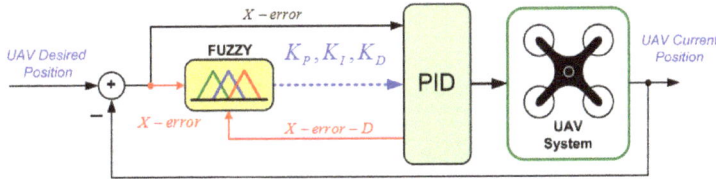

Figure 1. Scheme of fuzzy addition to the PID controller.

2.1. Cascaded-PID

The cascade strategy is a well-known scheme in control system design. It basically consists of the interaction between two control loops: an external one that is responsible for generating a reference signal and an internal, fed by the latter, which is responsible for computing the signals to drive the actuators.

The block diagram in Figure 2 illustrates the proposed simplified Cascaded-PID controller scheme. The external loop uses the position error to generate a setpoint reference value for the internal loop. Concerning the diagram, the internal control loop is responsible for controlling the UAV motors. The rotors' velocities are calculated after demonstrating that both roll and pitch angles converge to their (generated) reference values.

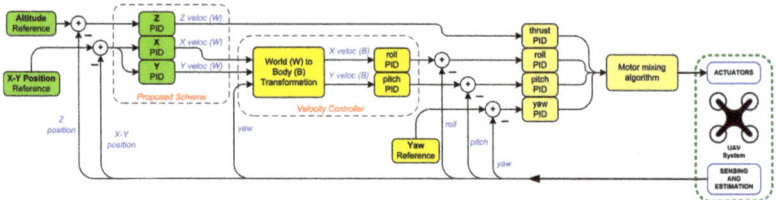

Figure 2. Schematic of the Simplified Cascaded-PID controller.

As can be seen in the diagram of Figure 2, the proposed scheme uses the altitude and position errors for generating and sending the velocity commands to the flight control unit controller.

2.2. ROS Integration

The general idea of the proposed methodology is illustrated in Figure 3. It consists of using a companion computer to perform the control of the UAV.

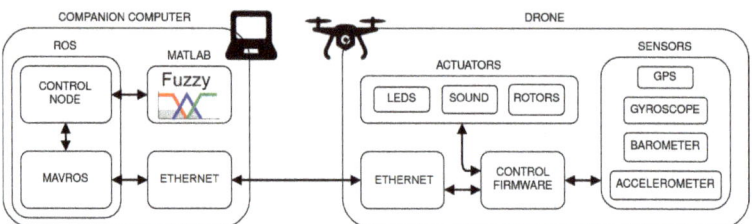

Figure 3. Signal flowchart and architecture.

This strategy allows a more straightforward implementation and testing tasks that are benefited from more advanced sensors, such as light detection and ranging and waypoint control integration. The Flight Management Unit (FMU) is responsible for acquiring information from the peripheral sensors and modifying these values in the actuators to which it has access. The primary sensor data are fed through the FMU to a specific ROS driver, the MAVROS. The data are then forwarded to a ROS node, in which the controller is implemented. The reverse-path takes place to perform motor driving at the end of the control loop.

The ROS interface with the UAV works as follows. The companion computer has the ROS core processing data in each of the code's pre-configured nodes. The computer receives the UAV state information from Ethernet using the MAVLink communication protocol. The UAV state is made available as a ROS topic by the MAVROS ROS driver. The calculations that enable the Cascaded-PID controller are performed in the main ROS node, which publishes the current PID gains and the errors to a topic that is used by the fuzzy logic module. The same ROS node subscribes to the messages with the new

PID gains sent by the fuzzy logic module. When it receives a message, it makes the PID calculations and transmits the target velocities to the FMU by publishing a MAVROS message. The gains are changed at a rate of 10 Hz, which is the best possible MAVLink rate.

2.3. Fuzzy Logic PID Tuner

In Figure 4, the MATLAB/Simulink block diagram with the Fuzzy Logic and the connection between Simulink and ROS are presented. It is shown, on the left side of the diagram, the subscription block that reads the information from the topics sent by ROS. The messages are divided in five variables, so they can be individually read by Simulink. The variable of the error derivative was suppressed, as it can be directly calculated from the error variable using Simulink, which guarantees that both have the same rate. Some displays are added only to keep track of evaluation during the tests. The the error and it's calculated derivative are sent to the Fuzzy Logic Controller With Ruleviewer block. This particular block is responsible for reading the fuzzy controller file and performing the fuzzy logic calculations. After that, the values from the fuzzy controller are sent to ROS, on the publish block on the bottom of the figure. The ROS messages of "fuzzy_values" and "defuzzy_values" were created for this work.

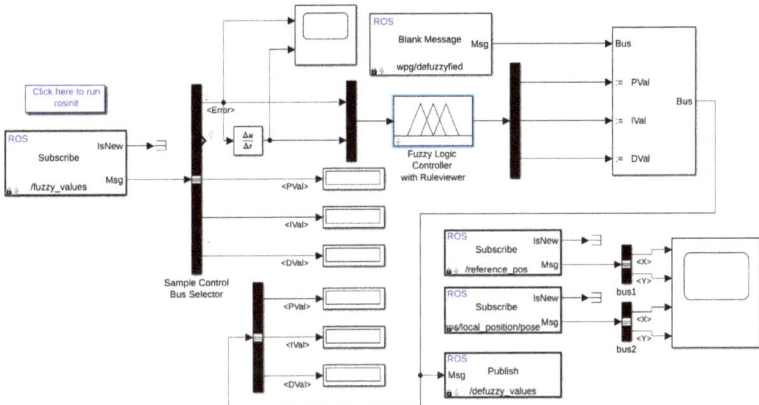

Figure 4. Simulink model for the fuzzy controller.

The Membership Functions types were chosen as Gaussian for the inputs and triangular for the outputs. This is the setup present in most of the literature cited by this work when it comes to PID gains tuning with fuzzy logic. The existing works were followed as a start point regarding the rules, and the rules were defined according to the UAV's desired behavior. For instance, if the position error X-error and its time derivative X-error-D are sufficiently small in the present approach, there is no need for large PID gain values. Since no "crisp functions" were used, final gains regularly tend to be slightly different from those in the center of the membership function. The fuzzy inference technique adopted in this work is based on the classical Mamdani [42] inference method and on the centroid defuzzification method.

The fuzzy rule Table 1 can be seen as the following. The symbol 'B' represents 'Big', 'M' for 'Medium', and 'L' for 'Low.' The outputs should be interpreted as K_P, K_D, and Alpha gains. Hence, the output notation 'LMB' should mean that the proportional (K_P) gain is Low, the derivative (K_D) gain is Medium, and Alpha gain is Big. For the integrative (K_I) gain, the formula below is used [43]. As the desired behavior of the controller is the same for when the position error is positive or negative, the absolute value of the error was used. Regarding the derivative, it was the derivative of the absolute value of the error.

$$K_I = K_P^2/(\alpha K_D) \qquad (1)$$

Table 1. Fuzzy Rule table.

		Error	
Error − D	**Low**	**Medium**	**Big**
Negative	LBB	MMM	BLL
Zero	BBM	BML	BLL
Positive	LBB	MMM	BLL

Selecting the Range of PID Gains for Horizontal Motion Control

The methodology to define the range of each of the PID gains for the horizontal motion control (x and y axis) followed [44], and [43] with a modification proposed by this work.

First, the ultimate gain (K_u) and the oscillation period (T_u) [45] are obtained. This is done by removing the derivative and integral parts of the PID and increasing the proportional gain until the point that the output of the control loop has stable and consistent oscillations. In this work, the measured ultimate gain was of 2.2 and the oscillation period of 4 s.

With the measured values of K_u and T_u, the Zhao/Larson gains $K_{P',min}$, $K_{P',max}$, $K_{D',min}$ and $K_{D',max}$ were calculated:

$$K_{P',min} = 0.32 K_u, \quad K_{P',max} = 0.6 K_u, \tag{2}$$

$$K_{D',min} = 0.08 K_u P_u, \quad K_{D',max} = 0.15 K_u P_u. \tag{3}$$

Provided that the traditional Ziegler-Nichols exhibits a high overshoot and that the overshoot is reduced but not removed by the above described method, this work proposes the following modification. The modification is to use a compression factor to reduce the fuzzy range of the PID gains, which will smooth the control signal and make it more suitable for UAV applications, where a fast convergence is desired but without a very high overshoot.

Therefore, the actual adopted gains $K_{P,min}$, $K_{P,max}$, $K_{D,min}$ and $K_{D,max}$ are obtained as follows:

$$K_{P,min} = K_{P',min} - \frac{K_{P',min}}{7}, \quad K_{P,max} = K_{P',min} + \frac{K_{P',max}}{7}, \tag{4}$$

$$K_{D,min} = K_{D',min} - \frac{K_{D',max}}{7}, \quad K_{D,max} = K_{D',min} + \frac{K_{D',max}}{7}. \tag{5}$$

Finally, the summary of the limits can be seen in Figure 5, where the left shows the two inputs and the right shows the three outputs.

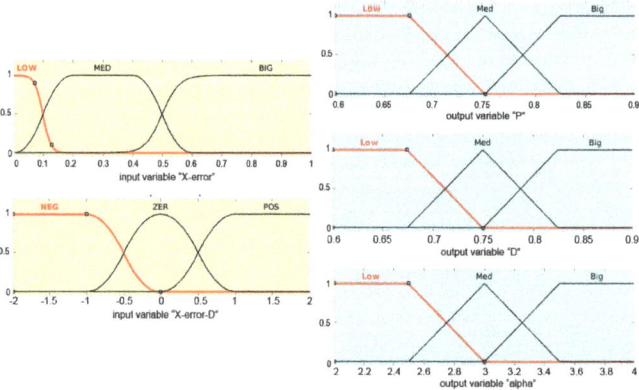

Figure 5. Fuzzy membership functions.

3. Results

3.1. Horizontal Motion Control Tests in the Simulation Environment

The proposed solution was tested in a simulation environment and compared with other solutions. The other solutions adopted for comparisons are:

- ArduPilot built-in GUIDED mode controller.
- Cascaded-PID solution tuned by the Ziegler-Nichols method.
- Cascaded-PID solution tuned by the methods in [44], which were detailed in [43].

The tests were performed in a simulation environment (Figure 6) that runs in Windows operating system with Windows Subsystem for Linux (WSL) Ubuntu 18.04.5 LTS 64 bits. The computer has an Intel i9-9900K 3600 MHz processor with an NVIDIA GeForce RTX 2080Ti graphics card and 48 GB RAM.

Figure 6. Simulation Environment provided by the software Unreal Engine 4 with AirSim.

The software Unreal Engine 4 (UE4) with the AirSim plugin [46] was used for the simulations trials. AirSim was developed by Microsoft, and is an abbreviation for Aerial Informatics and Robotics Simulation. This plugin program provided some important resources for this work, especially with regard to input/output signals, corresponding to the behavior of sensors and actuators, sent to and received from the UAV flight controllers. In the tests, the UAV was commanded to go from position 2 m at East from takeoff position to −2 m. Therefore, a total movement of 4 m was performed. The command to move was sent around 1.6 s after starting recording the logs and building the graphs.

Simulation Results

First, the three tuning methods for the Cascaded-PID were evaluated (Figure 7). It is possible to notice that, as already shown in [43,44], the Zhao/Larson method achieves the same convergence time as the Ziegler-Nichols but with lower overshoot and faster accommodation profile. In UAV applications, fast convergence is desirable, but the accommodation property is more important, as the UAV usually needs to acquire data with onboard sensors. Therefore, the solution proposed in this work is better suited for UAVs as it can achieve accommodation in a faster way with significantly less overshoot. By considering an acceptable error of 1 cm, the proposed solution reaches accommodation in 6.7 s, against 10.8 s achieved by using the Zhao/Larson tuning approach and more than 20 s achieved with the Ziegler-Nichols tuning method.

The proposed method was also compared with ArduPilot's built-in controller in GUIDED mode, in which ArduPilot accepts target position commands and updates the velocity control at a rate of 50 Hz. A ROS message of the target position was sent using MAVROS. The result is presented in Figure 8. From the curve, it is possible to conclude that the ArduPilot achieves a shorter time of accommodation. However, the proposed method performed very well to reach the desired position, being 1.5 s faster than the ArduPilot, with a small overshoot of less than 0.5 m (Figure 9).

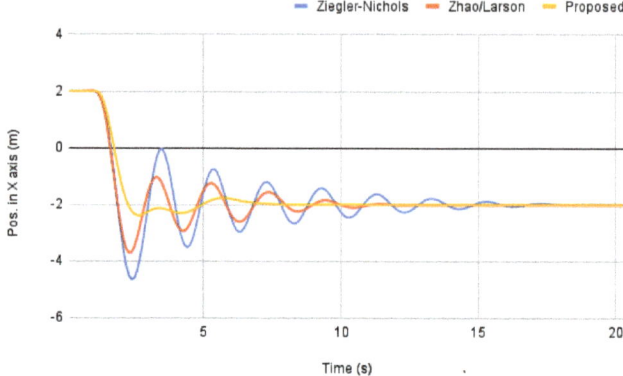

Figure 7. The comparison among Cascaded-PID tuning methods.

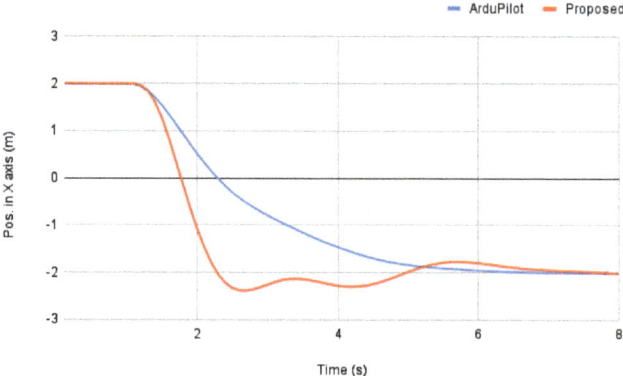

Figure 8. ArduPilot and Proposed method compared.

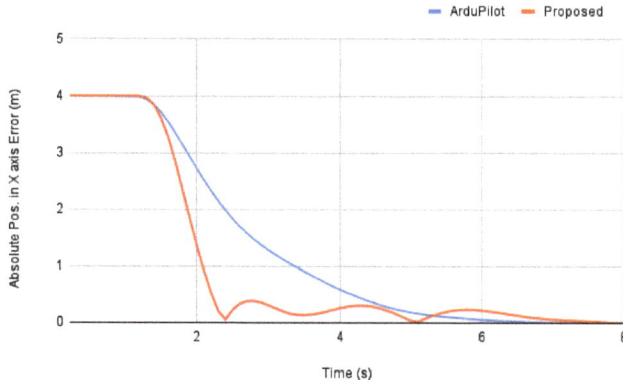

Figure 9. Absolute Error comparison between ArduPilot and the proposed method.

It is important to note that the ArduPilot controller is very advanced and has been developed for many years, counting with a big team of developers and contributors.

Furthermore, in comparison with ArduPilot's results, in Figure 10 the accumulated error can be analyzed. The accumulated error is a measure that shows how much the UAV was away from the desired position over time. The upper line shows the ArduPilot's accumulated error in terms of position in X axis, while the lower line represents the results of this proposed method. Both have crescent errors before two seconds, but it is shown that the ArduPilot stabilizes this error almost in 6 s, while the proposed method takes more than that to be a fairly horizontal line. However, the most meaningful information in this comparison is that the difference between both methods at the end of the experimentation is almost 25 m. This means that counting on every small error that both had on the trajectory tests, the proposed method shows itself significantly more accurate in terms of position in x.

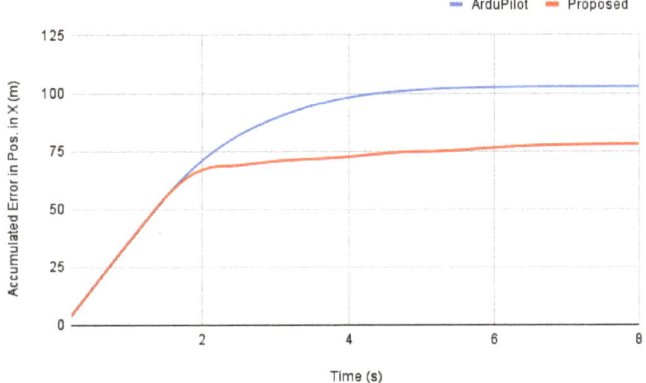

Figure 10. Accumulated Error comparison between ArduPilot and the proposed method.

Regarding the Fuzzy Logic PID gains used in the proposed method, they can be observed in Figure 11, by the side of a graph with the calculated error and error time derivative. The peak of the derivative error, in the beginning, is due to discontinuity generated by variable initialization and should not be considered. As explained in the previous section, the error curve is shown in absolute value.

The behavior of the PID gains variations shows that the Fuzzy Logic was properly configured as in the beginning when the error exhibits large values, the derivative gain has lower values but increases every time the error is approaching the setpoint. In addition, the derivative gain tends to reduce even more if the derivative of the error is negative, meaning that the UAV is progressing correctly. The integral gain tended to increase if there is an error for a long time, being smoothly reduced if the UAV is around the setpoint for some time. Regarding the proportional gain, it is possible to observe that it is basically directly following the error.

A fact that deserves to be highlighted in the graphs in Figure 11b are the oscillatory behaviors of the P, I and D gain estimates. These behaviors are probably caused by the measurement of the error derivative illustrated in Figure 12a, and are also propagated to the velocity command represented in Figure 12a. Although this type of oscillatory behavior is undesirable, it did not affect the actual measured velocity (Figure 12b). However, the issue regarding the chattering mitigation needs to be further investigated in future work.

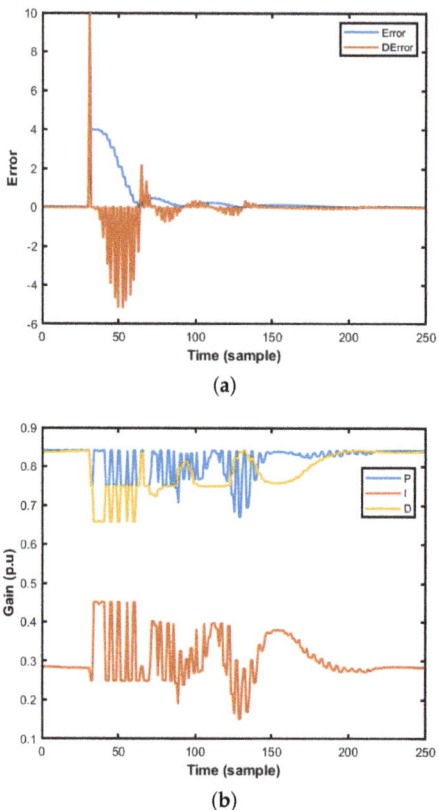

Figure 11. (**a**) Measured errors; and (**b**) Fuzzy PID gains.

Finally, the commanded velocity and the measured velocity can be compared in Figure 12. The sample times are different because the commands are sent at a rate 2.5 times superior than the measurements. Therefore, the graphs have the same time range in seconds.

It can be observed that the actual measured velocity of the UAV does not reach the commanded velocity. That happens because new commanded velocities are sent before the UAV can reach the previously sent one. In addition, the UAV has a limited maximum velocity, which was set to 3 m/s. Therefore, if the Fuzzy logic Cascaded-PID calculates a commanded velocity larger than 3 m/s, as it happened in the beginning as the UAV was steady and the error was big, the UAV is not able to immediately reach that velocity because of inertia.

In another perspective, in Figure 13, the proposed method is compared to fixed PID gains using the mean values of the Fuzzy Logic gains. This clarifies the importance of the adaptive controller for such purposes, while the mean values of fuzzy gains present still reasonable results, the variation of the fuzzy gains shows a smaller overshoot on the first attempt to stabilize, oscillates less than the mean-gains approach, and achieve the desired position in X axis before than the mean-values method. Still compared to a the method using fuzzy gains' mean values, the accumulated position error in X axis is slightly smaller for the proposed method at the end of the experimentation in Figure 14. The accumulated error from the proposed method is greater than the mean-values method only near 7 s, where it is trying to stabilize for the last time, with no further oscillations, while the mean-values method oscillates and thus briefly achieves the objective position.

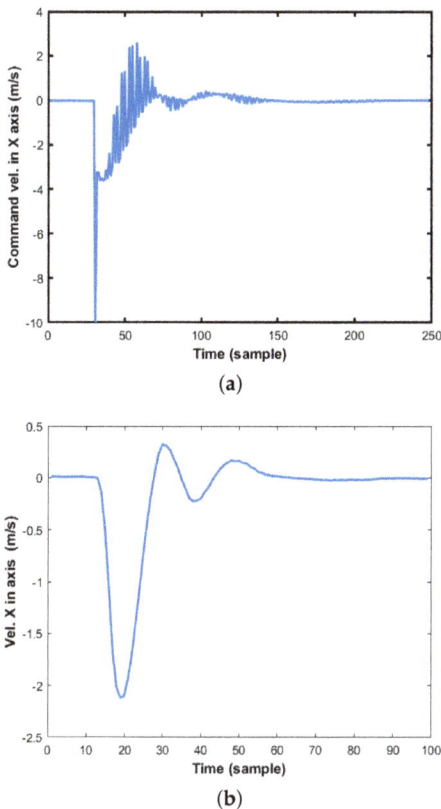

Figure 12. (**a**) Commanded; and (**b**) Measured velocities.

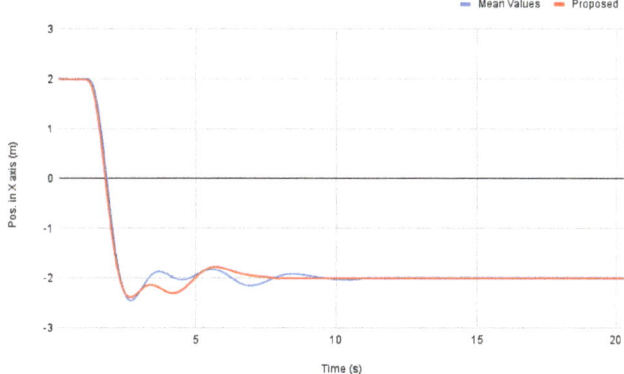

Figure 13. Comparison between the proposed method and using the mean values of the fuzzy logic gains.

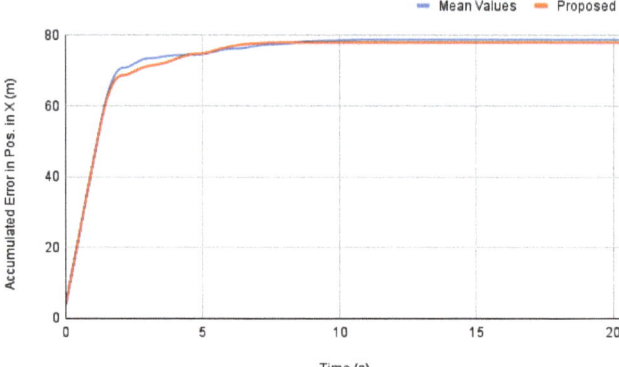

Figure 14. Accumulated Error comparison between the proposed method and using the mean values of the fuzzy logic gains.

Finally, a last simulation was performed, when a wind of 15 m per second was introduced. The results can be seen in Figure 15. It is possible to observe that the proposed method has a better performance than the approach using only the mean values. In addition, it achieves nearly the same time of accommodation than the Ardupilot solution, with faster convergence.

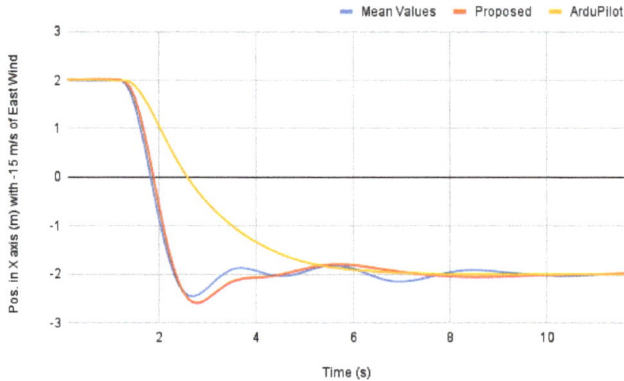

Figure 15. Comparison among different methods in the presence of 15 m/s wind to the West.

3.2. Altitude Control Tests in Real Flight

In the real implementation, the multirotor Parot bebop 2 was used. This UAV is shown in Figure 16. Its system provides a compatible network interface for ROS, by furnishing the required sensor information as well as the control interface. Any other compatible UAV can also be used, replacing only the proper interfaces.

Only height control is used in the test, while the original FMU algorithm still handles stabilization and position control. This ensures safety once the UAV is stable and locked into position. Figure 17 shows the UAV flight on a blue screen room during the test.

Figure 16. The Parrot Bebop 2.

Figure 17. The UAV height control flight test.

In order to avoid discontinuous movements, the method was tested using ramp profiles from one altitude setpoint to the next one. Figure 18 shows the control signal and the error measurements. The dotted line represents the overall behavior of the control signal that is applied in discrete time instants. Note that the error has a well-behaved time profile. The slightly oscillatory behavior is due to the effects of environmental conditions.

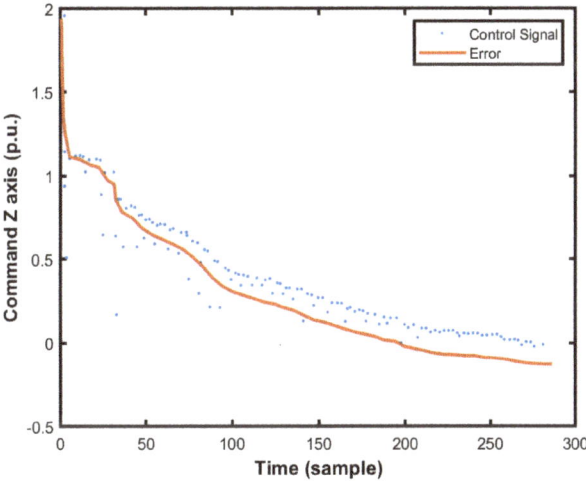

Figure 18. Control signal and error during the experimental trial.

The control PID gains estimated by the proposed fuzzy algorithm are shown in Figure 19. Their performances are within the expected patterns. It is also noticeable that the derivative estimation changes were more aggressive than the other two.

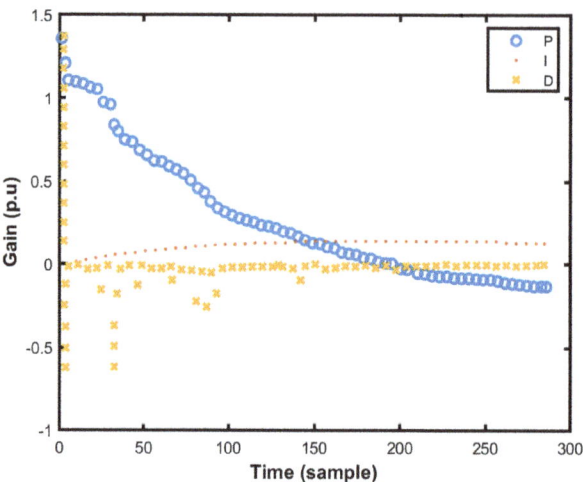

Figure 19. Individual control response during the experiment.

4. Discussion

In the literature, most of the Fuzzy Logic tuning solutions have rules defining the PID gains (K_P, K_D, and K_I) based on the output error and its time derivative. In this work, however, the absolute value of the output error is used to generate the Fuzzy rules. With this strategy, the deviation between the desired and current UAV position always assumes positive or null values, which facilitates the definition of "low", "medium" and "high" levels for the fuzzy algorithm. In the literature, it is common to find the same definitions of such levels for both the positive and negative excursions of the error time derivative, which is the same strategy adopted in fuzzy rules proposed in this work. Based on the observations made on the results obtained, we emphasize that the definition of different rules for the negative and positive excursions of the error derivative seems to influence the performance of the closed-loop and that, therefore, it should be an issue to be investigated in a future work.

In addition, one of the main challenges when using Fuzzy Logic for PID online tuning is to define a proper range of K_P, K_D, and K_I values. This work presents a method for defining a proper range for the horizontal motion PID gains specifically applied to UAVs based on the existing literature and with a novel modification that smooths the behavior. Regarding the altitude controller, the range of PID gains was chosen empirically. Similar tuning methods could have been used for this purpose, such as the Ziegler-Nichols ultimate gain and oscillation period method, or some variant tuning scheme as fast as the Zhao/Larson method, but with better settling time.

Regarding the results, some oscillations were verified on the commanded PID gains, which may be due to different rates among the different systems (ROS, FMU, MATLAB/Simulink). This affected the commanded velocity but did not affect the actual measured velocity. The MATLAB/Simulink was used for this solution's proof of concept version, and the Fuzzy Logic algorithm should be embedded in the ROS Python or C++ scripts in future work. By doing so, we expect to achieve a faster and more synchronized system.

In terms of evaluation, this research work opens up several future possibilities. For instance, it is expected the deployment of this method in a real UAV inspection mission.

Besides, the authors intend to insert the proposed control algorithm along with an optical flow algorithm for performing autonomous flight.

5. Conclusions

Nowadays, UAVs have been widely applied in diverse applications because of their flexibility of maneuvers, reducing risks to human life, ease to control, and cost-effectiveness. A drawback is the necessity of a platform with a high degree of stability to prevent the UAVs from falling and maintaining the desired orientation and path during the flight. Many solutions rely on PID controllers. However, its design and adjustment can be difficult. Therefore, this work proposed a novel approach composed of a Cascaded-PID Fuzzy Logic controller, where the system have adaptive capabilities engendered by the Fuzzy Logic part.

In this specific scenario, the fuzzy rules along with the membership functions play a crucial role in defining whether PID gains should prioritize stability. For instance, enhancing derivative gain in a constant position (zero error) scenario or prioritizing high gain PI for faster response when position error and its time derivative are both large.

Simulations and flight tests were conducted to demonstrate the effectiveness of the proposed method. As it can be seen from the results and discussions, when compared to the traditional Ziegler-Nichols method and another fuzzy method from the literature, the proposed solution provides much faster accommodation and smaller error amplitudes. Regarding the comparison with the built-in controller present in the ArduPilot flight control unit, the proposed solution achieves faster convergence and smaller accumulated error.

Author Contributions: Conceptualization, F.A.A.A. and M.F.P.; methodology, F.A.A.A. and I.P.G.; software, F.A.A.A., I.P.G. and G.F.C.; validation, M.F.P. and A.R.L.Z.; formal analysis, A.R.L.Z., L.F.A. and D.B.H.; investigation, F.A.A.A., M.F.P. and A.R.L.Z.; writing—original draft preparation, F.A.A.A., I.P.G., A.G.d.M. and G.F.C.; writing—review and editing, L.F.A., M.F.P., A.R.L.Z. and D.B.H.; visualization, F.A.A.A., A.G.d.M. and M.F.P.; supervision, F.A.A.A., M.F.P. and D.B.H.; project administration, F.A.A.A. and M.F.P.; funding acquisition, M.F.P., A.R.L.Z. and D.B.H. All authors have read and agreed with the submission of the current manuscript version.

Funding: The authors also would like to thank their home Institute CEFET/RJ, the federal Brazilian research agencies CAPES (code 001) and CNPq, and the Rio de Janeiro research agency FAPERJ, for supporting this work.

Institutional Review Board Statement: Not applicable.

Informed Consent Statement: Not applicable.

Data Availability Statement: The source-codes are openly available in https://github.com/piradata/wpg, accessed on 23 December 2021.

Conflicts of Interest: The authors declare no conflict of interest.

Abbreviations

The following abbreviations are used in this manuscript:

UAV	Unmanned Aerial Vehicle
ROS	Robot Operating System
GPS	Global Positioning System
IMU	Inertial Measurement Unit
PID	Proportional-Integral-Derivative
FMU	Flight Management Unit

References

1. Biundini, I.Z.; Pinto, M.F.; Melo, A.G.; Marcato, A.L.; Honório, L.M.; Aguiar, M.J. A Framework for Coverage Path Planning Optimization Based on Point Cloud for Structural Inspection. *Sensors* **2021**, *21*, 570. [CrossRef]
2. Pinto, M.F.; Honorio, L.M.; Melo, A.; Marcato, A.L. A Robotic Cognitive Architecture for Slope and Dam Inspections. *Sensors* **2020**, *20*, 4579. [CrossRef] [PubMed]

3. Kong, W.; Zhou, D.; Yang, Z.; Zhao, Y.; Zhang, K. UAV Autonomous Aerial Combat Maneuver Strategy Generation with Observation Error Based on State-Adversarial Deep Deterministic Policy Gradient and Inverse Reinforcement Learning. *Electronics* **2020**, *9*, 1121. [CrossRef]
4. Pinto, M.F.; Honório, L.M.; Marcato, A.L.; Dantas, M.A.; Melo, A.G.; Capretz, M.; Urdiales, C. ARCog: An Aerial Robotics Cognitive Architecture. *Robotica* **2021**, *39*, 483–502. [CrossRef]
5. Cabral, K.M.; dos Santos, S.R.B.; Givigi, S.N.; Nascimento, C.L. Design of model predictive control via learning automata for a single UAV load transportation. In Proceedings of the 2017 Annual IEEE International Systems Conference (SysCon), Montreal, QC, Canada, 24–27 April 2017; IEEE: Piscataway, NJ, USA, 2017; pp. 1–7.
6. Melo, A.G.; Pinto, M.F.; Marcato, A.L.; Honório, L.M.; Coelho, F.O. Dynamic Optimization and Heuristics Based Online Coverage Path Planning in 3D Environment for UAVs. *Sensors* **2021**, *21*, 1108. [CrossRef]
7. Fernández-Caramés, T.M.; Blanco-Novoa, O.; Froiz-Míguez, I.; Fraga-Lamas, P. Towards an autonomous industry 4.0 warehouse: A UAV and blockchain-based system for inventory and traceability applications in big data-driven supply chain management. *Sensors* **2019**, *19*, 2394. [CrossRef]
8. Pinto, M.F.; Melo, A.G.; Marcato, A.L.; Urdiales, C. Case-based reasoning approach applied to surveillance system using an autonomous unmanned aerial vehicle. In Proceedings of the 2017 IEEE 26th International Symposium on Industrial Electronics (ISIE), Edinburgh, UK, 19–21 June 2017; IEEE: Piscataway, NJ, USA, 2017; pp. 1324–1329.
9. Li, X.; Duan, H.; Li, J.; Deng, Y.; Wang, F.Y. Biological eagle eye-based method for change detection in water scenes. *Pattern Recognit.* **2022**, *122*, 108203. [CrossRef]
10. Cabassi, J.; Lazzaroni, M.; Giannini, L.; Mariottini, D.; Nisi, B.; Rappuoli, D.; Vaselli, O. Continuous and near real-time measurements of gaseous elemental mercury (GEM) from an Unmanned Aerial Vehicle: A new approach to investigate the 3D distribution of GEM in the lower atmosphere. *Chemosphere* **2022**, *288*, 132547. [CrossRef]
11. Bouguettaya, A.; Zarzour, H.; Taberkit, A.M.; Kechida, A. A review on early wildfire detection from unmanned aerial vehicles using deep learning-based computer vision algorithms. *Signal Process.* **2022**, *190*, 108309. [CrossRef]
12. Murray, C.C.; Raj, R. The multiple flying sidekicks traveling salesman problem: Parcel delivery with multiple drones. *Transp. Res. Part C Emerg. Technol.* **2020**, *110*, 368–398. [CrossRef]
13. Madridano, Á.; Al-Kaff, A.; Martín, D. 3d trajectory planning method for UAVs swarm in building emergencies. *Sensors* **2020**, *20*, 642. [CrossRef]
14. Wan, L.; Liu, R.; Sun, L.; Nie, H.; Wang, X. UAV swarm based radar signal sorting via multi-source data fusion: A deep transfer learning framework. *Inf. Fusion* **2022**, *78*, 90–101. [CrossRef]
15. Zapico, I.; Laronne, J.B.; Sánchez Castillo, L.; Martín Duque, J.F. Improvement of Workflow for Topographic Surveys in Long Highwalls of Open Pit Mines with an Unmanned Aerial Vehicle and Structure from Motion. *Remote Sens.* **2021**, *13*, 3353. [CrossRef]
16. An Unmanned Aerial Vehicle for Greenhouse Navigation and Video-Based Tomato Phenotypic Data Collection. In Proceedings of the 2021 ASABE Annual International Virtual Meeting, Anaheim, CA, USA, 11–14 July 2021; American Society of Agricultural and Biological Engineers: St. Joseph, MI, USA, 2021. [CrossRef]
17. Smrcka, D.; Baca, T.; Nascimento, T.; Saska, M. Admittance Force-Based UAV-Wall Stabilization and Press Exertion for Documentation and Inspection of Historical Buildings. In Proceedings of the 2021 International Conference on Unmanned Aircraft Systems (ICUAS), Athens, Greece, 15–18 June 2021; pp. 552–559. [CrossRef]
18. Casella, E.; Rovere, A.; Pedroncini, A.; Stark, C.P.; Casella, M.; Ferrari, M.; Firpo, M. Drones as tools for monitoring beach topography changes in the Ligurian Sea (NW Mediterranean). *Geo-Mar. Lett.* **2016**, *36*, 151–163. [CrossRef]
19. Pinto, M.F.; Marcato, A.L.; Melo, A.G.; Honório, L.M.; Urdiales, C. A framework for analyzing fog-cloud computing cooperation applied to information processing of UAVs. *Wirel. Commun. Mob. Comput.* **2019**, *2019*, 7497924. [CrossRef]
20. Addabbo, P.; Angrisano, A.; Bernardi, M.L.; Gagliarde, G.; Mennella, A.; Nisi, M.; Ullo, S.L. UAV system for photovoltaic plant inspection. *IEEE Aerosp. Electron. Syst. Mag.* **2018**, *33*, 58–67. [CrossRef]
21. Aghaei, M.; Bellezza Quater, P.; Grimaccia, F.; Leva, S.; Mussetta, M. Unmanned aerial vehicles in photovoltaic systems monitoring applications. In Proceedings of the European Photovoltaic Solar Energy 29th Conference and Exhibition, Amsterdam, The Netherlands, 22–26 September 2014; pp. 2734–2739.
22. Joyo, M.K.; Hazry, D.; Ahmed, S.F.; Tanveer, M.H.; Warsi, F.A.; Hussain, A. Altitude and horizontal motion control of quadrotor UAV in the presence of air turbulence. In Proceedings of the 2013 IEEE Conference on Systems, Process & Control (ICSPC), Kuala Lumpur, Malaysia, 13–15 December 2013; IEEE: Piscataway, NJ, USA, 2013; pp. 16–20.
23. Tanveer, M.H.; Hazry, D.; Ahmed, S.F.; Joyo, M.K.; Warsi, F.A.; Kamaruddin, H.; Razlan, Z.M.; Wan, K.; Shahriman, A. NMPC-PID based control structure design for avoiding uncertainties in attitude and altitude tracking control of quad-rotor (UAV). In Proceedings of the 2014 IEEE 10th International Colloquium on Signal Processing and its Applications, Kuala Lumpur, Malaysia, 7–9 March 2014; IEEE: Piscataway, NJ, USA, 2014; pp. 117–122.
24. Wang, B.; Zhang, Y. Adaptive Sliding Mode Fault-Tolerant Control for an Unmanned Aerial Vehicle. *Unmanned Syst.* **2017**, *5*, 209–221. [CrossRef]
25. Yuan, X.; Ren, X.; Zhu, B.; Zheng, Z.; Zuo, Z. Robust H Control for Hovering of a Quadrotor UAV with Slung Load. In Proceedings of the 2019 12th Asian Control Conference (ASCC), Kitakyushu, Japan, 9–12 June 2019; pp. 114–119.

26. Liu, D.; Liu, H.; Lewis, F.L.; Wan, Y. Robust Fault-Tolerant Formation Control for Tail-Sitters in Aggressive Flight Mode Transitions. *IEEE Trans. Ind. Inform.* **2020**, *16*, 299–308. [CrossRef]
27. Le Nhu Ngoc Thanh, H.; Hong, S.K. Quadcopter Robust Adaptive Second Order Sliding Mode Control Based on PID Sliding Surface. *IEEE Access* **2018**, *6*, 66850–66860. [CrossRef]
28. Eltag, K.; Aslamx, M.S.; Ullah, R. Dynamic stability enhancement using fuzzy PID control technology for power system. *Int. J. Control. Autom. Syst.* **2019**, *17*, 234–242. [CrossRef]
29. Hu, X.; Liu, J. Research on UAV Balance Control Based on Expert-fuzzy Adaptive PID. In Proceedings of the 2020 IEEE International Conference on Advances in Electrical Engineering and Computer Applications (AEECA), Dalian, China, 25–27 August 2020; IEEE: Piscataway, NJ, USA, 2020; pp. 787–789.
30. Ljubičić, R.; Strelnikova, D.; Perks, M.T.; Eltner, A.; Peña Haro, S.; Pizarro, A.; Dal Sasso, S.F.; Scherling, U.; Vuono, P.; Manfreda, S. A comparison of tools and techniques for stabilising unmanned aerial system (UAS) imagery for surface flow observations. *Hydrol. Earth Syst. Sci.* **2021**, *25*, 5105–5132. [CrossRef]
31. Megyesi, D.; Bréda, R.; Schrötter, M. Adaptive Control and Estimation of the Condition of a Small Unmanned Aircraft Using a Kalman Filter. *Energies* **2021**, *14*, 2292. [CrossRef]
32. Solihin, M.I.; Tack, L.F.; Kean, M.L. Tuning of PID controller using particle swarm optimization (PSO). In Proceeding of the International Conference on Advanced Science, Engineering and Information Technology, Bandar Baru Bangi, Malaysia, 14–15 January 2011; Volume 1, pp. 458–461.
33. Aly, A.A. PID parameters optimization using genetic algorithm technique for electrohydraulic servo control system. *Intell. Control Autom.* **2011**, *2*, 69. [CrossRef]
34. Doitsidis, L.; Valavanis, K.P.; Tsourveloudis, N.C.; Kontitsis, M. A framework for fuzzy logic based UAV navigation and control. In Proceedings of the IEEE International Conference on Robotics and Automation, 2004 (Proceedings. ICRA'04), New Orleans, LA, USA, 26 April–1 May 2004; IEEE: Piscataway, NJ, USA, 2004; Volume 4, pp. 4041–4046.
35. Dong, J.; He, B. Novel fuzzy PID-type iterative learning control for quadrotor UAV. *Sensors* **2019**, *19*, 24. [CrossRef] [PubMed]
36. Wang, Y.; Shi, Y.; Cai, M.; Xu, W.; Yu, Q. Optimization of air–fuel ratio control of fuel-powered UAV engine using adaptive fuzzy-PID. *J. Frankl. Inst.* **2018**, *355*, 8554–8575. [CrossRef]
37. Bucio-Gallardo, E.M.; Zavala-Yoé, R.; Ramírez-Mendoza, R.A. Mathematical Model and Intelligent Control of a Quadcopter, with Non-conventional Membership Functions. *J. Energy Power Eng.* **2016**, *10*, 634–642. [CrossRef]
38. Demaya, B.; Palm, R.; Boverie, S.; Titli, A. Multilevel qualitative and numerical optimization of fuzzy controller. In Proceedings of 1995 IEEE International Conference on Fuzzy Systems, Yokohama, Japan, 20–24 March 1995; IEEE, 1995; Volume 3, pp. 1149–1154.
39. Sun, C.; Liu, M.; Liu, C.; Feng, X.; Wu, H. An Industrial Quadrotor UAV Control Method Based on Fuzzy Adaptive Linear Active Disturbance Rejection Control. *Electronics* **2021**, *10*, 376. [CrossRef]
40. Peng-ya, X.; Yun-jie, W.; Jing-xing, Z.; Ling, C. Longitudinal attitude control of UAV based on fuzzy PID. In Proceedings of the 2018 IEEE CSAA Guidance, Navigation and Control Conference (CGNCC), Xiamen, China, 10–12 August 2018; IEEE: Piscataway, NJ, USA, 2018; pp. 1–5.
41. Carvalho, G.; Guedes, I.; Pinto, M.; Zachi, A.; Almeida, L.; Andrade, F.; Melo, A.G. Hybrid PID-Fuzzy controller for autonomous UAV stabilization. In Proceedings of the 2021 14th IEEE International Conference on Industry Applications (INDUSCON), São Paulo, Brazil, 15–18 August 2021; IEEE: Piscataway, NJ, USA, 2021; pp. 1296–1302.
42. Selvachandran, G.; Quek, S.G.; Lan, L.T.H.; Giang, N.L.; Ding, W.; Abdel-Basset, M.; Albuquerque, V.H.C. A new design of Mamdani complex fuzzy inference system for multi-attribute decision making problems. *IEEE Trans. Fuzzy Syst.* **2019**, *29*, 716–730. [CrossRef]
43. Larson, K. Fuzzy Logic Tuning of a Proportional-Integral-Derivative Controller. Ph.D. Thesis, California State Polytechnic University, Pomona, CA, USA, 2016.
44. Zhao, Z.Y.; Tomizuka, M.; Isaka, S. Fuzzy gain scheduling of PID controllers. *IEEE Trans. Syst. Man Cybern.* **1993**, *23*, 1392–1398. [CrossRef]
45. Ziegler, J.G.; Nichols, N.B. Optimum settings for automatic controllers. *Trans. ASME* **1942**, *64*, 759–765. [CrossRef]
46. Shah, S.; Dey, D.; Lovett, C.; Kapoor, A. AirSim: High-Fidelity Visual and Physical Simulation for Autonomous Vehicles. In *Field and Service Robotics*; Springer: Cham, Swotzerland, 2017.

Article

Recognition of Human Face Regions under Adverse Conditions—Face Masks and Glasses—In Thermographic Sanitary Barriers through Learning Transfer from an Object Detector [†]

Joabe R. da Silva [1], Gustavo M. de Almeida [1], Marco Antonio de S. L. Cuadros [1], Hércules L. M. Campos [2], Reginaldo B. Nunes [3], Josemar Simão [3] and Pablo R. Muniz [3,*]

1. Postgraduate Program in Control and Automation Engineering, Federal Institute of Espírito Santo, Serra 29173-087, Brazil; joabe.silva@estudante.ifes.edu.br (J.R.d.S.); gmaia@ifes.edu.br (G.M.d.A.); marcoantonio@ifes.edu.br (M.A.d.S.L.C.)
2. Institute of Health and Biotechnology, Federal University of Amazonas, Coari 69460-000, Brazil; herculeslmc@ufam.edu.br
3. Department of Electrical Engineering, Federal Institute of Espírito Santo, Vitoria 29040-780, Brazil; regisbn@ifes.edu.br (R.B.N.); josemars@ifes.edu.br (J.S.)
* Correspondence: pablorm@ifes.edu.br
† This paper is an extended version of our paper published in" Joabe Ruella da Silva; Yngrith Soares da Silva; Felipe de Souza Santos; Natália Queirós Santos; Gustavo Maia de Almeida; Josemar Simão; Reginaldo Barbosa Nunes; Marco Antonio de Souza Leite Cuadros; Hércules Lázaro Morais Campos; Pablo Rodrigues Muniz. Utilização da transferência de aprendizado no detector de objetos para regiões da face humana em imagens termográficas de barreiras sanitárias." In Proceedings of the 2021 14th IEEE International Conference on Industry Applications, São Paulo, Brazil, 15–18 August 2021.

Abstract: The COVID-19 pandemic has detrimentally affected people's lives and the economies of many countries, causing disruption in the health, education, transport, and other sectors. Several countries have implemented sanitary barriers at airports, bus and train stations, company gates, and other shared spaces to detect patients with viral symptoms in an effort to contain the spread of the disease. As fever is one of the most recurrent disease symptoms, the demand for devices that measure skin (body surface) temperature has increased. The thermal imaging camera, also known as a thermal imager, is one such device used to measure temperature. It employs a technology known as infrared thermography and is a noninvasive, fast, and objective tool. This study employed machine learning transfer using You Only Look Once (YOLO) to detect the hottest temperatures in the regions of interest (ROIs) of the human face in thermographic images, allowing the identification of a febrile state in humans. The algorithms detect areas of interest in the thermographic images, such as the eyes, forehead, and ears, before analyzing the temperatures in these regions. The developed software achieved excellent performance in detecting the established areas of interest, adequately indicating the maximum temperature within each region of interest, and correctly choosing the maximum temperature among them.

Keywords: COVID-19; thermography; fever; computer vision; intelligent systems; artificial intelligence

Citation: da Silva, J.R.; de Almeida, G.M.; Cuadros, M.A.d.S.L.; Campos, H.L.M.; Nunes, R.B.; Simão, J.; Muniz, P.R. Recognition of Human Face Regions under Adverse Conditions—Face Masks and Glasses—In Thermographic Sanitary Barriers through Learning Transfer from an Object Detector. *Machines* 2022, *10*, 43. https://doi.org/10.3390/machines10010043

Academic Editors: Marcos de Sales Guerra Tsuzuki, Marcosiris Amorim de Oliveira Pessoa and Alexandre Acássio

Received: 23 November 2021
Accepted: 24 December 2021
Published: 7 January 2022

Publisher's Note: MDPI stays neutral with regard to jurisdictional claims in published maps and institutional affiliations.

Copyright: © 2022 by the authors. Licensee MDPI, Basel, Switzerland. This article is an open access article distributed under the terms and conditions of the Creative Commons Attribution (CC BY) license (https://creativecommons.org/licenses/by/4.0/).

1. Introduction

Coronaviruses (CoVs) are viruses that cause respiratory infections in animals such as birds and mammals, including humans. There have been seven recorded CoVs that have caused serious harm to human health, with two of them responsible for the epidemics that emerged in Hong Kong in 2003 and Saudi Arabia in 2012 [1]. In December 2019, a new CoV called SARS-CoV-2 (the virus that causes the disease COVID-19) emerged in the city of Wuhan, China. In the first part of 2020, this virus spread to virtually every country in the world. On 30 January 2020, the World Health Organization (WHO) declared the COVID-19 pandemic an international emergency [1].

As the transmission rate increased, many studies developing detection and diagnostic technologies for people infected with COVID-19 have emerged. Mild cases of COVID-19 present with symptoms such as cough, fever, runny nose, sore throat, and difficulty breathing, whereas severe cases can cause pneumonia [2]. As fever is one of the most recurrent symptoms of the infection, some countries perform temperature measurements on people at airports, bus and train stations, company gates, and other shared and public spaces. These actions sought to detect symptomatic patients to contain the spread of the virus [3–5].

Currently, there are several ways to measure temperatures using devices or infrared thermometers. However, these methods do not guarantee measurement reliability, as incorrect use may lead to measurements of a person or an object with significant errors. Furthermore, as it is a manual process, the collection and recording of temperatures are subject to human error in addition to intrinsic measurement uncertainties, causing many screening errors in barriers currently in use [6–8].

In health, noninvasive measuring and intervention equipment can prevent germ spread and contribute to the practicality of the diagnosis process [9]. Furthermore, the high demand for fever measuring devices has caused many companies to enter the field without understanding the mechanism of human temperature measurement, employing technology that may incorrectly measure body temperature [10].

Thermal imaging cameras, or thermal imagers, utilize infrared thermography to measure temperature; this is a noninvasive, fast, and objective technique. All objects with temperatures above 0 K emit infrared radiation, and the amount of radiation emitted increases with temperature. Therefore, thermography can measure the surface temperature of the bodies [11]. Thus, it is possible to develop innovative technologies and automate the measurement processes by applying computer vision algorithms to thermographic images [12].

Computer vision, a subfield of artificial intelligence, can help provide solutions to many complex problems in health; it can also assist in the diagnosis and spread prevention of COVID-19 [13]. Additionally, Ulhaq et al. [13] presented computer vision techniques to control the virus spread, including infrared thermography processing. In this case, the algorithms detect regions of interest (ROIs) extracted from the original image.

In [14], the authors applied a series of machine learning (ML) algorithms to different tasks related to processing thermographic facial images. This study also presents a method to estimate head position to increase the ability to detect reference points in nonfrontal faces. These techniques are essential for improving the accuracy of the detection algorithm by capturing the face at a more appropriate angle. However, for Wang et al. [9], the facial temperature measurement should not be used only in a small area and ignore other sectors because the facial thermal image can show specific thermal characteristics among different regions.

Thermographic cameras also have significant potential for use in measuring the temperature of the human body surface, i.e., skin temperature [15]. However, the literature indicates inconsistent diagnostic performance, possibly due to wide variations in the implemented methodologies. This study evaluated the effectiveness of fever diagnosis and the effect of measuring temperatures in 17 facial regions; it contributes to the elucidation of the impact that location has on facial temperature measurement and other issues regarding the performance of febrility detection methods.

Propelled by the lack of published studies and the urgent need for sanitary barriers for screening people who may have COVID-19, this study aims to develop a solution for screening without physical contact or holding people and with only minimal interference of the flow of the site where the barrier is established. This work was supported by the Government of the State of Espírito Santo through the Foundation for Support to Research and Innovation of Espírito Santo (FAPES).

This paper presents an intelligent system using the Transfer Learning technique of a Deep Learning network trained to detect objects. The algorithm analyzes thermographic images to quickly detect the subject's face, forehead, eyes, and ears, as these areas present the highest temperatures in the frontal and lateral regions of the head. The algorithm can then analyze the temperature of the ROI and, subsequently, estimate body surface temperature more accurately and efficiently than manual measurement methods. Additionally, the algorithm can incorporate suitable diagnostic criteria for the different ROIs with different febrility thresholds.

This article is an expanded article from a conference paper presented at the 14th IEEE/IAS International Conference on Industry Applications (Induscon) [16], whose theme was 'Innovation in the time of COVID-19'. This version introduces more details on human infrared thermography, presents more tests with volunteers, and applies Optical Character Recognition (OCR) technology to identify maximum and minimum temperatures in thermographs. These add-ons improved the work previously carried out in the automatic detection of febrile people at sanitary barriers, which is very relevant in this phase of the COVID-19 pandemic, where new variants of the virus are emerging.

2. Materials and Methods

This section presents the main steps in developing an automatic system for measuring the human temperature at sanitary barriers by combining thermography and computer vision technologies.

The Research Ethics Committee of the Federal Institute of Espírito Santo, linked to the National Research Ethics Commission of the Ministry of Health of Brazil, approved this research under the Certificate of Presentation and Ethical Appreciation (CAAE) 33502120.2.0000.5072, opinion number 4.180.201, on 29 July 2020.

The volunteers who participated in this research were informed about the objectives, the scope of their participation, the confidential treatment of their data, and the consolidated statistically grouped method of disclosing data. All participants provided written consent.

The inclusion criterion considered the volunteers 18 years old or older and the signature on the consent term of free participation without any burden or bonus for the volunteer or researchers, with the possibility of withdrawing from the study at any time.

2.1. Fever and Human Thermography

Fever occurs when there is an increase in the body's thermal threshold, usually maintained at around 37 °C, triggering metabolic responses of heat production and conservation, for example, tremors and peripheral vasoconstriction. These responses help to raise the body temperature to the new threshold. After fever is resolved or treated, threshold returns to baseline and heat loss processes begin, e.g., peripheral vasodilation and sweating [17].

The surface temperature threshold for determining whether a patient is in a febrile state varies among different authors. However, the most adopted thresholds are 37.5 °C ([18–20]) and 38.0 °C ([21–23]).

However, the surface temperature of the human body is different from the core temperature, which is the gold standard for diagnosing fever. The surface temperature presents different and typically lower values from the core temperature for the different regions of the face, a surface commonly inspected in sanitary barriers. Different face regions can vary the nonfever temperature from 32.3 °C up to 35.9 °C [24–26]. Therefore, properly identifying the region of interest (ROI) on the human face where the temperature is being measured and applying an adequate threshold leads to a more accurate diagnosis than measuring the maximum face temperature without considering which region of the face is being treated.

For this reason, this work adopts the following ROIs: medial palpebral commissure (eyes region), temporal (forehead region), and external acoustic meatus (ear region). These regions are recommended by literature [25,27,28].

2.2. Infrared Thermography

In physics, waves are periodic disturbances that maintain their shape as they propagate through space as a function of time. The literature describes visible light, ultraviolet radiation (UV), and infrared radiation (IR) specifically as types of electromagnetic (EM) waves. The spatial periodicity, or the interval between two wave peaks, is called the wavelength, λ, and is given in meters, nanometers, or micrometers. The temporal periodicity, or the time interval between two wave peaks, is denoted as the oscillation period, T, and is given in seconds or its submultiples. The frequency, ν, is the inverse of period T, with the unit as Hertz.

Figure 1 presents an overview of the most common characteristics of EM waves. Visible light, defined by the range over which the light receptors of human eyes can detect, covers a small range within the spectrum, with wavelengths ranging from 380 nm to 780 nm [11].

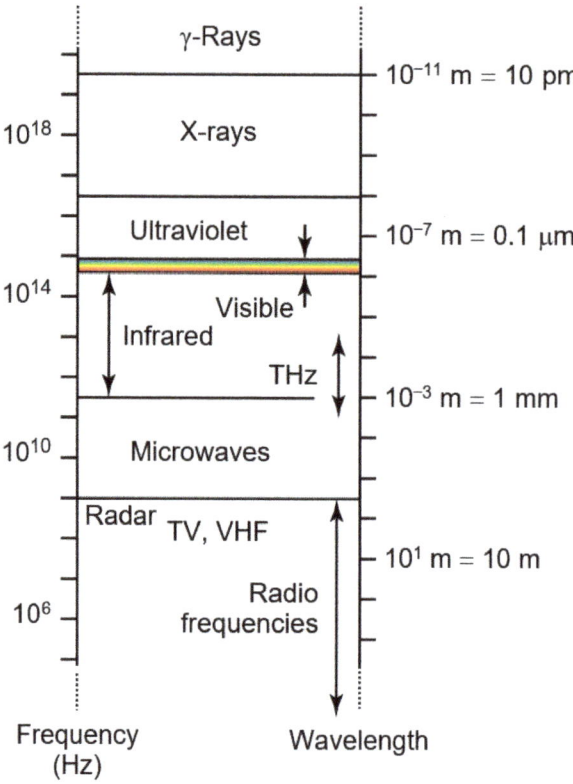

Figure 1. Overview of the electromagnetic wave spectrum. Adapted from [11].

The spectral region with wavelengths in the range of 0.7–1000 μm is generally called the infrared region, which is the focus of this work [11]. Infrared radiation is invisible to the human eye and has a long wavelength and low energy [29]. Anybody with a temperature above absolute zero (0 K, −273.15 °C) emits infrared radiation perceived as heat. The amount of radiation emitted by a body depends on the temperature and properties of the material [11].

Within the infrared spectrum, some bands exhibit particular characteristics, which affect their applications. The main ranges are near-infrared (NIR, from 0.7–1 μm), short-wave infrared (SWIR, 1–2.5 μm), mid-wave infrared (MWIR, 3–5 μm), long-wave infrared (LWIR, 7.5–14 μm), and very long-wave infrared (VLWIR, from 14 to 1000 μm) [29]. Figure 2 depicts these ranges.

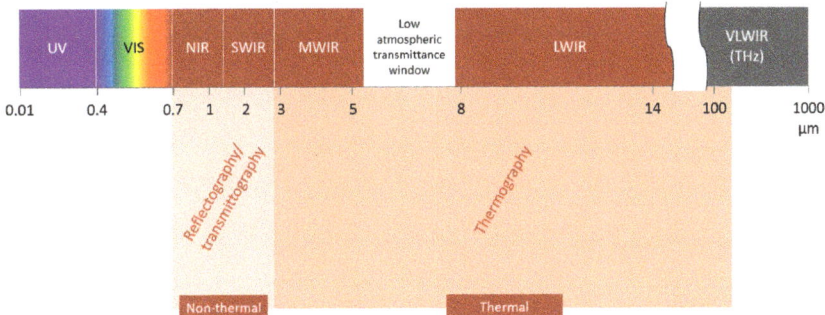

Figure 2. Infrared spectral regions. Adapted from [29].

The infrared radiation spectrum bands generally applied in technologies involving thermographic images are MWIR and LWIR [29].

Infrared thermal imaging, also called infrared thermography (IRT), is a rapidly evolving technology. Currently, researchers are applying IRT to intelligent solutions in different fields, including condition monitoring, predictive maintenance, and gas detection. Medicine is another area that has benefited from this technology, employing IRT in oncology (breast, skin, etc.), surgery, medication effectiveness monitoring, and, more recently, for acute respiratory syndrome testing applications [30].

Technologies based on IRT can detect the intensity of thermal radiation emitted by objects since the bodies transmit, radiate, and reflect infrared radiation. Radiation transmission, or transmissivity, is the ability of a material to allow infrared radiation to pass through it. Emissivity is the capacity of a material to emit infrared radiation. Finally, reflectivity is the capability of the material/object surface to reflect radiation, that is, temperature reflected from the object.

2.3. Machine Learning

With the high volume of data generated by devices, sensors, and users, machines capable of identifying patterns and assisting in making decisions have become essential, with supervised learning and unsupervised machine learning being the most widely adopted methods. Reinforcement and semisupervised learning are other methods that may be used [31].

Deep learning is a set of machine learning technologies that utilize algorithms to detect, recognize, and classify objects and text in images or other documents. One of the leading deep learning architectures is the convolutional neural network (CNN), which is used to solve most image analysis problems [32].

2.3.1. Convolutional Neural Networks

CNNs have been widely applied in image classifiers. They excel in analyzing images and learning abstract representations. A typical CNN has an input layer, an output layer, and several hidden layers. The hidden layers of a CNN generally consist of a series of convolutional layers. The first convolution layer learns to identify the simple features. The following layers learn to detect more significant and complex characteristics. Other operations include the rectified linear unit (ReLU), grouping, fully connected, and normalizing layers. Finally, backpropagation is used for error distribution and weight adjustment [33,34].

Digital images can be represented by a matrix in which each pixel contains one or more values. First, a CNN trains and tests each input image with the pixel values going through a series of convolution operations with filters (kernels). Then, the results are grouped (pooling) to reduce the matrix dimensions and generate a new, simplified matrix. These operations complete the feature-extraction step. Then, a vector is created from the feature map, which is used to feed the input layer of a multilayer neural network (fully connected, FC) [35]. Figure 3 presents a simplified diagram of a CNN [35,36].

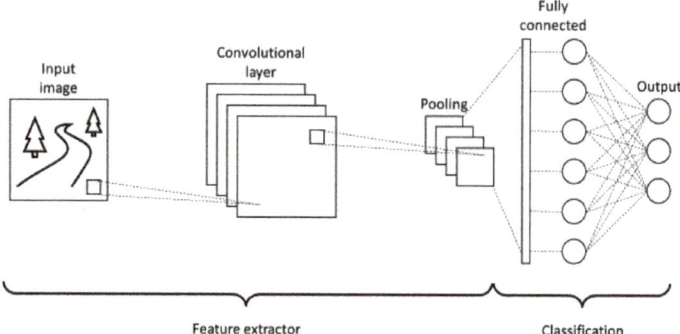

Figure 3. Simplified diagram of the structure of a CNN.

2.3.2. Region Based Convolutional Neural Networks

Region-based CNNs (R-CNNs) emerged as an improvement of CNNs. They are able to detect and locate specific objects in an image. The architecture of an R-CNN is similar to that of a CNN. However, an added step of extracting the region containing the object to be detected is included [37]. Figure 4 presents a simplified diagram of an R-CNN [36,37].

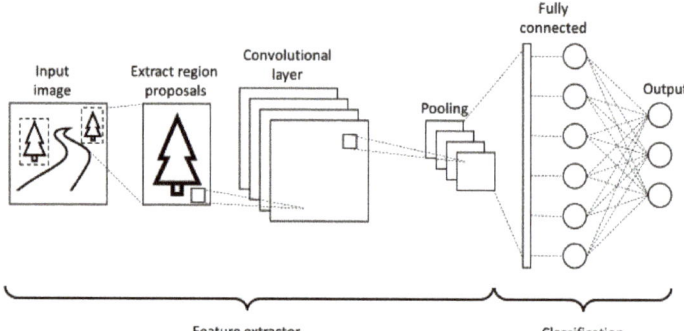

Figure 4. Simplified diagram of the structure of an R-CNN.

The R-CNN detector consists of four main steps: candidate box generation, resource extraction, classification, and regression. For candidate box generation, approximately 2000 boxes are determined in the image using the selective search method. For resource extraction, the CNN extracts the resources of each candidate box. In the third step, a classifier determines whether the extracted features belong to a specific class. Finally, the regression step adjusts the position of the bounding box by referring to a particular resource [38,39].

2.3.3. You Only Look Once Network

According to [36], many improved algorithms have emerged from proposals of R-CNN models, all providing different degrees of improvement in the detection performance compared to the original R-CNN.

The You Only Look Once (YOLO) network, proposed by [40], is a pretrained object detector in the common objects in context (COCO) image dataset, with RGB (red, green, and blue) images of various object classes. Its main contribution is real-time image detection. Additionally, unlike other object detection algorithms, the YOLO network input is an entire image. It performs object detection through a fixed-grid regression consisting of

24 convolutional layers and two multilayer neural networks. The network can process images in real-time at 45 frames per second (FPS). Furthermore, YOLO produces fewer false positives than other similar architectures [41]. In this study, YOLO was used to apply the transfer of learning in the training of a specific dataset.

The learning transfer is a technique that takes advantage of a structure of a pre-trained CNNs structure for a given application as a starting point for a new task that is, until then, unknown. Thus, the structure of convolutional layers and filters in the feature extraction stage are reused for a new application. Afterward, changes are made in the FC layer, where the classes of the pre-trained network can be removed and/or new classes can be added to meet the new application. After the FC changes, only this layer needs to be retrained, drastically reducing the effort of training a complete CNN, which demands a high computational cost and requires a large amount of training data to achieve high performance. In this work, a pre-trained structure with a dataset of 998 images will be used to recognize volunteers' faces. The aim is for the new structure to be able to detect the ROIs in human faces, with difficulties not originally imposed: volunteers wearing semifacial masks and glasses.

2.3.4. Optical Character Recognition

Optical character recognition (OCR) is a technology that allows the recognition and extraction of characters in image files to generate analyzable, editable, and searchable data [42]. This technology uses image and natural language processing to solve different challenges [43].

Tesseract is a free open-source OCR software, originally developed at Hewlett-Packard Laboratories Bristol and Hewlett-Packard Co., Greeley, Colorado, between 1985 and 1994. From 2006 to 2018, Google improved the software, and it is currently available on GitHub [44]. It can recognize texts in over 100 languages.

In this study, Tesseract was used to identify the minimum and maximum temperature values in the temperature scale of the analyzed thermographs.

2.4. Dataset

The dataset used in this study is publicly available in [45]. The authors used a FLIR Vue Pro camera to capture thermographic images for the dataset. During image capture, participants looked at a fixed point while the camera was moved to nine equidistant positions, forming a semicircle around the volunteer. Thus, the dataset contained nine thermographic images of the face of each participant. Figure 5 displays examples of photos that comprise the dataset.

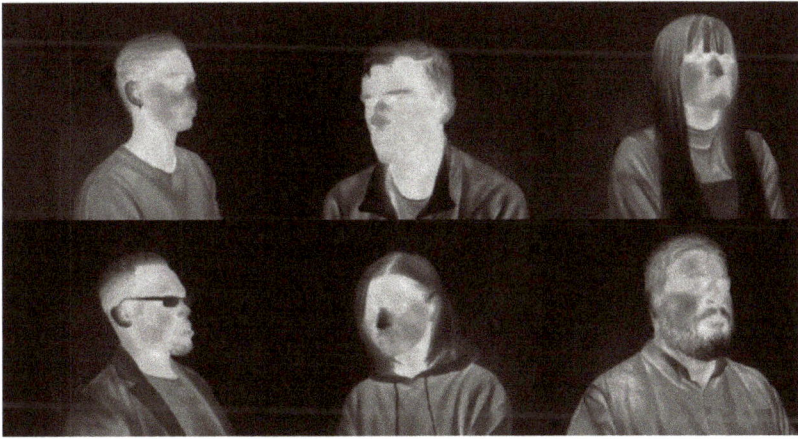

Figure 5. Examples of dataset images [45].

The complete dataset contained 998 images from 111 participants. However, to work on a balanced dataset concerning volunteers' gender, only 781 images were used. Of these, 658 were used for training and 123 for validation of transfer learning by the YOLO network. Aiming to evaluate the performance of YOLO for object detection, it is necessary to label each image with the annotations of their respective bounding boxes.

The face ROIs are the ear, eye, forehead, and whole face. These areas have known temperature thresholds for febrility and can be see directly; thus, they are suitable for screening febrile people using thermography [46]. However, not all regions are constantly visible on the person owing to the use of glasses, face masks, hair over the forehead or ear, and others.

LabelImg software, a free graphical tool for image and video annotation [47], was used to label all images used in this study. Figure 6 shows the graphical interface of the LabelImg software.

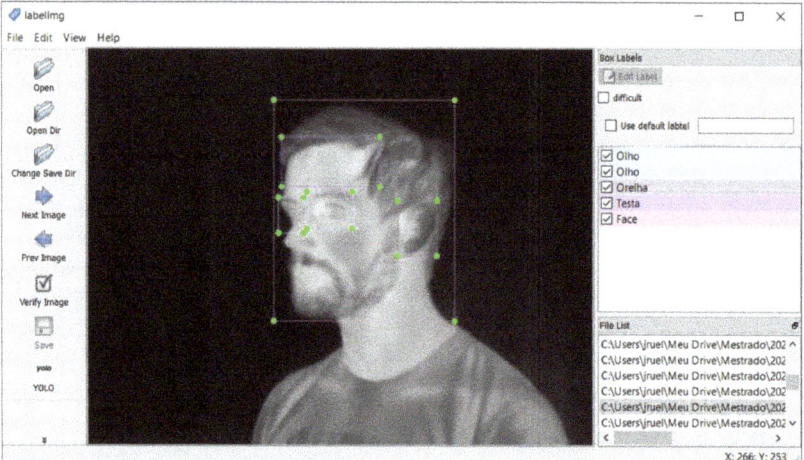

Figure 6. LabelImg software graphical interface.

3. Results

The training of the object detector was performed on Google Collaboratory, a computational environment that runs in the cloud and requires no configuration. It allows the writing and executing of code directly in the browser.

For the R-CNN training assessment, it was necessary to quantify the prediction accuracy by comparing the prediction made by this model with the real object location in the image. Thus, the mean average precision (mAP), which is one of the most common metrics for determining the accuracy of object detectors, was employed [48].

Other methods to evaluate the performance of the trained network were precision (P), recall (R), and F1-score (F1). For these metrics, a higher value indicates a better result. Additionally, the values of true positives (TP), false positives (FP), and false negatives (FN) were employed as performance metrics.

The resultant metric values of the trained R-CNN, with a confidence limit of 25% (conf_threshold = 0.25), were as follows: TP = 452, FP = 46, FN = 9, P = 0.91, R = 0.98, and F1 = 0.94. The mAP with an intersection over union (IoU) greater than 50%, also known as mAP@0.50, was 0.97. From Table 1, it is possible to evaluate the performance of the model for each class.

Tests were carried out with photos of six volunteers, different from those present in the training dataset, to evaluate the prediction accuracy of new images. A Testo-885 thermographic camera captured the new images. After transfer learning, the object detector algorithm analyzed these images and detected all ROIs, even for volunteers wearing masks, caps, and with long hair. Figure 7 displays some of these images.

Table 1. Performance for each ROI during validation.

Class Name	Precision mAP	True Positives—TP	False Positives—FP
Face	97.38%	120	15
Eye	98.61%	161	15
Forehead	97.25%	111	7
Ear	94.54%	60	9

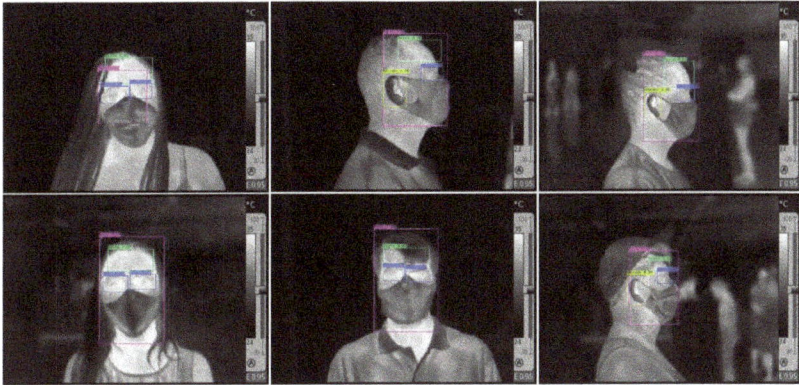

Figure 7. Images produced using the Testo-885 camera and analyzed by YOLO.

When identifying an object, the YOLO detector provides the coordinates, width, and height of the bounding boxes. This allows delimiting ROIs where the temperature is analyzed. From each ROI, the values of the pixels with the highest temperatures were extracted. Thus, the algorithm discards regions covered by hair, sweat, and fabric, which are generally at lower temperatures. The higher temperatures are shown as the lightest colors in Figure 8.

Figure 9 displays a boxplot of the pixel values in each ROI, as depicted in Figure 8. The distribution of the pixels in the forehead region displays lower values than those in the eye regions, indicating that the eyes are at a higher temperature than the forehead.

According to [25], the maximum or mean temperatures of ROIs can be adopted to assess human body surfaces. However, the segmentation of ROIs performed in this paper may include background images, parts of the surfaces of glasses, masks, and hair, decreasing the mean temperature of the ROI. Therefore, to avoid this issue, the maximum temperature for each ROI was adopted.

The temperature scale on the right side of Figure 8 indicates that: darker colors are close to a temperature of 24 °C and lighter colors approach 35 °C. In the thermal imager standard operating mode, these values are automatically generated by the camera's operating software, where the highest value indicates the maximum temperature of the objects in the thermal imager's field of view and the lowest value indicates the minimum temperature of the objects.

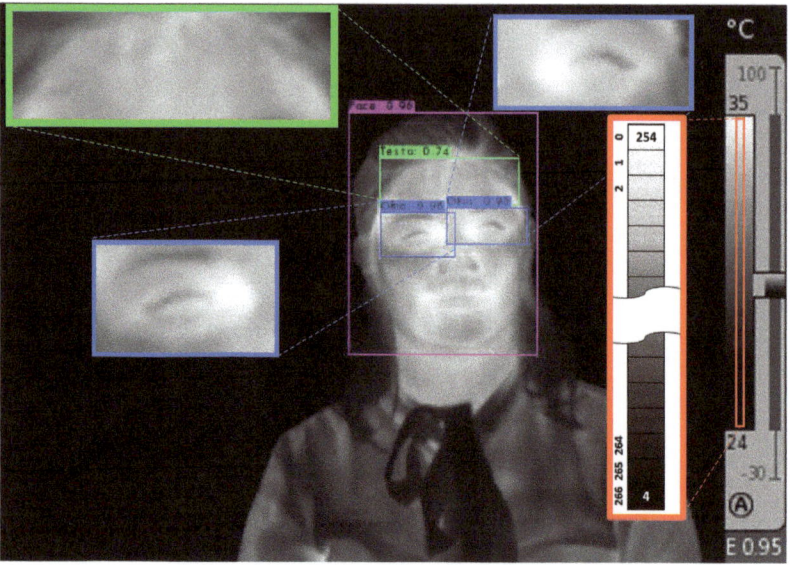

Figure 8. Extraction of ROIs to be analyzed.

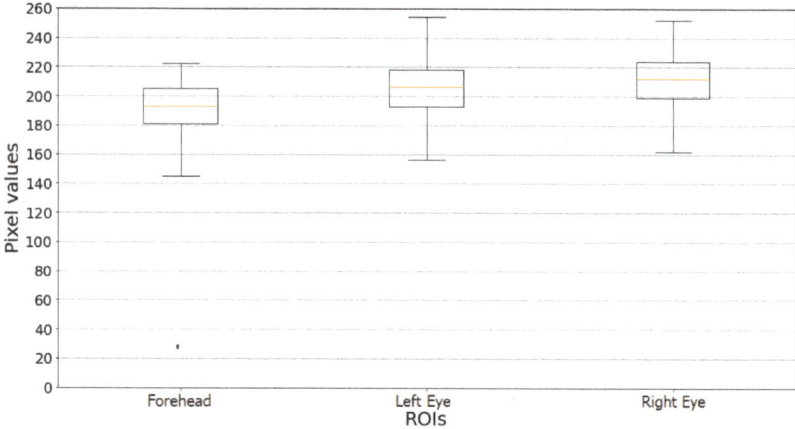

Figure 9. Distribution of pixel values of the ROIs depicted in Figure 8.

Thermogram radiometric output is not always available, depending on the imager manufacturer. Thus, a method for reading temperatures in the region of interest directly in the thermal image was developed, so that the method can be widely used.

Along the temperature scale, there are 267 pixels, where the first one, pixel zero, has a value of 254. The last pixel of the scale has a value of 4. Figure 10 presents the relationship between the pixel values and their respective positions on the scale as a dashed line (in green). Equation (1) shows a first-order linear proportionality relation, the first-order polynomial, obtained through linear regression, with a coefficient of determination (R^2) of 0.9941. The solid line (in red) on the graph shows the behavior of the equation of the straight line that describes this relationship.

$$i = -1.064v + 270.256, \tag{1}$$

where v is the pixel value and i is the position on the temperature scale.

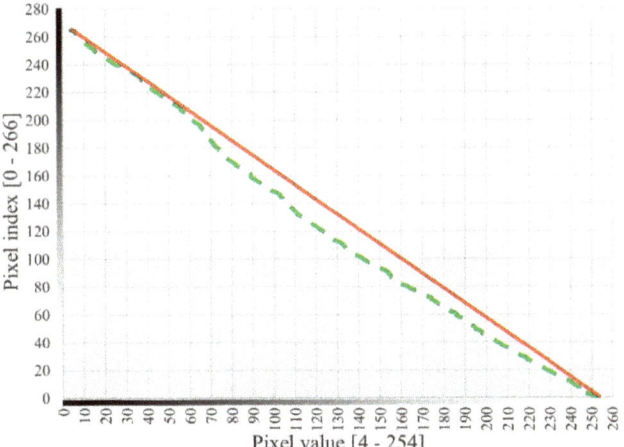

Figure 10. Pixel value as a function of position on the temperature scale. The dashed green line depicts the pixel values extracted from the image. The solid red line displays the behavior of the equation obtained by linear regression.

As Equation (1) indicates a first-order linear proportionality between pixel position on the scale and temperature, higher temperatures will produce higher pixel values. Thus, the pixels positioned at the beginning of the scale represent the highest temperatures, and pixels at the end indicate the lowest. Figure 11 depicts the relationship between the pixel positions and the respective temperatures of the scale, as shown in Figure 8. Equation (2) presents a straight line that describes this relationship.

$$T = [(y_2 - y_1)/266]\, i + y_1, \qquad (2)$$

where i is the pixel position, obtained using Equation (1), y_1 is the highest value recorded on the temperature scale, y_2 is the lowest value recorded on the temperature scale, and T is the temperature (in °C) of the analyzed pixel.

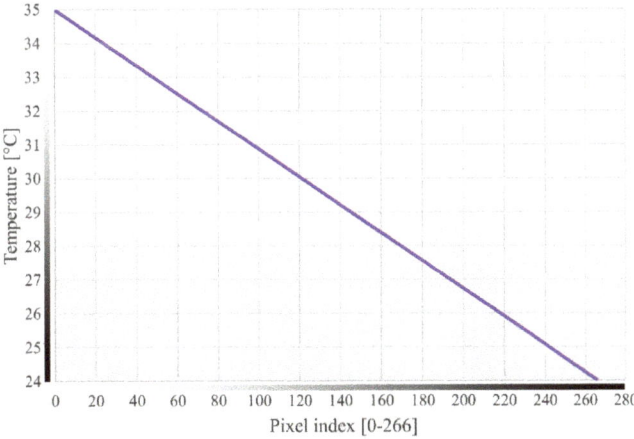

Figure 11. Temperature as a function of pixel position.

Through algebraic manipulation of Equations (1) in (2), it is possible to obtain the value of the temperature T of the analyzed pixel, through the value, v, of the pixel, the highest value recorded on the temperature scale, y_1, and the lowest value, y_2. Equation (3) displays the algebraic manipulation:

$$T = 1/266 \, [y_1(1.064v - 4.256) - y_2(1.064v - 270.256)]. \qquad (3)$$

After obtaining the highest temperatures of each image ROI, the highest value among these temperatures represents the final temperature of the volunteer. Figure 12 shows images of 24 volunteers, and Table 2 lists the highest temperature recorded in each ROI for each person.

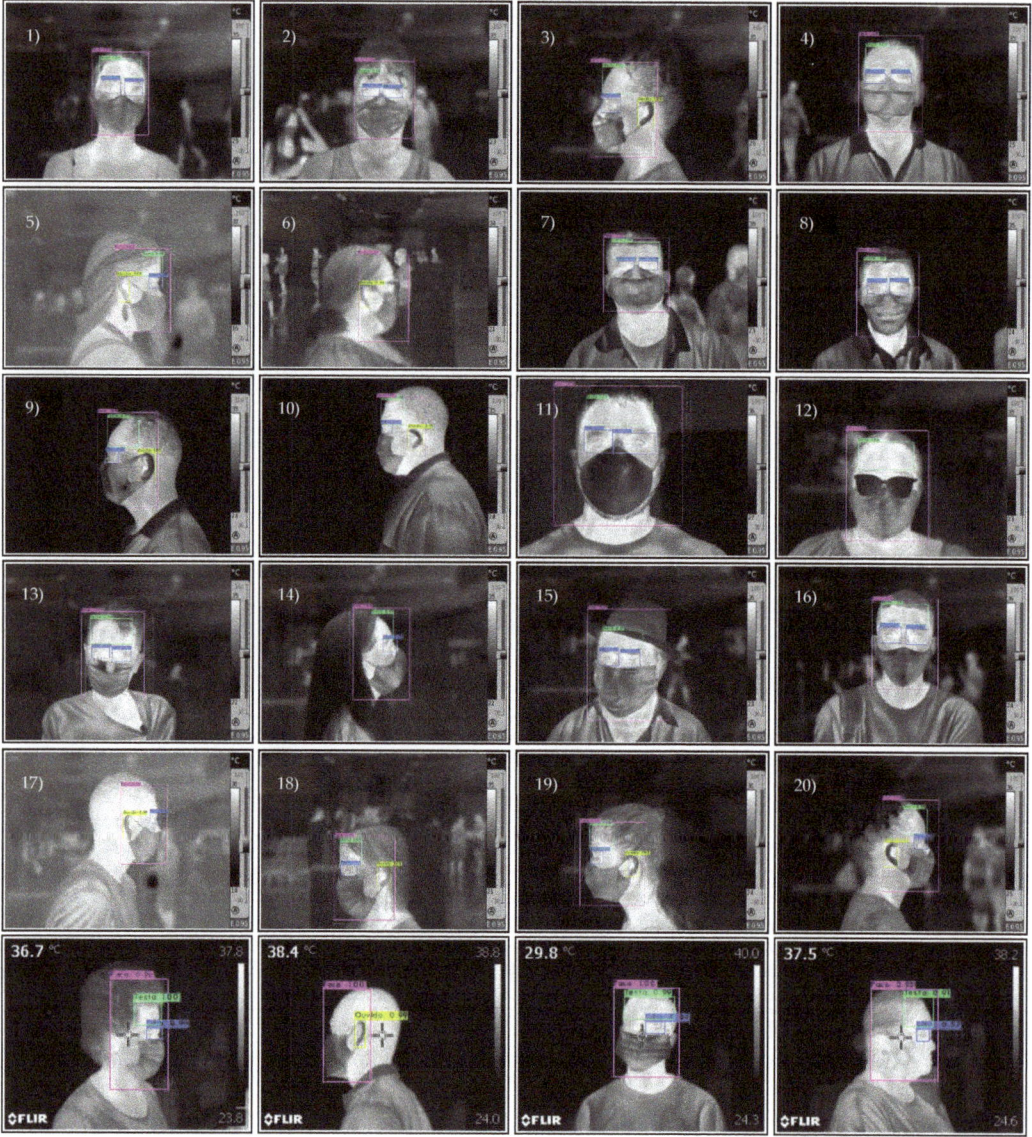

Figure 12. Images produced using the Testo-885 camera and analyzed by YOLO, with temperature estimated using Equation (3).

Table 2. Estimated Volunteer Temperatures.

Volunteer	Right Eye [°C]	Left Eye [°C]	Forehead [°C]	Ear [°C]
Nonfebrile volunteers				
01	34.9	34.9	34.9	-
02	34.9	34.9	34.1	-
03	-	34.5	34.5	35.0
04	35.0	34.8	33.8	-
05	35.1	-	35.6	36.9
06	-	-	-	34.0
07	35.0	35.1	33.4	-
08	34.6	34.8	34.5	-
09	-	33.3	33.8	35.0
10	-	33.9	34.4	34.6
11	34.7	34.5	35.0	-
12	-	-	34.9	-
13	35.0	35.0	34.7	-
14	34.3	-	34.9	-
15	35.0	34.9	34.3	-
16	34.9	35.0	34.4	-
17	33.7	-	-	34.9
18	-	33.2	34.6	35.1
19	-	34.3	34.3	35.0
20	34.2	-	34.6	-
Nonfebrile volunteers (Mean ± Std. Dev.)	34.7 ± 0.4	34.5 ± 0.6	34.5 ± 0.5	35.1 ± 0.8
Febrile Volunteers				
21	-	37.2	37.1	-
22	-	-	-	37.9
23	-	39.7	39.2	-
24	37.2	-	37.3	-

Table 2 initially shows that there are small variations in temperature (less than 1 °C) among most volunteers. However, considering that the human being is homeothermic, this shows that the surface temperature undergoes variations not experienced by the body temperature. Furthermore, it is confirmed that face surface temperature is predominantly lower than body temperature, as shown by the mean temperature values of non-febrile volunteers.

One can note that for volunteers 6 and 22, only one ROI was visible in the image, permitting to obtain only one temperature for the analysis. The identification of only one ROI in volunteers 6 and 22 is an example of both the limitation of the intelligent system and that its objective was achieved. Volunteers were not in adequate direct sight of the thermal imager, in a way that would allow the identification of more ROIs. However, the system managed to identify at least one ROI, enabling the person's temperature analysis.

Unfortunately, only four febrile volunteers were obtained in the image production campaign with volunteers, who were previously diagnosed as feverish by a health team by checking body temperature. Their measured temperatures are identified as 21, 22, 23, and 24 on Table 2. Despite the low sampling of febrile people, it is noted that: the maximum temperature detected in the region of the face of the volunteers 21 and 24 (37.1 °C and 37.3 °C, respectively) did not exceed the usual fever threshold for central temperature (37.5 °C or 38.0 °C); the temperatures of different facial regions of volunteer 23 showed a 0.5 °C discrepancy, which is significant for the diagnosis of fever. This supports the hypothesis that the febrile diagnostic criteria of core body temperature (37.5 °C or 38.0 °C) should not be applied to human facial temperature.

4. Conclusions

This study employed a transfer deep learning method to detect and recognize ROIs, including the face, forehead, eyes, and ears, in thermographic images using the YOLO object detector. After training a CNN from a dataset made available by other researchers, images of new volunteers obtained in the laboratory served as input to the CNN to evaluate the detection performance of the ROIs.

Tests verified that ROI detection was feasible even with the use of masks, caps, helmets, or with features hidden by hair.

As displayed in Figure 9, there were variations in temperature among ROIs. The criterion of adopting the highest temperature within each ROI proved to be efficient, as areas without a direct target, such as those covered by hair, are disregarded.

This study presents a simple system for obtaining temperature values directly from thermographic images without significant computational processing. These improvements in detecting the maximum and minimum temperatures of ROIs can provide better results for identifying febrile people.

As infrared thermography measures surface (skin) temperature and not the core temperature of the human body, future work will apply adequate criteria to analyze the febrility of the individuals from the temperatures of the ROIs. This avoids the use of a single temperature threshold to indicate a feverish state for all regions of the human face. Screening people with fevers through infrared thermography should apply a different and adequate threshold temperature for each face region, typically higher than the body threshold temperature (37.5 °C). Additionally, expanding the dataset will improve the detection of ROIs and allow more reliable screening of febrile people.

Finally, other deep learning algorithms will be applied, evaluated, and compared to the results presented in this work.

Author Contributions: Conceptualization, R.B.N., J.S. and P.R.M.; data curation, J.R.d.S.; formal analysis, J.R.d.S., R.B.N. and J.S.; funding acquisition, M.A.d.S.L.C. and P.R.M.; investigation, J.R.d.S., G.M.d.A. and M.A.d.S.L.C.; methodology, J.R.d.S., G.M.d.A., M.A.d.S.L.C., H.L.M.C., R.B.N., J.S. and P.R.M.; project administration, P.R.M.; resources, M.A.d.S.L.C. and P.R.M.; software, J.R.d.S., G.M.d.A., M.A.d.S.L.C., R.B.N. and J.S.; supervision, G.M.d.A., M.A.d.S.L.C. and P.R.M.; validation, J.R.d.S., R.B.N. and J.S.; visualization, R.B.N.; writing–original draft, J.R.d.S.; writing–review & editing, R.B.N., J.S. and P.R.M. All authors have read and agreed to the published version of the manuscript.

Funding: This research was funded by FAPES (Espírito Santo Research and Innovation Support Foundation), grant numbers 14/2019 (Master's scholarship), 03/2020 (Induced Demand Assessment–COVID-19 Project), and 04/2021 (Research Support). APC and text review were partially funded by the Federal Institute of Espírito Santo through the Institutional Scientific Diffusion Program (PRODIF).

Institutional Review Board Statement: The study was conducted according to the guidelines of the Declaration of Helsinki, and approved by the Ethics Committee of Federal Institute of Espírito Santo - Brazil (protocol code CAAE 33502120.2.0000.5072, approved on 29 July 2020).

Informed Consent Statement: Informed consent was obtained from all subjects involved in the study.

Acknowledgments: The authors thank Masterplace Mall, Construtora Paulo Octávio, Sagrada Família Church in Jardim Camburi, and the Energy Laboratory of Ifes Campus Vitória for collaborating in the field research with volunteers. This work was also supported by the Federal Institute of Espírito Santo and the National Council for Scientific and Technological Development (CNPq).

Conflicts of Interest: The authors declare no conflict of interest.

References

1. Lana, R.M.; Coelho, F.C.; Gomes, M.F.d.C.; Cruz, O.G.; Bastos, L.S.; Villela, D.A.M.; Codeço, C.T. Emergência do novo coronavírus (SARS-CoV-2) e o papel de uma vigilância nacional em saúde oportuna e efetiva. *Cad. Saude Publica* **2020**, *36*, e00019620. [CrossRef] [PubMed]
2. Lima, C.M.A.d.O. Information about the new coronavirus disease (COVID-19). *Radiol. Bras.* **2020**, *53*, V–VI. [CrossRef] [PubMed]

3. Willingham, R. Victorian Students in Coronavirus Lockdown Areas to Get Daily Temperature Checks on Return to Classrooms. Available online: https://www.abc.net.au/news/2020-07-09/victorian-school-kids-to-get-coronavirus-temperature-checks/12438484 (accessed on 22 January 2021).
4. Thiruvengadam, M. You May Have Your Temperature Checked before Your Next Flight at JFK. Available online: https://www.travelandleisure.com/airlines-airports/jfk-airport/honeywell-temperature-monitoring-device-jfk-airport (accessed on 22 January 2021).
5. Kekatos, M. FAA Opens the Door to Pre-Flight COVID-19 Screenings as a Small Airport in Iowa Rolls Out Temperature Checks and Questionnaires That Could Eventually Expand to the Nation's more than 500 Airports. Available online: https://www.dailymail.co.uk/health/article-9169095/Small-airport-Iowa-rollout-temperature-checks-questionnaires.html (accessed on 22 January 2021).
6. Jung, J.; Kim, E.O.; Kim, S.-H. Manual Fever Check Is More Sensitive than Infrared Thermoscanning Camera for Fever Screening in a Hospital Setting during the COVID-19 Pandemic. *J. Korean Med. Sci.* **2020**, *35*, e389. [CrossRef]
7. Normile, D. Airport screening is largely futile, research shows. *Science* **2020**, *367*, 1177–1178. [CrossRef]
8. Mekjavic, I.B.; Tipton, M.J. Myths and methodologies: Degrees of freedom—limitations of infrared thermographic screening for Covid-19 and other infections. *Exp. Physiol.* **2021**, EP089260. [CrossRef] [PubMed]
9. Wang, Z.H.; Horng, G.J.; Hsu, T.H.; Chen, C.C.; Jong, G.J. A Novel Facial Thermal Feature Extraction Method for Non-Contact Healthcare System. *IEEE Access* **2020**, *8*, 86545–86553. [CrossRef]
10. Beall, E.B. Infrared Fever Detectors Used for COVID-19 Aren't as Accurate as You Think. Available online: https://spectrum.ieee.org/news-from-around-ieee/the-institute/ieee-member-news/infrared-fever-detectors-used-for-covid19-arent-as-accurate-as-you-think (accessed on 22 January 2021).
11. Vollmer, M.; Möllmann, K.P. *Infrared Thermal Imaging: Fundamentals, Research and Applications*; Wiley-VCH: Weinheim, Germany, 2010; ISBN 9783527407170.
12. Bilodeau, G.A.; Torabi, A.; Lévesque, M.; Ouellet, C.; Langlois, J.M.P.; Lema, P.; Carmant, L. Body temperature estimation of a moving subject from thermographic images. *Mach. Vis. Appl.* **2012**, *23*, 299–311. [CrossRef]
13. Ulhaq, A.; Born, J.; Khan, A.; Gomes, D.P.S.; Chakraborty, S.; Paul, M. COVID-19 Control by Computer Vision Approaches: A Survey. *IEEE Access* **2020**, *8*, 179437–179456. [CrossRef]
14. Kopaczka, M.; Kolk, R.; Schock, J.; Burkhard, F.; Merhof, D. A Thermal Infrared Face Database with Facial Landmarks and Emotion Labels. *IEEE Trans. Instrum. Meas.* **2019**, *68*, 1389–1401. [CrossRef]
15. Zhou, Y.; Ghassemi, P.; Chen, M.; McBride, D.; Casamento, J.P.; Pfefer, T.J.; Wang, Q. Clinical evaluation of fever-screening thermography: Impact of consensus guidelines and facial measurement location. *J. Biomed. Opt.* **2020**, *25*, 097002. [CrossRef]
16. da Silva, J.R.; da Silva, Y.S.; de Souza Santos, F.; Santos, N.Q.; de Almeida, G.M.; Simao, J.; Nunes, R.B.; de Souza Leite Cuadros, M.A.; Campos, H.L.M.; Muniz, P.R. Utilização da transferência de aprendizado no detector de objetos para regiões da face humana em imagens termográficas de barreiras sanitárias. In Proceedings of the 2021 14th IEEE International Conference on Industry Applications (INDUSCON), São Paulo, Brazil, 15–18 August 2021; pp. 475–480. [CrossRef]
17. Longo, D.; Fauci, A.; Kasper, D.; Hauser, S.; Jameson, J.; Loscalzo, J. (Eds.) *Harrison's Manual of Medicine*, 18th ed.; McGraw-Hill: New York, NY, USA, 2014; ISBN 007174519X.
18. Ring, E.F.J.; Jung, A.; Zuber, J.; Rutkowski, P.; Kalicki, B.; Bajwa, U. Detecting Fever in Polish Children by Infrared Thermography. In Proceedings of the 2008 International Conference on Quantitative InfraRed Thermography, Krakow, Poland, 2–5 July 2008. QIRT Council.
19. Sun, G.; Saga, T.; Shimizu, T.; Hakozaki, Y.; Matsui, T. Fever screening of seasonal influenza patients using a cost-effective thermopile array with small pixels for close-range thermometry. *Int. J. Infect. Dis.* **2014**, *25*, 56–58. [CrossRef]
20. Tay, M.R.; Low, Y.L.; Zhao, X.; Cook, A.R.; Lee, V.J. Comparison of Infrared Thermal Detection Systems for mass fever screening in a tropical healthcare setting. *Public Health* **2015**, *129*, 1471–1478. [CrossRef]
21. Chiu, W.; Lin, P.; Chiou, H.Y.; Lee, W.S.; Lee, C.N.; Yang, Y.Y.; Lee, H.M.; Hsieh, M.S.; Hu, C.; Ho, Y.S.; et al. Infrared Thermography to Mass-Screen Suspected Sars Patients with Fever. *Asia Pac. J. Public Health* **2005**, *17*, 26–28. [CrossRef]
22. Nishiura, H.; Kamiya, K. Fever screening during the influenza (H1N1-2009) pandemic at Narita International Airport, Japan. *BMC Infect. Dis.* **2011**, *11*, 111. [CrossRef] [PubMed]
23. Silvino, V.O.; Gomes, R.B.B.; Ribeiro, S.L.G.; Moreira, D.D.L.; Santos, M.A.P. Dos Identifying febrile humans using infrared thermography screening: Possible applications during COVID-19 outbreak. *Rev. Context. Saúde* **2020**, *20*, 5–9. [CrossRef]
24. Brioschi, M.; Teixeira, M.; Silva, M.T.; Colman, F.M. *Medical Thermography Textbook: Principles and Applications*, 1st ed.; Andreoli: São Paulo, Brazil, 2010; ISBN 978-85-60416-15-8.
25. Haddad, D.S.; Brioschi, M.L.; Baladi, M.G.; Arita, E.S. A new evaluation of heat distribution on facial skin surface by infrared thermography. *Dentomaxillofacial Radiol.* **2016**, *45*, 20150264. [CrossRef] [PubMed]
26. Haddad, D.S.; Oliveira, B.C.; Brioschi, M.L.; Crosato, E.M.; Vardasca, R.; Mendes, J.G.; Pinho, J.C.G.F.; Clemente, M.P.; Arita, E.S. Is it possible myogenic temporomandibular dysfunctions change the facial thermal imaging? *Clin. Lab. Res. Dent.* **2019**. [CrossRef]
27. Ferreira, C.L.P.; Castelo, P.M.; Zanato, L.E.; Poyares, D.; Tufik, S.; Bommarito, S. Relation between oro-facial thermographic findings and myofunctional characteristics in patients with obstructive sleep apnoea. *J. Oral Rehabil.* **2021**, *48*, 720–729. [CrossRef] [PubMed]
28. Derruau, S.; Bogard, F.; Exartier-Menard, G.; Mauprivez, C.; Polidori, G. Medical Infrared Thermography in Odontogenic Facial Cellulitis as a Clinical Decision Support Tool. A Technical Note. *Diagnostics* **2021**, *11*, 2045. [CrossRef] [PubMed]

29. Liu, Z.; Ukida, H.; Ramuhalli, P.; Niel, K. (Eds.) *Integrated Imaging and Vision Techniques for Industrial Inspection*; Advances in Computer Vision and Pattern Recognition; Springer: London, UK, 2015; ISBN 978-1-4471-6740-2.
30. Diakides, M.; Bronzino, J.D.; Peterson, D.R. (Eds.) *Medical Infrared Imaging*; CRC Press: Boca Raton, FL, USA, 2012; ISBN 9780429107474.
31. Ongsulee, P. Artificial intelligence, machine learning and deep learning. In Proceedings of the 2017 15th International Conference on ICT and Knowledge Engineering (ICT&KE), Bangkok, Thailand, 22–24 November 2017; pp. 1–6.
32. Sultana, F.; Sufian, A.; Dutta, P. Advancements in image classification using convolutional neural network. In Proceedings of the 2018 4th IEEE International Conference on Research in Computational Intelligence and Communication Networks, ICRCICN, Kolkata, India, 22–23 November 2018.
33. Andrearczyk, V.; Whelan, P.F. Deep Learning in Texture Analysis and Its Application to Tissue Image Classification. In *Biomedical Texture Analysis*; Elsevier: Amsterdam, The Netherlands, 2017; pp. 95–129. ISBN 9780128121337.
34. Ren, J.; Green, M.; Huang, X. From traditional to deep learning: Fault diagnosis for autonomous vehicles. In *Learning Control*; Elsevier: Amsterdam, The Netherlands, 2021; pp. 205–219.
35. Phung, V.H.; Rhee, E.J. A High-accuracy model average ensemble of convolutional neural networks for classification of cloud image patches on small datasets. *Appl. Sci.* 2019, *9*, 4500. [CrossRef]
36. Zhao, Z.Q.; Zheng, P.; Xu, S.T.; Wu, X. Object Detection with Deep Learning: A Review. *IEEE Trans. Neural Netw. Learn. Syst.* 2019, *30*, 3212–3232. [CrossRef]
37. Girshick, R.; Donahue, J.; Darrell, T.; Malik, J. Rich feature hierarchies for accurate object detection and semantic segmentation. In Proceedings of the IEEE Computer Society Conference on Computer Vision and Pattern Recognition, Columbus, OH, USA, 24–27 June 2014.
38. Yang, J.; Li, S.; Wang, Z.; Dong, H.; Wang, J.; Tang, S. Using Deep Learning to Detect Defects in Manufacturing: A Comprehensive Survey and Current Challenges. *Materials* 2020, *13*, 5755. [CrossRef]
39. Wang, J.; Zhang, T.; Cheng, Y.; Al-Nabhan, N. Deep Learning for Object Detection: A Survey. *Comput. Syst. Sci. Eng.* 2021, *38*, 165–182. [CrossRef]
40. Redmon, J.; Divvala, S.; Girshick, R.; Farhadi, A. You only look once: Unified, real-time object detection. In Proceedings of the IEEE Computer Society Conference on Computer Vision and Pattern Recognition, Las Vegas, NV, USA, 27–30 June 2016.
41. Ivašić-Kos, M.; Krišto, M.; Pobar, M. Human detection in thermal imaging using YOLO. In Proceedings of the ACM International Conference Proceeding Series, Daejeon, Korea, 10–13 November 2019.
42. Memon, J.; Sami, M.; Khan, R.A.; Uddin, M. Handwritten Optical Character Recognition (OCR): A Comprehensive Systematic Literature Review (SLR). *IEEE Access* 2020, *8*, 142642–142668. [CrossRef]
43. Islam, N.; Islam, Z.; Noor, N. A Survey on Optical Character Recognition System. *arXiv Preprint* 2017, arXiv:1710.05703.
44. Indravadanbhai Patel, C.; Patel, D.; Patel Smt Chandaben Mohanbhai, C.; Patel, A.; Chandaben Mohanbhai, S.; Patel Smt Chandaben Mohanbhai, D. Optical Character Recognition by Open source OCR Tool Tesseract: A Case Study. *Artic. Int. J. Comput. Appl.* 2012, *55*, 975–8887. [CrossRef]
45. Panetta, K.; Samani, A.; Yuan, X.; Wan, Q.; Agaian, S.; Rajeev, S.; Kamath, S.; Rajendran, R.; Rao, S.P.; Kaszowska, A.; et al. A Comprehensive Database for Benchmarking Imaging Systems. *IEEE Trans. Pattern Anal. Mach. Intell.* 2020, *42*, 509–520. [CrossRef] [PubMed]
46. de Souza Santos, F.; Da Silva, Y.S.; Da Silva, J.R.; Simao, J.; Campos, H.L.M.; Nunes, R.B.; Muniz, P.R. Comparative analysis of the use of pyrometers and thermal imagers in sanitary barriers for screening febrile people. In Proceedings of the 2021 14th IEEE International Conference on Industry Applications (INDUSCON), São Paulo, Brazil, 15–18 August 2021; pp. 1184–1190. [CrossRef]
47. Lin, T. LabelImg. Available online: https://github.com/tzutalin/labelImg (accessed on 24 January 2021).
48. Padilla, R.; Netto, S.L.; Da Silva, E.A.B. A Survey on Performance Metrics for Object-Detection Algorithms. In Proceedings of the International Conference on Systems, Signals, and Image Processing, Niteroi, Brazil, 1–3 July 2020.

Article

A Model-Based and Goal-Oriented Approach for the Conceptual Design of Smart Grid Services [†]

Miguel Angel Orellana [1], Jose Reinaldo Silva [2,*] and Eduardo L. Pellini [3]

[1] Electronic and Electric Engineering Department, Universidade do Estado do Amazonas, Manaus 69050-025, Brazil; mpostigo@uea.edu.br
[2] Mechatronics Department, Escola Politécnica, Universidade de São Paulo, São Paulo 05508-900, Brazil
[3] Energy and Automation Engineering Department, Escola Politécnica, Universidade de São Paulo, São Paulo 05508-900, Brazil; elpellini@usp.br
* Correspondence: reinaldo@usp.br; Tel.: +55-11-99485-5734
[†] This paper is an extended version of our paper published in Postigo, M.; Pellini, E.; Silva, J. Proposta de método sistêmico baseado em modelos para Smart Grid. In Proceedings of the 14th IEEE International Conference on Industry Applications (INDUSCON), São Paulo, Brazil, 16–18 August 2021.

Abstract: A solid demand to integrate energy consumption and co-generation emerged worldwide, motivated, on one hand, by the need to diversify and enhance energy supply, and, one the other hand, by the pressure to attend to the requirements of a heterogeneous class of users. The coupling between energy service provision and final users also includes balancing user needs, eliminating excesses, and optimizing energy supply while avoiding blackouts. Another motivation is the challenge of having sustainable sources and many adapted to the user ecosystem. Altogether, these motivations lead to more abstract design approaches to co-generation-distributed systems, such as those based on goal-oriented requirements used to model smart grids. This work considers the available design practices and its difficulties in proposing a new method capable of producing a flexible requirement model that could serve for design and maintenance purposes. We suggest coupling the approach based on goal-oriented requirements with model-based engineering to support such a model. The expected result is a sound and flexible requirements model, including a model for the interaction with the final user (now being considered a producer and consumer simultaneously). A case study is presented, wherein a small energy service system in an isolated community in the Amazon rain forest was designed.

Keywords: microgrid model-based systems engineering; service systems; goal-oriented requirements engineering

1. Introduction

Sustainable smart grid (SG) systems are a viable alternative to face the demand of improving and diversifying energy supply services, relying on distributed systems that are also adaptable to specific consumers.

The design challenge is to integrate legacy energy supply; to fit all requirements emanated from hybrid consumers who are now producers (or prosumers); and to simultaneously maintain energy quality. The resulting system is open since it is possible to introduce or remove new prosumers. New design approaches are a demand to face this challenge [1].

Traditional distribution energy supply relies on a general method related to matching theory (even if the designers explicitly use this association) [2]. A "well-posed" problem P related to energy supply should match known systems and equipment or previously studied and practiced methods. Typically, the demand is to organize and optimize distribution from a centralized energy provider and user demands are not personalized. Users are classified into generic classes.

Automation is a control or optimized control problem for traditional energy systems, admitting a small degree of "intelligence". For distributed production-consuming flexible energy services, as described above, design automation is a collaborative multiagent system, which could also be intelligent–coupling with the consumer results from personal requirements and no longer from generic classification. Design methods should face all these demands.

In a single word, we must say that the demand is now for a distributed prosumer service system, where the term "service" carries the usual meaning provided by the Service Science [3]. Figure 1 illustrates this change [4].

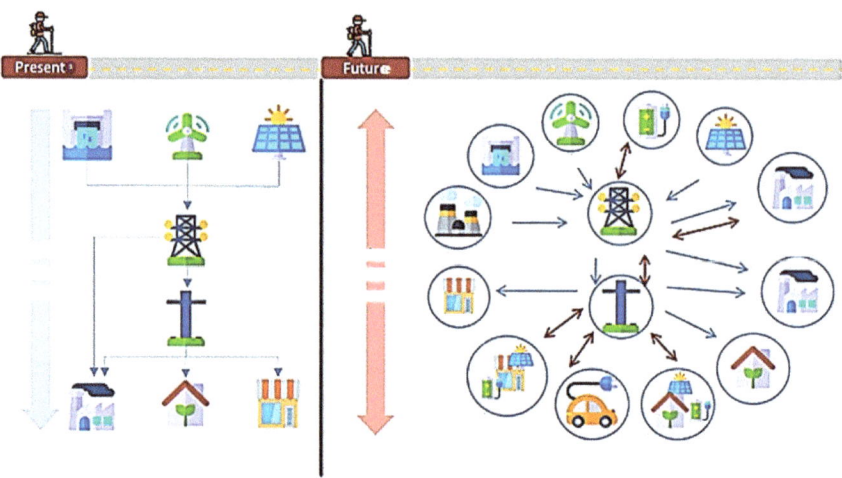

Figure 1. Smart grid present and future. (Reprinted with permission from Ref. [5]. Copyright 2021 IEEE).

We claim that new design methods for energy service distribution must be interdisciplinary, integrating traditional methods and practices, and others typically used in ICT (Information and Communication Technology) or in multiagent systems. Since we are now dealing with services, instead of direct provision, another demand is to introduce Service Engineering Design (SED).

The initial requirements phase acquires importance once it should model the integration of distributed agents in a service ecosystem merged by a central energy provider. The coupling with the consumer (who is also a provider) must be included in this model, which was never done by traditional methods [6].

Requirements validation is not straightforward for the described ecosystem. Actually, for automated systems, formal representation and verification are key issues, leading to a cycle composed of modeling, formalization, and verification. Existing methods consider this cycle a preliminary stage, where the problem P is modeled after some refinements, leading to a model-based requirements approach [7–9].

This article presents a proposal for the formal requirements modeling cycle, relying on a goal-oriented approach customarily used for automated systems including hardware and software, as in metro systems. We anticipate formalization to the requirements phase using a preliminary representation in a schematic language called KAOS (Knowledge Acquisition and Object Specification) [10], fitting into model-based requirements engineering (MBRE) [11]. Therefore, requirements are captured in a diagramatic model (KAOS), starting with a preliminary requirement specification and a refinement cycle composed of analysis and formalization in a formal language (Petri Nets); followed by a property analysis and (formal) verification; and a post-modeling analysis, which points to further improvements. The cycle iterates up to a final model that feeds the design of solutions, formalized in the same language (Petri Nets). In the design phase, developers should look for matching between the problems specified and the available (or new) solutions to produce energy to store [12] or insert other sustainable alternatives as wind turbines [13].

This article is organized as follows: Section 2 presents the background and state of the art. Section 3 provides a conceptual description of the proposed method. Section 4 presents its application in a case study and the results. Finally, a concluding section overviews the contributions of this work and suggests perspectives on future work.

This article is an extended version of the paper presented in the 14th International Conference of *Industrial Applications*, INDUSCON2021, in Portuguese [5].

2. Design Perspectives to Energy Systems

Smart grid (SG) ecosystem design brings new demands that insert this process in the category of complex energy systems due to its connections with different areas of knowledge (e.g., physics, engineering, communication, and even human–systems interaction) [14].

However, from a systems engineering perspective, the most significant change from traditional methods is treating generation and distribution environments as systems rather than arrays of products or devices [15]. Attempts to modify and modernize the conventional design approach by considering the introduction of ICT or even software methods such as the "V" model, also used by the manufacturing industry, have been employed. The "V" model derived from the structured approach for software was adapted to model systems, including software, hardware, and interactions with users. Essentially, the "V" is a "top-down" method, where one part of the "V" goes down from the requirements to prototyping and goes through both system analysis and design. The other part of the "V" goes up from prototyping to system delivering, going through local tests, integration tests, and final deployment. In attempting to adapt this method to the design of electrical energy systems, Roboam [15] proposed an extended model, including some recursion and interlaced multiples Vs. Figure 2 depicts the essence of this hierarchical view.

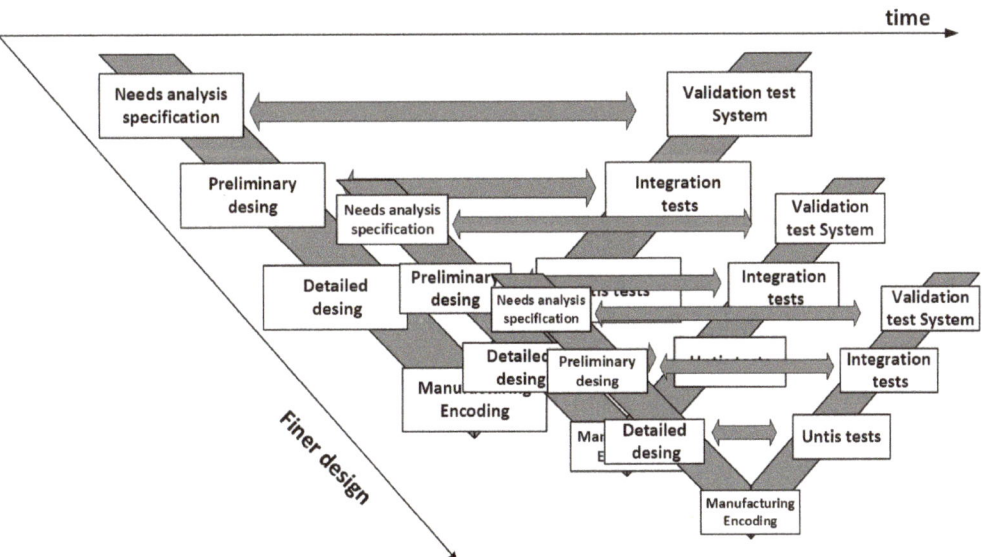

Figure 2. Design V Cycle. (Based on [15], Reprinted with permission from Ref. [5]. Copyright 2021 IEEE).

Roboam [15] was already concerned with presenting a systemic proposal, even if it still followed the basic steps functionally, as in the classic "V". The novelty relies on the proposition of an extended life cycle, starting with a requirements phase, which is not particularly present in the conventional design of energy systems. The drawback is the difficulty of adapting this proposal for distributed systems. An alternative was to use an iteration of "V" cycles to recover flexibility. Requirements were represented in UML (Unified Modeling Language) and also used in several reference models for energy systems design.

Another perspective is the use of the Life Cycle Assessment (LCA) using multi-objective optimization techniques [14]. LCA also stresses the importance of the early phase and it is an interdisciplinary approach, integrating concepts from engineering design, ecology, sustainability, and economic and thermodynamic aspects. Figure 3 presents the basics for this proposal. Information is organized and structured using concepts of system modeling, also redirecting energy distribution projects towards systems design engineering.

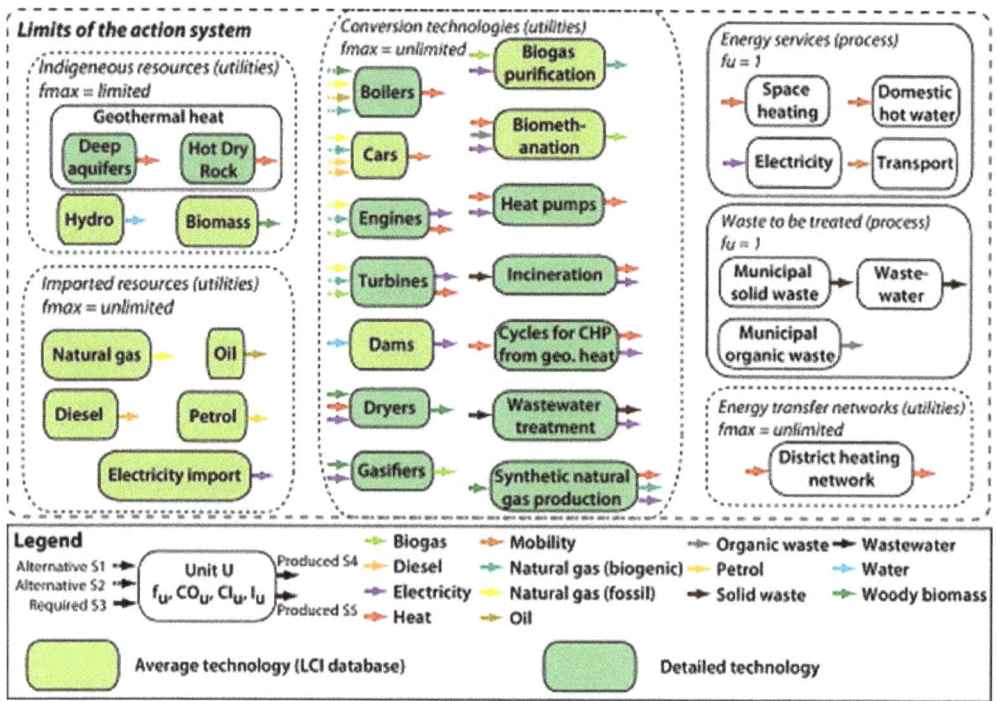

Figure 3. LCA metodology (Adapted with permission from Ref. [14], Copyright 2021 Elsevier)

LCA is a very precise technique that goes into operational details since the initial stages and, for this exact reason, does not allow for a broader and systemic view. This leads to the anticipation of decision-making, which is a very inconvenient feature in complex projects that also requires flexibility. Additionally, it does not consider domain restrictions from the local environment where the project should be deployed, which are very important to reinforce sustainability and user interactions. Therefore, system maintenance could become more complicated. Furthermore, the proposed method is mainly oriented towards systems with a centralized provisioning of energy.

Another perspective brings a specification formalism using a requirements technique based on graph theory for modeling and verification. This reinforces the need to formalize the project from the very beginning: the requirements phase. The work was presented by Frangopoulos [16] and proposed a method for solving energy problems with a large number of non-linear and complex degrees of freedom.

Frangupoulos presented a proposal directly based on the analysis and formal verification of requirements. However, he pointed to a design concerned exclusively with technical aspects without considering the application domain, as required in systemic approaches. Therefore, although there is a good improvement associated with the formal modeling and capture of workflow, using graph theory, aspects related to user interaction and regional insertion are not covered. It would be necessary to recover all external domain restrictions and also user interactions (as consumer and producer) from parallel documentation, either formal or informal, to complete the modeling based on graphs, as shown in Figure 4.

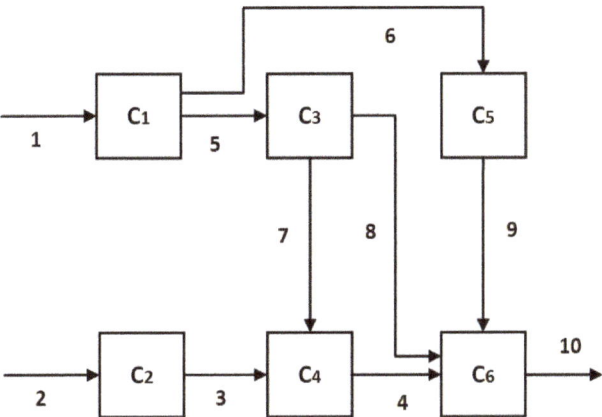

Figure 4. Graph showing a "mental map" (requirements) for the connections between energy components (Reprinted with permission from Ref. [16]. Copyright 2021 Christos Frangopoulos).

We advocate the importance of a plain life cycle to energy systems, particularly related to the production/consumer distributed architecture based on smart grids. The life cycle should be based on the anticipation of a formal representation of requirements. Still, it should also allow for process analysis, domain coupling (covering both interactions and restrictions), and proper modeling of transactions with the prosumer. To avoid losing connections with the conventional methods and practitioners in the area, we also should consider a possible coupling with reference models as proposed by reliable institutions responsible for maintaining regular standards, simplifying the design process.

The Reference Models for Smart Grid Systems

There is an international effort to develop a model or architecture that is globally recognized and used by the leading players in the electricity sector which allows for the unification of methods and criteria, including the development of projects including SGs [17].

The IEEE 2030 standard is considered the most significant recent effort to standardize architectures for the development of SG systems [17]. In this context, the classic "V" model used in the systems area received the additional contribution of a semi-formal functional approach based on UML [18,19], introducing the need to insert visual diagrams from the elicitation stage and requirements analysis.

We identify current methodological trends for the power system design:

- Methods are generally prescriptive, based on methodological or local regulations, but the domain (the region where the electric system will be implemented) is not included as an essential element of design;
- Most methods are primarily functional, that is, they prioritize system functionality and general requirements are restricted to that, generating a severe risk of consistency due to lack (or incompleteness) of non-functional requirements;

- Many methods are device-oriented, that is, they are centered on the matching to specific devices probably to facilitate the integration with legacy distribution systems, regional conditions, or users;
- Many methods consider SG as a product or an arrangement of products composing the electrical system, with few or no consideration to their interaction with users;
- Some approaches consider centralized master–slave systems, where SGs are individual resources tightly coupled to a master system that centralizes all control or information, basically eliminating flexibility;
- In many conventional methods, the evolution of the system is not a design concern, as in systems design, which makes it difficult to introduce changes or innovations (or even to meet emerging requirements); and
- Despite all the effort to standardize and build project repositories for reuse, reuse is not part of design methods.

Therefore, we conclude that a modern approach to designing energy systems requires a requirements cycle that must be closed (as a model should be), combining hierarchy with distributed system concepts. It should also be formal. Based on this, we present a proposal for a design cycle for energy multiagent-distributed systems.

Reference models have been developed to orient the design of electric systems. In the following section, we will describe only principals.

EPRI/NIST (Electric Power Research Institute/National Institute of Standards) was sponsored by IntelliGrid, while the European Commission (EC) developed the SGAM (smart grid architecture model) architecture. They are the strongest reference architectures for SG design. These architectures are based on open standards, allowing for interoperability between products and systems. Additionally, they embrace a systematic approach for addressing multidisciplinary distributed systems-based SGs, starting with a UML requirement specification. Reusability was also considered using a IEC/PAS (International Electrotechnical Commission/Public Available Specification) 62559-based repository of use-case model histories [7,8,19,20]. IntelliGrid and SGAM architectures dominated the scene both in the academy and market. Still, we should point out that they consist of a purely functional approach based on use case, as conceived for requirements elicitation rather than analysis.

SGAM architecture has a method for mapping SG use-case information through top-down layered development [8]. Layers are derived from the use-case analysis, starting with the component, business, functional, information, and communication layers.

Basic reference models also rely on functionality and are suitable for merging conventional methods, reinforcing a longer life cycle and making the requirements analysis difficult.

3. A New Proposed Approach for SG Systems Design

The most improving design feature is to treat SG as a system, including model-based systems engineering (MBSE). The International Council on Systems Engineering defines MBSE as the formal application of modeling to support system requirements, design, analysis, verification, and validation activities, starting from the conceptual design phase and continuing throughout the development as well as later life cycle phases [21–23]. Additionally, MBSE has several frameworks for examining different corporate system views: business view, system, technology, operations, and service [24].

From the systems design perspective, an "ideal" design must be correct and complete, for which formalization is a crucial issue [11]. The formal approach can clarify objectives and provide both uniqueness and closeness in the different phases of design:

- In requirement (defining the what and why);
- In design (defining how to satisfy the requirements efficiently with controlled costs); and
- In implementation (deploying a prototype that interacts with the end-user and works properly up to decommission).

Frequently, a new system is required to "replace" an existing one that has become obsolete. The process should start by modeling the "System As-Is", that is, the legacy system. "New requirements" should then detach the differences between the "System As-Is" and the target "System To-Be", as shown in Figure 5.

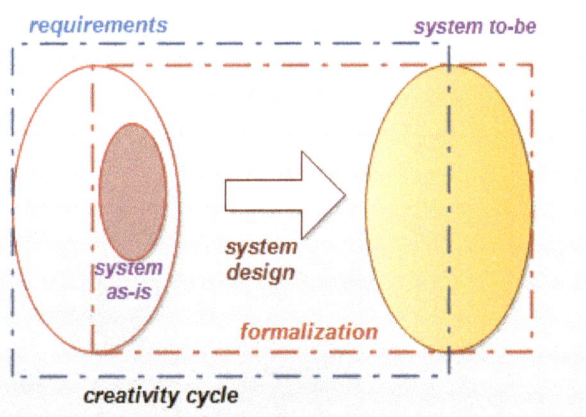

Figure 5. Creativity cycle based on (Reprinted with permission from Ref. [25]. Copyright 2021 Brazilian Society of Automation).

Another important feature is related to the paradigm shift that moves "product" systems into service systems, leading to thinking in models rather than prototypes, aiming to disconnect from the legacy functionality of their parts [26]. Therefore, structuring a complex system means using the concept of hierarchy, that is, to divide the primary system into parts recursively until the level where components are very simple or already developed is reached.

However, a simple hierarchical approach does not help to understand a complex system. Integration is challenging and emerging requirements could appear after deployment (sometimes when the system is already being used), forcing a re-design. The hierarchy should combine loosely coupling components or services facilitating the reuse.

In this scenario, the concept of System of Systems (SoS) emerges. SoS is defined as a set of constituent systems that exist and are developed to perform specific services or tasks independently from the legacy system they work with [27].

3.1. Smart Grid as a Distributed Arrangement of Services

The NIST Framework and Roadmap for Smart Grid Interoperability Standards, Release 1.0, was adapted to include an SoS approach for SG. Figure 6 shows a diagrammatic view that connects different systems (and services), implying some responsibility in the consumption or provision and including the final user in the design process. It could be interpreted as a domain functional representation, relating to operations, provision, destination, and an implicit agent responsibility.

Analyzing SG systems as an SoS [28] leads to a model where the SG can be thought of as a "company providing energy services to different markets". Therefore, this system's design is hardly conceivable without strategic planning and service control management to maximize performance and optimize its functioning. Thus, technological, environmental, and socioeconomic aspects must be considered in the same perspective and be reflected upon in a knowledge model related to requirements. It necessarily points to a requirements model that transcends functionality.

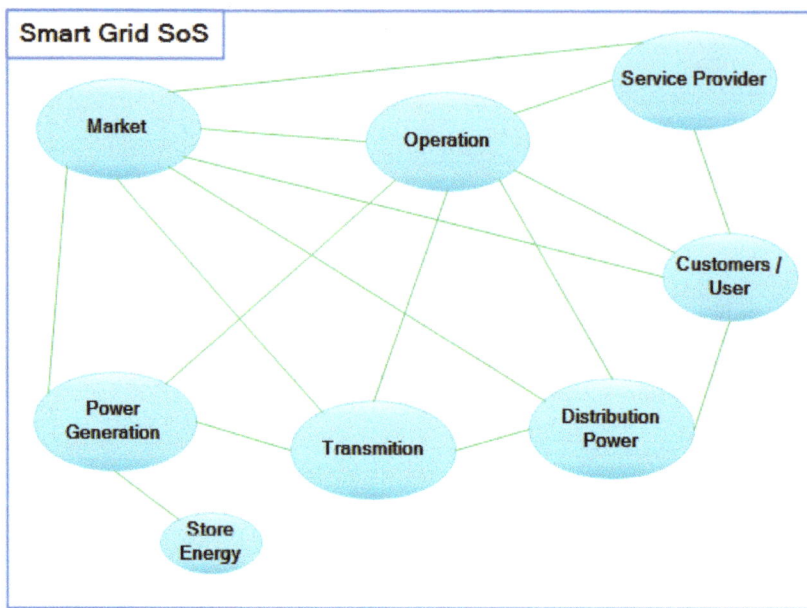

Figure 6. Smart grid SoS.

Designing SG as a distributed arrangement of services in an SoS perspective requires an MBSE approach, starting with the requirements phase and going through both modeling and formalization. Such an approach should include and transcend functionality, looking for alternative methods to requirements engineering that relate to agents and map responsibility, and represent user coupling [29].

Therefore, considering the importance of the requirements stage in SG design, we propose extending the strictly functional approach to goal-oriented requirements Engineering (GORE), exploring the possibility of providing a model-based requirements cycle.

3.2. The GORE Method for SG Systems Design

The GORE approach aims to eliminate the dichotomy between functional and non-functional requirements [10]. Although working with functionalities sounds more intuitive, as a one-way delivering action, objectives are related to quality, user satisfaction, and feedback. Therefore, it is suitable for the user-coupling proposed in service engineering. Objectives represent the most stable information in the system and are problem-oriented, while other functional techniques are solution-oriented. Thus, objectives become an excellent communication tool with stakeholders regarding a particular problem solution.

An objective is a statement of intent that the system must satisfy through cooperation between agents, active components of the system, and those responsible for the operations behind the actions and processes [30]. In the last decade, the GORE method's popularity has increased, reaching a higher level of maturity [31,32]. The main reason for this is its capability to ease decision-making and lead to optimized solutions, different from the purely functional method, which depends on the completeness after including the non-functional requirements set.

For these reasons, we chose a goal-oriented approach for requirements elicitation, modeling, and analysis, composing a model-based requirements cycle. We also use a diagrammatic semi-formal language which is easily understood by the stakeholders [33].

Goal-Oriented Elicitation and Modeling

Goal-oriented elicitation and modeling use a semi-formal visual language called KAOS (Knowledge Acquisition in Automated Specification) to identify business requirements, build a network of objectives and operations, and attribute responsibility to agents, associating them with goals and requirements. KAOS generates a causal visual diagram with a suit transference to a formal LTL (linear temporal logic) specification. It was developed by the University of Oregon (USA) and the University of Louvain (Belgium) in 1990 [34].

Thus, we can construct formal LTL specifications from KAOS requirement models. Objectives are defined at different levels of abstraction, from high-level goals related to organizational, strategy, and diffusely specified objectives, to lower-level objectives with more technical, detailed, and system design-related specifications [10]. Figure 7 presents the schematic diagram of such a model.

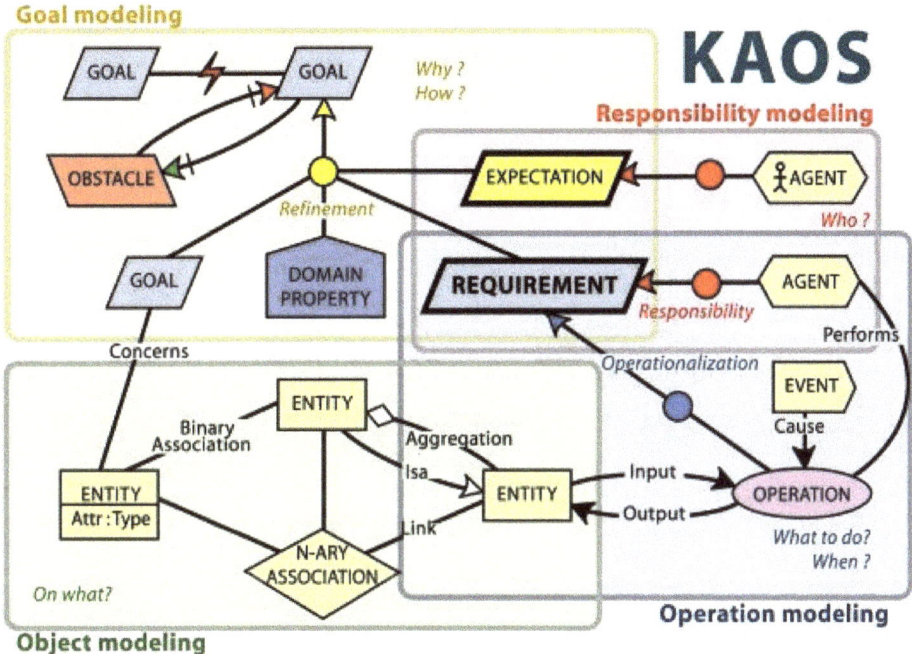

Figure 7. KAOS diagrams with goal-oriented elements (Reprinted with permission from Ref. [35]. Copyright 2021 Respect-IT).

A tree represents KAOS objectives in which the nodes represent objectives and edges represent relations (composition, refinement, dependency, and constraint).

The main objective (matching the "system goal") abstracts the design problem, while sub-objectives represent compositional refinements or sub-goals. Expectations or intentions are related to objectives, which would not be possible in a purely functional method. Agents would stand by active objects that could change the state of the system. Figure 8 shows the basic elements for KAOS diagrams.

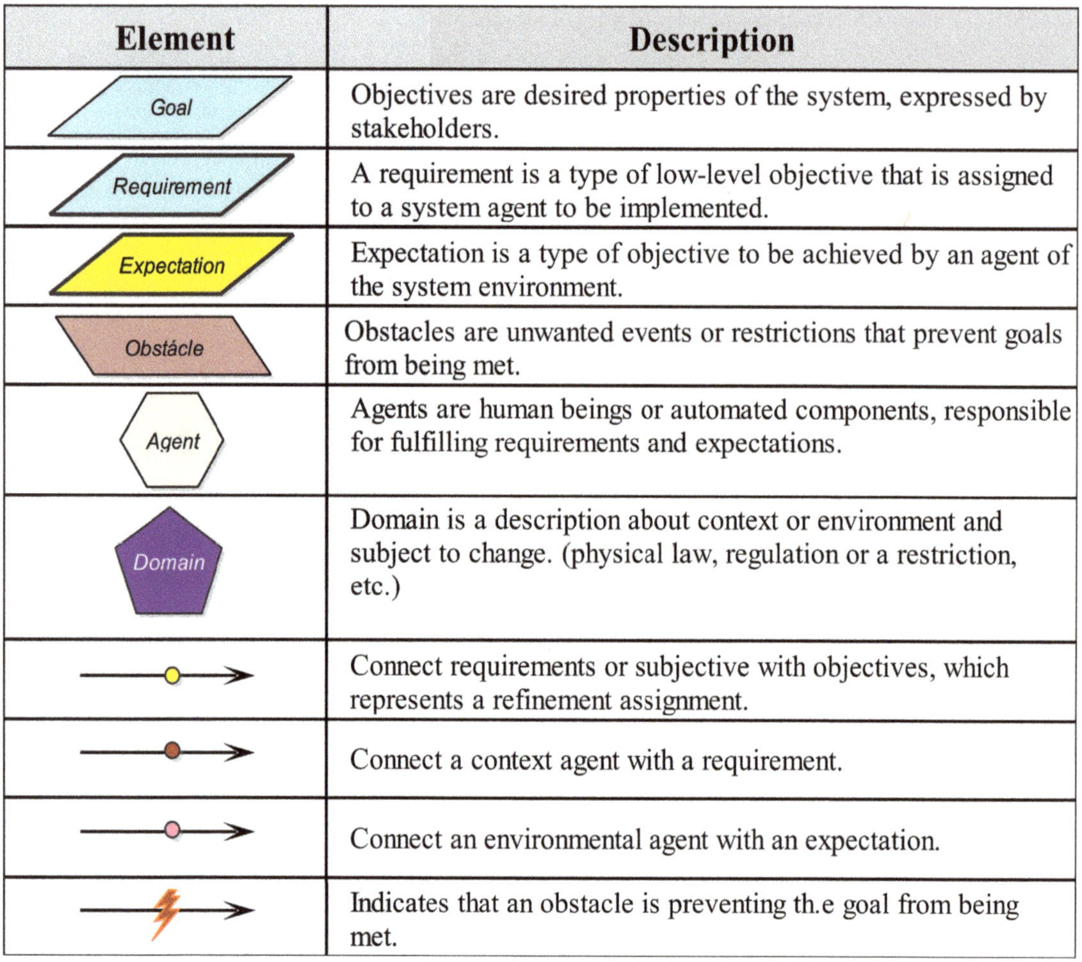

Element	Description
Goal	Objectives are desired properties of the system, expressed by stakeholders.
Requirement	A requirement is a type of low-level objective that is assigned to a system agent to be implemented.
Expectation	Expectation is a type of objective to be achieved by an agent of the system environment.
Obstácle	Obstacles are unwanted events or restrictions that prevent goals from being met.
Agent	Agents are human beings or automated components, responsible for fulfilling requirements and expectations.
Domain	Domain is a description about context or environment and subject to change. (physical law, regulation or a restriction, etc.)
—o→	Connect requirements or subjective with objectives, which represents a refinement assignment.
—●→	Connect a context agent with a requirement.
—o→	Connect an environmental agent with an expectation.
—⚡→	Indicates that an obstacle is preventing th.e goal from being met.

Figure 8. KAOS elements (Reprinted with permission from Ref. [5]. Copyright 2021 IEEE).

The main difference between a KAOS and UML diagram is that while KAOS integrates the requirements model in four diagrams, UML requires up to thirteen structural and twelve behavior diagrams. Although it is unnecessary to use all UML diagrams to model requirements, it raises the problem of finding a minimal set.

Regarding requirements model formalization, there is a direct algorithm to transfer KAOS diagrams' linear temporal logic (LTL) [36]. It is instrumental in requirements analysis, but to model automated systems, it is also necessary to use processes and a formal language capable of representing workflow, such as Petri Nets.

Some of authors worked to introduce process and workflow analyses in the requirements cycle, proposing algorithms to transfer UML [37] or KAOS diagrams into Petri Nets [38]. Thus, we can have a requirement cycle using UML to fit some of the reference models or use KAOS diagrams. In this work, we will explore the second option.

3.3. Verification of Requirements Based on Petri Nets

Verification is crucial to the proposed method for designing a power electronic mesh of sustainable resources integrated into a conventional electrical distribution system. The basic assumption is that requirement specifications must be formalized, analyzed, and verified before searching for an effective solution.

As with any automated system, integrated electronic power plants have to guarantee the proper convergence of objectives towards the satisfaction of user needs (as services) while also avoiding undesirable states and obstacles. This could be obtained from the modeling and verification of requirements, which would lead to the design of solutions that could also be verified using the same formalism, increasing traceability in the whole design process.

Verification could be done by generating the state space or by analyzing workflows (of control and items) using property analysis in Petri Nets, or both. Temporal and hierarchical extensions are also welcome.

4. A New Requirement Cycle for Microgrid Systems Design

We propose for a requirements cycle to be inserted in a design approach for microgrids that have the following characteristics:

- Systemic vision, that is, a holistic approach where the microgrid is a service system component and therefore should be considering since the early design phase to work in the context of a more extensive and evolving system that will enforce flexibility and reusability;
- The microgrid service system is built based on system goals and in a legacy system (the system-as-is);
- The approach should be based on objectives, especially in the early phase (goal-oriented requirements);
- The method follows service-oriented engineering, which means that the proper coupling between system and user is part of the design;
- The process is recursive and distributed, allowing for the use of subsystems already developed or specifying components that will be developed later; and
- Formalization should be present since the requirements phase using LTL and Petri Nets.

In this work, we will focus on the early phase. MBSE will be the overall approach to propose a requirements cycle that can specifically sustain electrical systems based on the interaction of different subsystems, namely a systems of systems (SoS). Subsystems could also be microgrids, generators, or primary power sources. Requirements are adapted to fit user profiles and preferences, and to anticipate emerging requirements and new demands raised after implementation.

We start by representing the existing system (the System As-Is) and using it as a reference to define new goals, expecting to convert it into the System To-Be. Suppose there is no system-as-is (in which case the challenge is to make a design from scratch). In that case, a model is derived from the requirements elicitation, capturing the intentions of stakeholders and final users.

The distributed system arrangement is treated as a service. Using GORE techniques will help to overcome the so-called *functional dilemma*: the balance between functional and non-functional requirements. The goal approach potentializes design reuse, flexibility, and maintenance while anticipating formalization accuracy.

Figure 9 shows the process, which starts by collecting data from the domain where the system is supposed to be used (including user profiles). Once the system-to-be goals are derived and synthesized in a diagrammatic model, they should be formalized in LTL or Petri Nets.

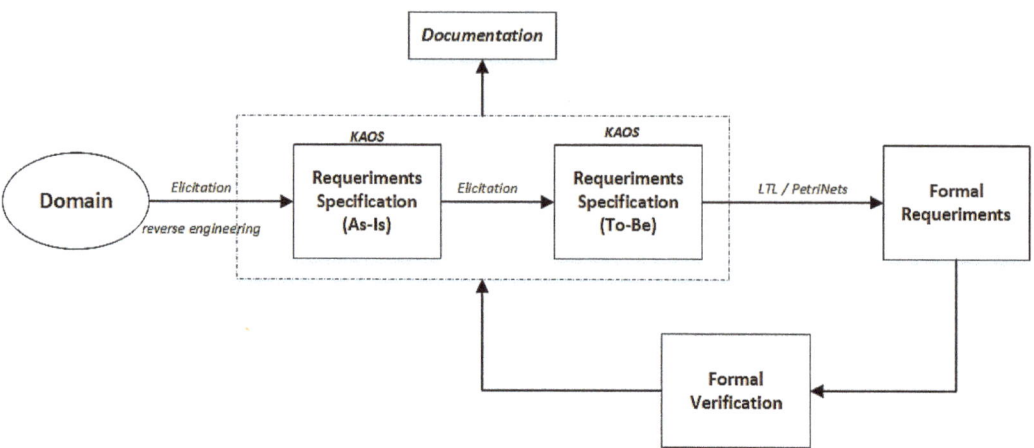

Figure 9. The requirements life cycle (Reprinted with permission from Ref. [5]. Copyright 2021 IEEE).

Petri Nets structural and behavioral properties can analyze the dynamic process and workflow [37]. Verification should certify the work done and also provide directions for the new requirements cycling.

In the following, we will show the requirement cycle in a case study for a power supply system (a smart grid) installed in an isolated Amazon forest community.

5. Case Study: Energy Supply in an Isolated Community

The proposed method was applied in a case study based on the R&D Project "Minigrid with intermittent sources to serve isolated areas", which was a project conceived by a research group and implemented in an isolated community in 2012 [39]. The project's main goal was to develop and implement a solar–diesel battery microgrid, which we take as the system-as-is. The implementation domain was a small community in Laranjal do Jarí, located in Amapá province, covered by the Brazilian Amazon forest (Figure 10). In this region, many county communities are energetically isolated from the conventional hydroelectric provider.

The project was revisited to provide maintenance, that is, to attend to missed or emergent requirements and to improve automation. We detected some emerging requirements using the goal-oriented modeling built over the legacy system-as-is used to design the new system. Another goal of this maintenance process was to check and enhance sustainability and compatibility between the system and the local environment. Finally, the user coupling (originally only consumer) was also a goal, especially in terms of exploring this user's possibility to produce energy.

The microgrid mixed both AC and DC architecture, as shown in Figure 11. The legacy system (system-as-is) combined intermittent solar energy with battery sources and diesel generation units in a single automated generation system. The diesel generator was a backup unit to guarantee operation when the solar source was not producing the expected charge.

A data acquisition system (DAS) for control, monitoring, and storage was embedded in the network of inverters used to provide information about the system's behavior, such as the amount of energy generated and specific load consumption.

Figure 10. Location of the small villages where the project was implemented. Thet are close to a large hydroelectric facility but did not have energy.

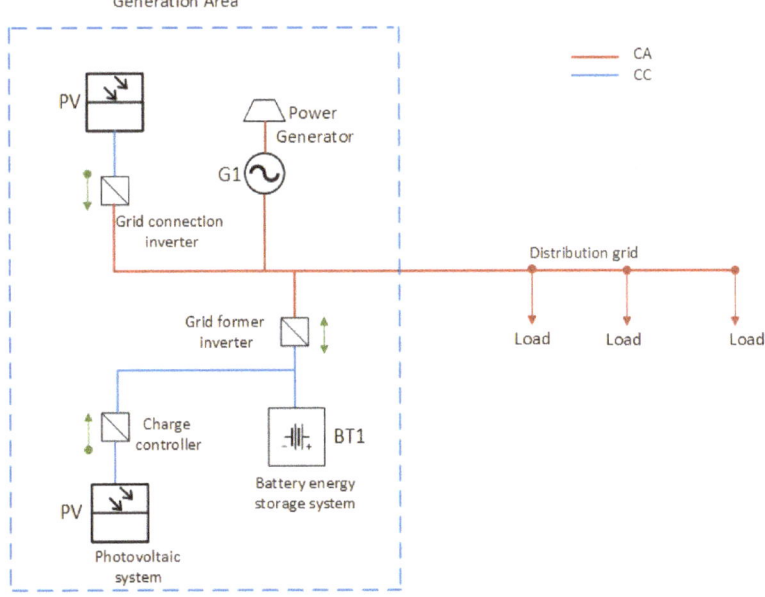

Figure 11. Microgrid legacy system. There were two sources (PV): the microgrid provided an alternate current (CA) and the photovoltaic system provided power associated with a battery BT1, which provided a continuous current (CC). (Reprinted with permission from Ref. [25]. Copyright 2021 Sociedade Brasileira de Automática).

Energy meters monitor each customer's consumption and show when they overtake pre-established values. The monthly energy consumption of the units was not monitored by energy meters, as the purpose was not billing.

5.1. Modeling the System-As-Is

Following the requirement cycle of Figure 9, we started by organizing and classifying information, as shown in Table 1. This information was used to identify and justify objectives.

Table 1. Project data.

Data	Description
Technical data Demand study Availability of daily energy to be granted Availability of the energy matrix Dimensioning of the photovoltaic system Dimensioning of the battery bank Dimensioning of the auxiliary generation system Dimensioning of electrical connections Characteristic of the existing electric grid Energy management system	Project data
Operational data Monitoring of operational variables Maintenance program for microgrid systems Microgrid data and communication storage system System security procedures	Microgrid operation and control
Financial data Budget spreadsheets Acquisition of materials and labor Cost of transport logistics	Financial and economic analysis of the project
Legal/social data Regulatory standard Socioeconomic survey Perspective of population increase	Environmental licenses
Environmental data Survey of environmental restrictions	Environmental impact survey worksheet

Some revisiting points emerged from the analysis. Most of them were derived from the emerging requirements or changes that appearred after implementation either in the environment or in the consumer's attitude. Such requirements would not be identified using a strictly functional approach, which justifies choosing a goal-oriented method.

The objective model for the legacy system's (system-as-is) objectives is shown in Figure 12. Basic information derived from the original project's documentation but additional expectations and constrains (and conflicts) emerged from the analysis of Table 2.

Table 2. Possible new goals to the project.

List of Possible New Goals
• Excessive energy consumption coupled with the presence of clandestine load connections not foreseen in the basic project, thus can increase consumption by approximately 240% • Increase in diesel fuel consumption, overtaking the subsidized quota due to increased energy demand • Abnormal operating conditions of the battery banks, working below the state of charge (SoC) of 50% • The increase in energy consumption also caused an overload in transformers and power cables • There is no monitoring control in the energy consumption • There are no penalties due to energy theft and due to the lack of energy contracts • The "culture of wasting energy" was not disciplined or prohibited

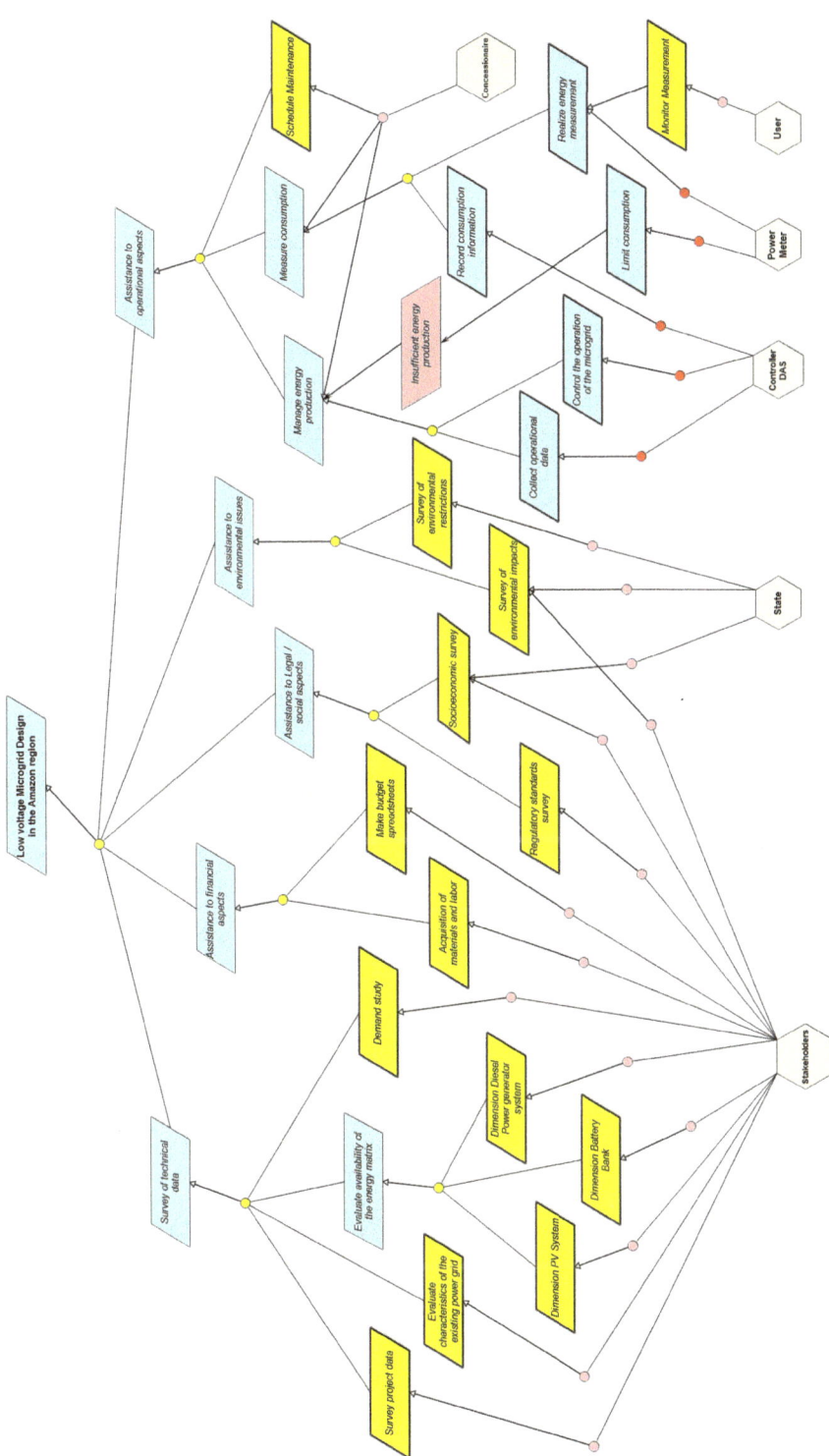

Figure 12. Objetives for System-As-Is (Adapted with permission from Ref. [25]. Copyright 2021 Brazilian Society of Automation).

The main goal for the system-as-is was to "Provide a low voltage microgrid" and it was refined in the sub-goals successively. For instance, "Meeting operational requirements" depends on the sub-objectives "Managing energy production", "Measuring consumption", and the expectation "Schedule maintenance". Requirement expectations are linked to the agents concerned with them: the Concessionaire, responsible for maintenance; the User; the Power Meter (a passive interface in the legacy system); the data acquisition system (DAS) Controller; or the financial provider (the Government). Notice that in the original project, the agent identified as "User" is responsible only for "Monitor Measurement", that is, this agent is not a consumer (recall that the system does not charge its consumers). Consequently, the project does not identify the system and its components as a service.

5.2. Modeling the System-To-Be

We first need to identify the target problem and its relation to the execution environment to model the system-to-be. From the analysis of the system-as-is, we understand that a possible problem is related to the excessive energy consumption due to the advantages that electricity offers rather than to the technical energy losses related to the end-users' alienation from the energy provision. Energy stealing was also a problem, motivated by changes in the user attitude.

The system's environment does not a general provider, turning it into an isolated microgrid very dependent on the energy it could generate and responsible for its own backup to avoid interruptions.

We propose changing the domain from an isolated microgrid to a distributed microgrid (on-grid) with a higher degree of automation to solve these problems. Domain changes, such as automated measurement and new communication technologies, appear as new objectives. In addition to technical changes, the new system must be safe and reliable, and must meet all interested parties' requirements (such as the utility, financial, and environmental agencies). In the new scenario, the user becomes a prosumer and should provide information feedback to optimize the microgrid system.

The new system can no longer be considered a "product". Therefore, it will be necessary to prioritize the requirements of the system that provides this service.

The primary (system) goal of the system-to-be is *to improve the low voltage legacy microgrid design in the Amazon region*. Still, now the objective is to enhance the automation level. Emerging requirements anticipated in the system-as-is modeling appear in expectations, obstacles, or restrictions, and lead to new requirements.

Thus, to achieve the System To-Be's objectives, we propose a SoS+ architecture that supports a new arrangement of services (subsystems) where agents are dynamic elements. The final user is one of these dynamical elements and has a complex structure of resources, as shown in Figure 13. One of these constituent systems is the smart meter (SM).

The system-as-is goal showed that the power supply would be blocked whenever the user consumption exceeds a pre-established limit of 100 kWH. There is no consideration of historical consumer data.

Figure 14 shows the objective diagram for the system-to-be, in which the main objective is to "Improve the low Voltage legacy Microgrid in the Amazon region". The sub-objectives "Improving Strategic Planning", "Optimize energy production", and "Improve operational management" should contribute to the achievement of this primary goal. These sub-objectives can be successively refined.

The analysis of the objective diagrams points to the need for a service system that supports a higher level of control and interaction with the prosumer. Parameters such as the energy production and consumption in each unit should be monitored, as well as its balance between provision and consumption. All data should be directed to an advanced metering infrastructure (AMI) which acts as a service provider instead of a passive supervisor. The detection of electricity stealing is a goal. Naturally, the SM should process information and commands to activate or deactivate the dealership's energy provision.

The goal of "Reducing energy consumption" would fail if the user does not take the actions required. These actions should range from energy-saving acts to the production of energy as a prosumer. The adequate coupling with the end-user is ensured by monitoring the transaction communication.

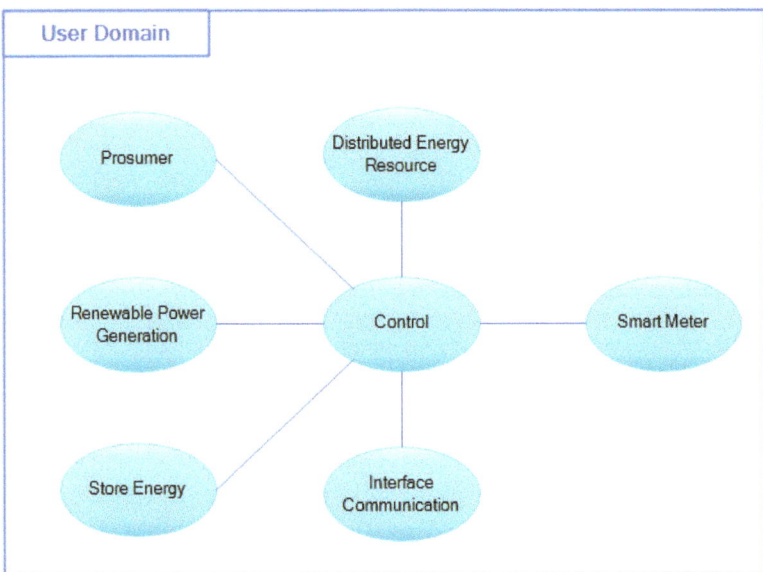

Figure 13. User domain.

5.2.1. The Object Model

An object model should describe all the agents that contribute to the functioning of the system service, be it humans or machines. Object modeling is also practiced in other representations, including UML, and does not constitute a novelty. It is also essential for the goal-oriented approach to support the interaction between the energy system service and its prosumers. We will further illustrate the modeling of the role of a machine, namely the smart meter.

The prosumer characteristic appears in the goal-oriented model, including the same individual prosumer as an external user (and therefore a consumer) and as part of the system (an energy service provider).

Agent modeling will also be used to support responsibility diagrams.

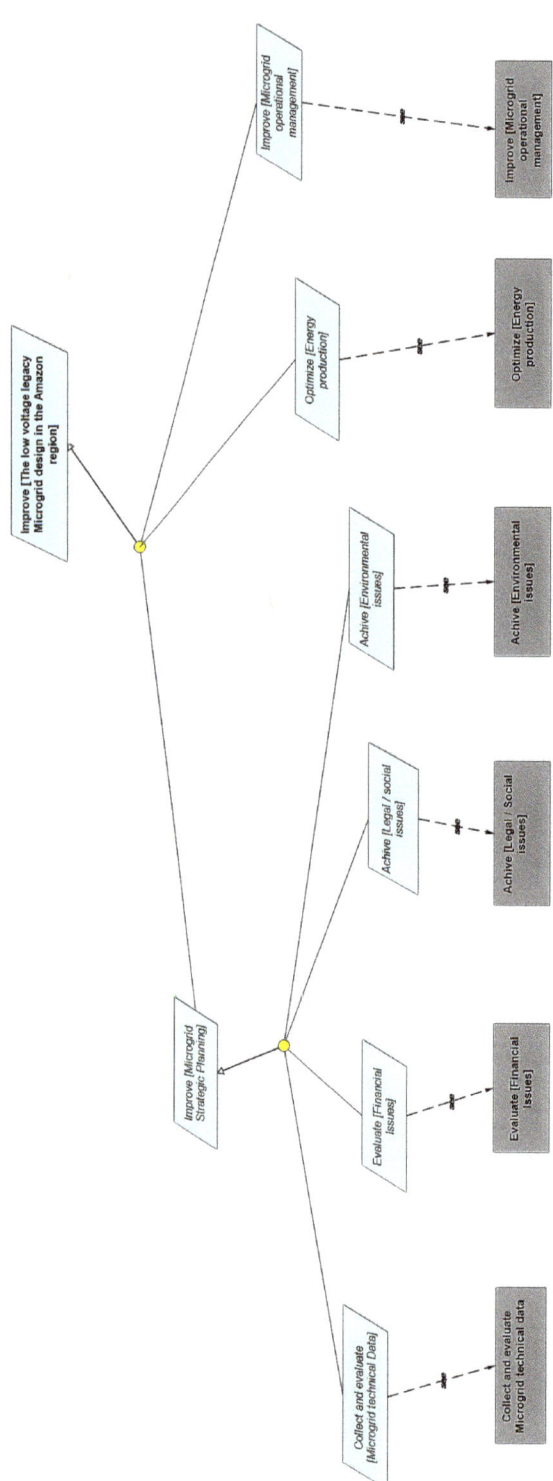

Figure 14. Goal diagram for the system-to-be.

5.2.2. The Responsibility Model

The responsibility model maps the relationship between agents and their role in requirements satisfaction and restrictions. Building a responsibility model means analyzing different requirements and expectations, and assigning agents responsible for launching, performing, or receiving their results.

The responsibility model denotes relations where an agent can be the service provider or service consumer for modeling services. A provider or consumer relation was detailed only in the Petri Nets model since it does not belong to the current goal-oriented method.

Figure 15 shows the responsibility diagram for our case study and describes both the new requirements and expectations assigned to the end user and SM agents, respectively.

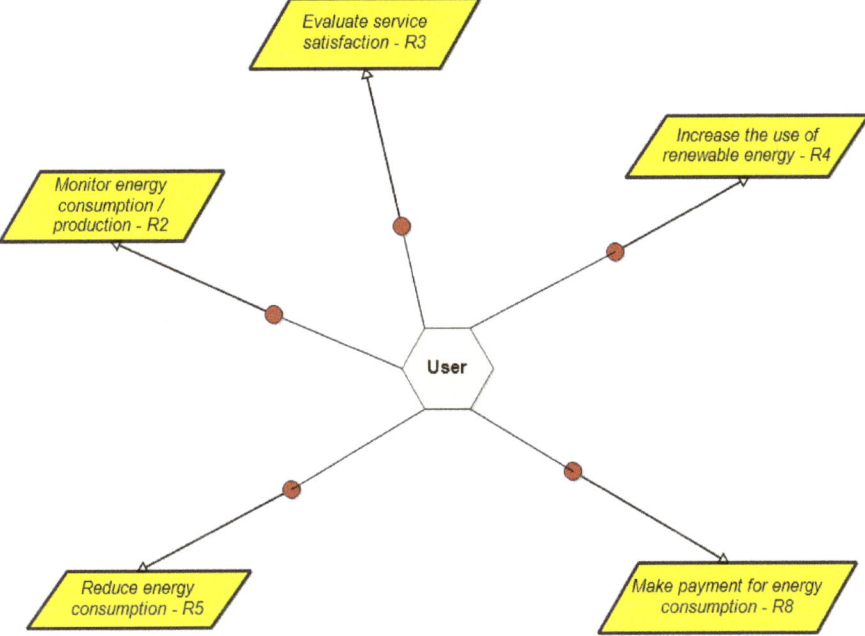

Figure 15. Agent user.

5.2.3. The Operation's Diagram

The operation's diagram describes agents' behavior for meeting requirements. For instance, the SM agent must fulfill the requirement "Receive information of energy consumed and produced" and the expectation "Inform the user about Energy consumed and produced". In further maintenance, any change in the communication with the user would lead to an investigation, reprogramming, or modification of this agent.

Figure 16 shows a definition for the object Smart Meter ("SM") and Figure 17 shows its behavior carrying out operations that affect the "Consumption report" and cause the activation of the "Register consumption" operation. It is verified that this operation has an entry called the "Measurement data received" event and the corresponding output is represented by the "Consumption report" object. The "Recorded record" event represents the end of the operation.

5.2.4. The Operation's Diagram

The operation's Diagram describes agents' behavior for meeting requirements. For instance, the SM agent must fulfill the requirement "Receive information of energy consumed and produced" and the expectation "Inform the user about Energy consumed and produced", respectively. In further maintenance, any change in the communication with the user would lead to an investigation, reprogramming, or modification of this agent.

Figure 17 also shows the "SM" behavior carrying out operations that causes the activation of the "Register consumption" operation. It is verified that this operation has an entry called "Measurement data received" event, which corresponding output is represented by the "Consumption report" object. The "Recorded record" event represents the end of the operation.

Figure 16 presents the "SM" agent's requirements and expectations.

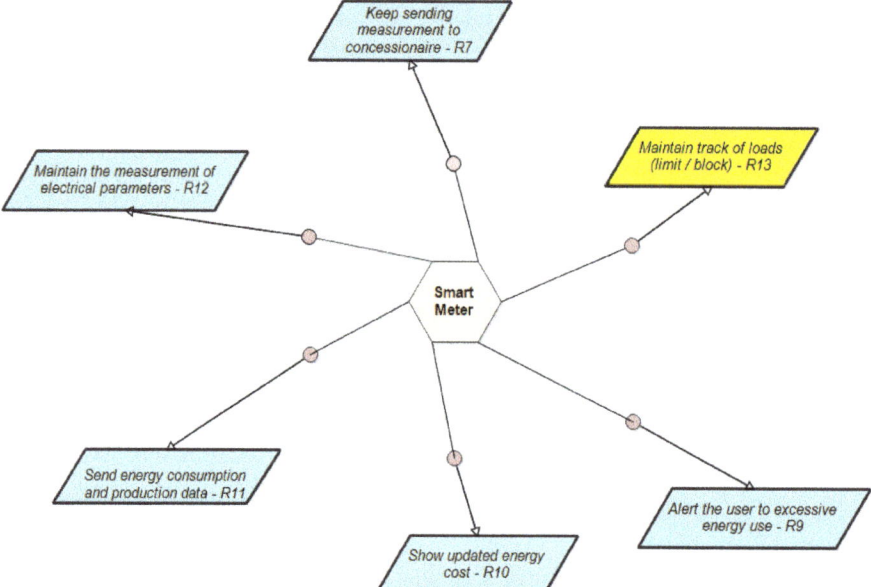

Figure 16. Agent SM.

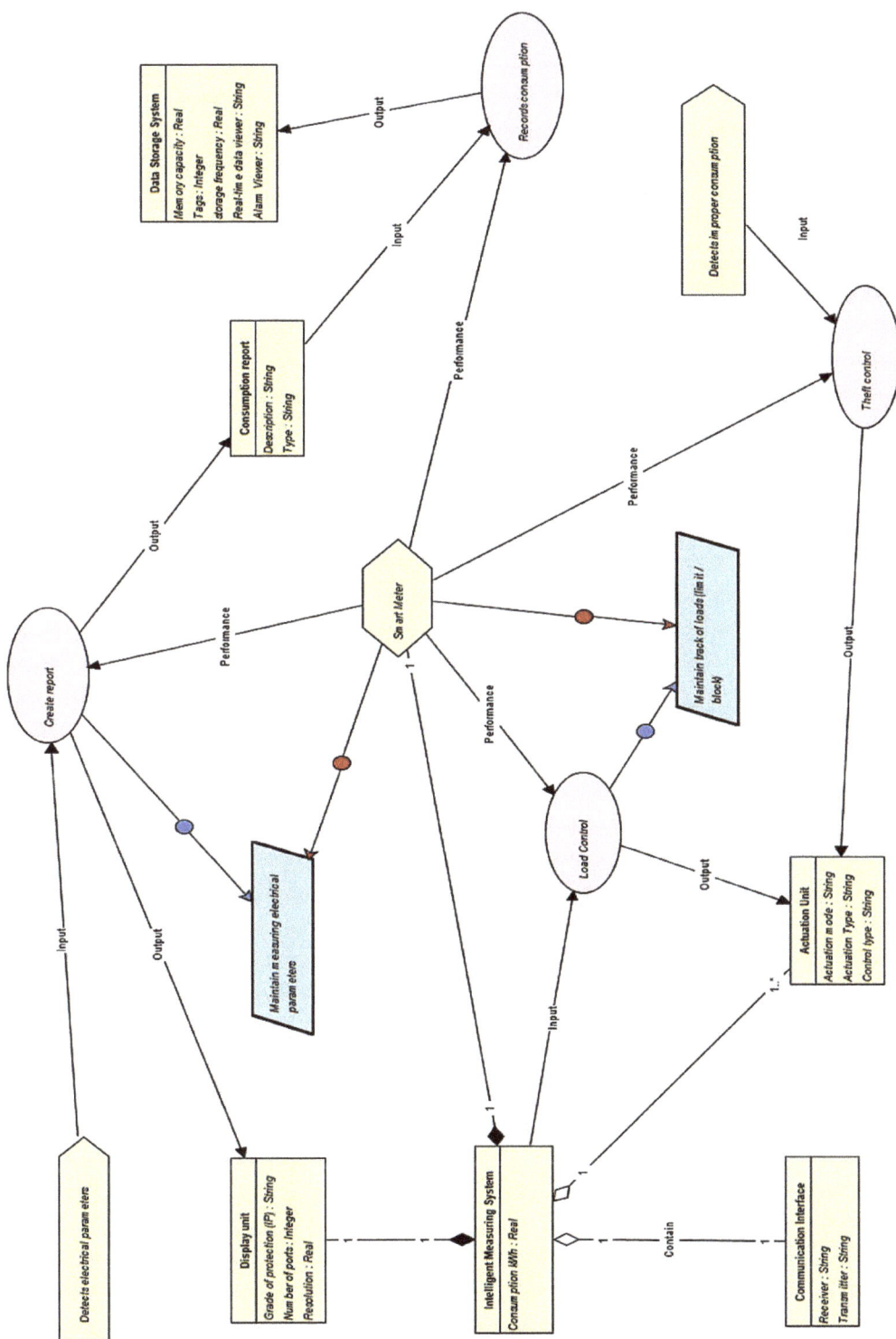

Figure 17. SM operation to monitor consumption.

6. Formalizing the Requirements Model

Finally, the requirements model, composed of the objective, object, operation, and responsibility diagrams, could be transferred to a formal representation in linear temporal logic (LTL) or Petri Nets. Since we intend to work with automated processes, we will introduce a formal representation in Petri Nets, following the standard IEC 61850.

7. Analysis of the Partial Results

The outcomes of the proposed methods appear in the design process of the case study presented.

The proposed requirements cycle starts modeling the system-as-is (if a legacy system exists) to understand and analyze the current design situation. The system-as-is is scrutinized to find flaws or improvements to be inserted in the system-to-be. This part of the method is already proposed in existing material for KAOS [35,36,40].

We should also point out that we applied the goal-oriented approach, delaying the expression of discrimination between functional and non-functional aspects, and always relying on goals to collapse or hide this need. We avoided the pressure to balance functional and non-functional requirements at the beginning of the process while dealing with the problem definition (not with solutions) and with incomplete knowledge. Non-functional requirements will be included as domain restrictions later.

The service design appeared in the process only heuristically and although very useful, it still needs to be formalized in further work. For instance, the current model admits direct formal modeling for users' classes but the coupling between the service provided and the user agents was not formalized. In other words, it is essential to highlight the importance of value co-creation modeling to support (human) user interaction with the service. A formal approach for that is being developed by the authors and will be published later.

The Petri Net in Figure 18 can be analyzed using available environments to evaluate its transition graph (as an automaton) or to perform property analysis. Either way, we can extract and verify the desired properties or detect some unexpected behavior that should be removed from the control/service system. A Petri Net is a formal schema that captures the behavioral and structural properties of a target system. It is well used in modeling automated dynamical systems. The ISO/IEC 15.909 standard provides a unified definition of the formalism used by the market and academy.

The proposed method follows a cycle of model-driven requirements development for a distributed energy service system, mapping the interaction between all agents, be it humans or machines, and its relationship with the surrounding environment. This is very helpful to ensure sustainability and human well-being as part of the design process.

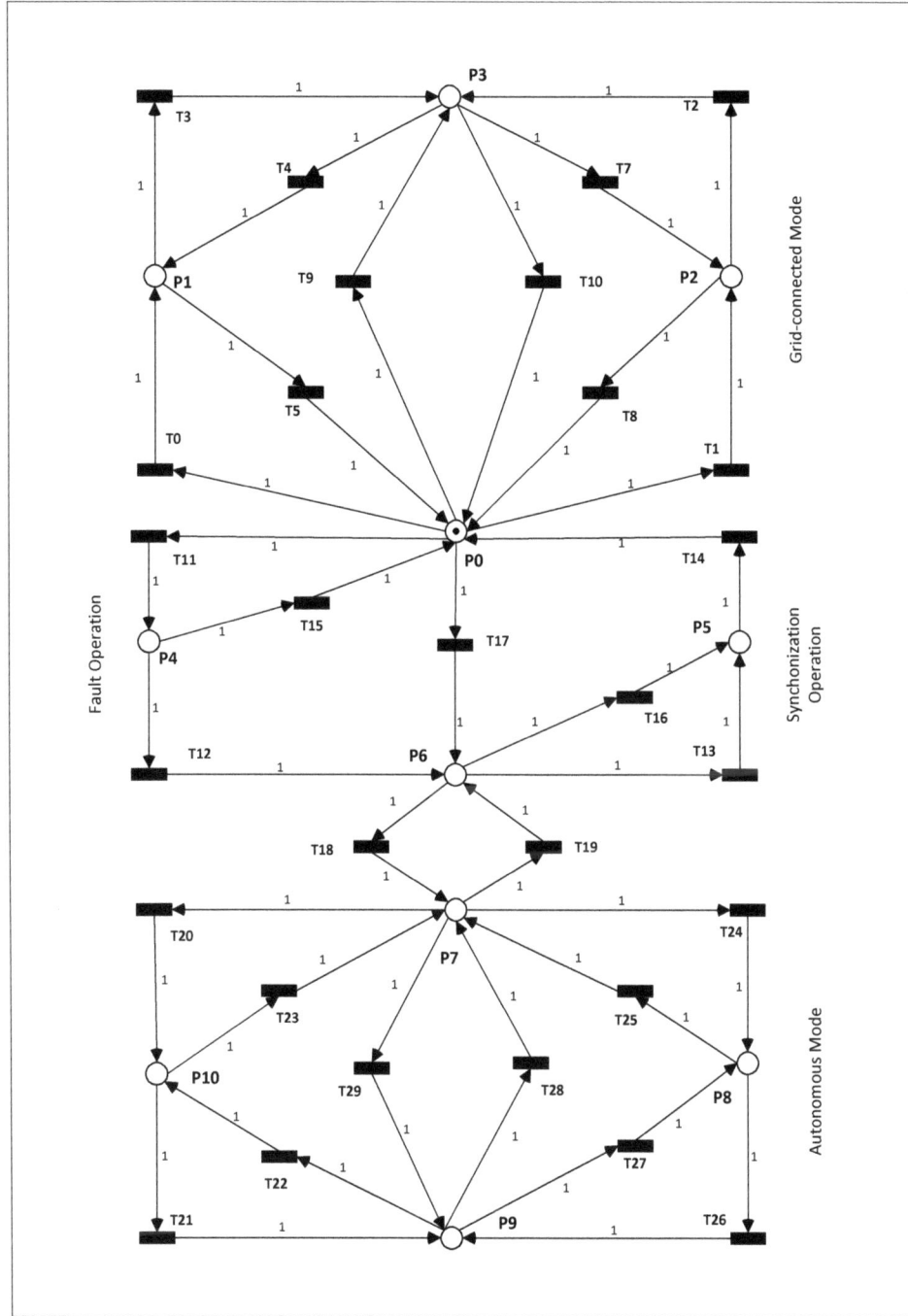

Figure 18. Petri Net model IEC for the system service operation (Reprinted with permission from Ref. [5]. Copyright 2021 IEEE).

8. Conclusions

This paper's contribution is a requirement cycle that faces the growing complexity of microgrid design in sustainable environments. In such cases, the tendency is to couple the low-voltage distribution of energy [41] with a sustainable network of prosumer units, reducing harmful interference in the environment while integrating the energy service with its prosumers. The system is open, meaning that it will be frequently used for designing, implementing, and integrating (or removing) new prosumer arrangements on top of a legacy system that minimizes long-wire transmissions, especially inside forests.

To fit this demand, we advocate that:

- Distributed microgrid systems should be modeled and designed as systems or systems of systems (SoS);
- Distributed microgrid systems should be designed as services or service of services (SoS+);
- Distributed microgrid systems design should be modeled and formalized, allowing for a sound verification and coupling with the legacy system; and
- Distributed microgrid clusters must be easily adapted and reconfigured.

To achieve these goals, we proposed a model-driven requirements life cycle based on a goal-oriented approach to model services, stressing the communication and coupling with the end-useR, which is a critical issue for the service approach. Dynamic requirements synthesized by this approach are invariant and will be preserved both in the design process and even after, up to the system elimination. Formalization as well as property and process analysis are performed in Petri Nets.

Once the focus is on the life cycle, we prefer to avoid a tedious mathematical explanation of all aspects. Instead, we chose to show the method's practical applicability, preserving soundness. A case study based on a real application provided a suitable example: a microgrid design that needs to be expanded to a distributed smart grid service to provide energy in a small community in a tropical forest with substantial sustainability restrictions. We used its implementation as system-as-is and revised the design, leading to emergent requirements and a missing coupling with the final user. Integrating a service-oriented component with goal-oriented requirements is the main contribution of this paper, reinforcing environmental restrictions and (human) user integration.

The state-transition direct approach to represent dynamics—the main characteristic of an automated process—was introduced using a Petri Net formal representation. However, the transference from KAOS modeling to Petri Nets was not fully automated. A new development concerns creating an XML representation for KAOS that could match PNML (Petri Nets Markup Language) as defined by ISO/IEC 15.909 (which defines the Petri Net approach). In further work, we plan to introduce this automated transfer system.

A possible drawback of the proposed method is related to scalability. From one side, the intense use of hierarchy in systems design is a way to deal with scalability and the SoS (and SoS+) approach could add the possibility to analyze components separately or to reuse some of them. From the other side, exploring XML to export models and integrate documentation in XML (actually in KML, which is a Kaos Markup Language) should make this problem slightly better than in UML, even if not solving it.

Finally, some communication problems detected will be treated using the standard IEC 61850 [42], wherein reference models must be represented appropriately in goal-oriented diagrams and translated to Petri Nets. This development was slightly shown in the case study but is not fully automated yet. The authors are also developing a software tool to do this.

Author Contributions: Conceptualization, M.A.O. and J.R.S.; data curation, M.A.O., J.R.S. and E.L.P.; formal analysis, M.A.O. and J.R.S.; investigation, M.A.O. and J.R.S.; methodology, M.A.O. and J.R.S.; supervision, J.R.S. and E.L.P.; validation, M.A.O. and J.R.S.; data analysis, M.A.O., J.R.S. and E.L.P.; visualization, J.R.S. and E.L.P.; writing—original draft preparation, M.A.O., J.R.S. and E.L.P. All authors have read and agreed to the published version of the manuscript.

Funding: Funded by the Graduate Program in Mechanical Engineering, Escola Politécnica, Universidade de São Paulo.

Institutional Review Board Statement: Not applicable.

Informed Consent Statement: Not applicable.

Acknowledgments: The authors acknowledge their respective institutions for supporting their research activities: UEA, USP, and the support given by FAPEAM (Amazon Research Foundation) to the first author.

Conflicts of Interest: The authors declare no conflict of interest.

Abbreviations

The following abbreviations are used in this manuscript:

SG	smart grids
KAOS	Knowledge Acquisition and Object Specification
MBRE	model-based requirements engineering
EPRI/NIST	Electric Power Research Institute/National Institute of Standards
SGAM	smart grid architecture model
UML	Unified Modeling Language
IEC	International Electrotechnical Comission
MBSE	model-based systems engineering
SoS	systems of systems
SoS+	service of services
GORE	goal-oriented requirements engineering
LTL	linear temporal logic
SM	smart meter

References

1. de Oliveira, V.C.; Silva, J.R. A service-oriented framework to the design of information system service. *J. Serv. Sci. Res.* **2015**, *7*, 55–96. [CrossRef]
2. Han, Z.; Gu, Y.; Saad, W. *Matching Theory for Wireless Networks*; Springer: Berlin/Heidelberg, Germany, 2017.
3. Spohrer, J.; Anderson, N.; Pass, N.; Ager, T.; Gruhl, D. Service Science. *J. Grid Comput.* **2008**, *6*, 313–324. [CrossRef]
4. Nyso. *Distributed Energy Resources Roadmap for New York's Wholesale Electricity Markets: A Report by the New York Independent System Operator*; New York Independent System Operator: Nwe York, NY, USA, 2017.
5. Postigo, M.; Pellini, E.; Silva, J. Proposta de método sistêmico baseado em modelos para Smart Grid. In Proceedings of the 14th IEEE International Conference on Industry Applications (INDUSCON), São Paulo, Brazil, 15–18 August 2021; pp. 41063–41070.
6. Kendal, S.L.; Creen, M. *An Introduction to Knowledge Engineering*; Springer: Berlin/Heidelberg, Germany, 2007.
7. Rohjans, S.; Dänekas, C.; Uslar, M. Requirements for smart grid ICT-architectures. In Proceedings of the 2012 3rd IEEE PES International Conference and Exhibition, Innovative Smart Grid Technologies (ISGT Europe), Berlin, Germany, 14–17 October 2012; pp. 1–8.
8. Uslar, M.; Specht, M.; Dänekas, C.; Trefke, J.; Rohjans, S.; González, J.M.; Rosinger, C.; Bleiker, R. *Standardization in Smart Grids: Introduction to IT-Related Methodologies, Architectures and Standards*; Springer Science & Business Media: Berlin/Heidelberg, Germany, 2012.
9. Postigo, M.A.O.; Silva, J.R. Modeling in Petri Nets for micro smart grid operation based on IEC 61850 architecture. In Proceedings of the 2018 Simposio Brasileiro de Sistemas Eletricos (SBSE), Niteroi, Brazil, 12–16 May 2018; pp. 1–6.
10. Van Lamsweerde, A. *Requirements Engineering: From System Goals to UML Models to Software*; John Wiley & Sons: Chichester, UK, 2009; Volume 10.
11. Cretu, L.G.; Dumitriu, F. *Model-Driven Engineering of Information Systems: Principles, Techniques, and Practice*; CRC Press: Boca Raton, FL, USA, 2014.
12. Zaman, M.; Halder, R.; Buakharl, S.; Ashraf, H.; Kim, C.H. Impacts of Responsive Loads and Energy Storage System on Frequency Response of a Multi-Machine Power System. *Machines* **2019**, *7*, 34. [CrossRef]
13. Ju, X.; Liu, F.; Wang, L.; Lee, W.J. Wind farm layout optimization based on support vector regression guided genetic algorithm with consideration of participation among landowners. *Energy Convers. Manag.* **2019**, *196*, 1267–1281. [CrossRef]
14. Gerber, L. A systematic methodology for the environomic design and synthesis of energy systems combining process integration, Life Cycle Assessment and industrial ecolog. *Comput. Chem. Eng.* **2013**, *59*, 2–16. [CrossRef]
15. Roboam, X. *Systemic Design Methodologies for Electrical Energy Systems: Analysis, Synthesis and Management*; John Wiley & Sons: Hoboken, NJ, USA, 2012.

16. Frangopoulos, C.A.; Von Spakovsky, M.R.; Sciubba, E. A brief review of methods for the design and synthesis optimization of energy systems. *Int. J. Thermodyn.* **2002**, *5*, 151–160.
17. Nafi, N.S.; Ahmed, K.; Gregory, M.A.; Datta, M. A survey of smart grid architectures, applications, benefits and standardization. *J. Netw. Comput. Appl.* **2016**, *76*, 23–36. [CrossRef]
18. Ranganathan, P.; Nygard, K.E.; Magel, K. UML Design Patters in Smart Grid. In Proceedings of the Computers and their Application CATA, New Orleans, LA, USA, 23–25 March 2011; pp. 114–119.
19. IEC. *Intelligrid Methodology for Developing Requirements for Energy Systems*; IEC: Geneva, Switzerland, 2008.
20. Gottschalk, M.; Uslar, M.; Delfs, C. *The Use Case and Smart Grid Architecture Model Approach: The IEC 62559-2 Use Case Template and the SGAM Applied in Various Domains*; Springer: Berlin/Heidelberg, Germany, 2017.
21. INCOSE. *Systems Engineering Vision 2020*; INCOSE-TP-2004-004-02; INCOSE: San Diego, CA, USA, 2007.
22. INCOSE. *SEBOK v.2.2*; INCOSE: San Diego, CA, USA, 2020.
23. INCOSE. *Guide for Writing Requirements*; Prepared by: Requirements Working Group; INCOSE: San Diego, CA, USA, 2015.
24. Jamshidi, M. *Systems of Systems Engineering: Principles and Applications*; CRC Press: Boca Raton, FL, USA, 2008.
25. Postigo, M.A.O.; Silva, J.R. Microgrid System Design Based on Model Based Systems Engineering: The Case Study in the Amazon Region. *Simpósio Brasileiro de Sistemas Elétricos-SBSE* **2020**, *1*. [CrossRef]
26. Silva, J.R.; Nof, S.Y. Manufacturing service: From e-work and service-oriented approach towards a product-service architecture. *IFAC-PapersOnLine* **2015**, *48*, 1628–1633. [CrossRef]
27. Buede, D.M.; Miller, W.D. *The Engineering Design of Systems: Models and Methods*; John Wiley & Sons: Hoboken, NJ, USA, 2016.
28. Lopes, A.J.; Lezama, R.; Pineda, R. Model based systems engineering for smart grids as systems of systems. *Procedia Comput. Sci.* **2011**, *6*, 441–450. [CrossRef]
29. FitzPatrick, G.J.; Wollman, D.A. NIST interoperability framework and action plans. In *IEEE PES General Meeting*; IEEE: Piscataway, NJ, USA, 2010; pp. 1–4.
30. van Lamsweerde, A. Goal-oriented requirements engineering: From system objectives to UML models to precise software specifications. In Proceedings of the 25th International Conference on Software Engineering, Portland, OR, USA, 3–10 May 2003; pp. 744–745.
31. Mavin, A.; Teufl, S.; Femmer, H.; Eckardt, J.; Mund, J. Does Goal-Oriented Requirements Engineering Achieve its Goal? In Proceedings of the IEEE 25th International Conference on Engineering Requirements, Lisbon, Portugal, 4–8 September 2017; pp. 174–183.
32. Parveen, S.; Imam, A. Analysis of different techniques of GORE (Goal oriented requirement engineering). *Glob. Sci-Tech* **2017**, *9*, 22–36. [CrossRef]
33. Lapouchnian, A. *Goal-Oriented Requirements Engineering: An Overview of the Current Research*; Technical Report; University of Toronto: Toronto, ON, Canada, 2005.
34. Ponsard, C.; Darimont, R.; Michot, A. Combining Models, Diagrams and Tables for Efficient Requirements Engineering: Lessons Learned from the Industry. In Proceedings of the INFORSID, Biarritz, France, 26–29 May 2015; pp. 235–250.
35. Respect, I. A Kaos Tutorial, 2007. Available online: http://www.objectiver.com/fileadmin/download/documents/KaosTutorial.pdf (accessed on 3 November 2021).
36. Horkoff, J.; Aydemir, F.B.; Cardoso, E.; Li, T.; Maté, A.; Paja, E.; Salnitri, M.; Piras, L.; Mylopoulos, J.; Giorgini, P. Goal-oriented requirements engineering: An extended systematic mapping study. *Requir. Eng.* **2019**, *24*, 133–160. [CrossRef] [PubMed]
37. Salmon, A.; del Foyo, P.; Silva, J. Verification of Automated systems Using Invariants. In Proceedings of the Brazilian Congress of Automation, Belo Horizonte, Brazil, 20–24 September 2014; pp. 3511–3518.
38. Silva, J.M.; Silva, J.R. Combining KAOS and GHENeSys in the requirement and analysis of service manufacturing. *IFAC-PapersOnLine* **2015**, *48*, 1634–1639. [CrossRef]
39. Martinez, J.R.; Saidel, M.A.; Fadigas, E.A. Influence of non-dispatchable energy sources on the dynamic performance of MicroGrids. *Electr. Power Syst. Res.* **2016**, *131*, 96–104. [CrossRef]
40. Nwokeki, J.; Clark, T.; Barns, B. Towards a Comprehensive Meta-model for KAOS. In Proceedings of the IEEE Workshop on Model-Driven Requirements Engineering Workshop—MoDRE, Rio de Janeiro, Brazil, 15 July 2013; pp. 30–39.
41. Gomes, R.; Costa, C., Jr.; Silva, J.; Sicchar, J. SmartLVGrid Platform—Convergence of Legacy Low-Voltage Circuits toward the Smart Grid Paradigm. *Energies* **2019**, *12*, 2590. [CrossRef]
42. Deng, W.; Pei, W.; Shen, Z.; Zhao, Z.; Qu, H. Adaptive micro-grid operation based on IEC 61850. *Energies* **2015**, *8*, 4455–4475. [CrossRef]

Article

Blending Colored and Depth CNN Pipelines in an Ensemble Learning Classification Approach for Warehouse Application Using Synthetic and Real Data [†]

Paulo Henrique Martinez Piratelo [1,2], Rodrigo Negri de Azeredo [1], Eduardo Massashi Yamao [1], Jose Francisco Bianchi Filho [2,3], Gabriel Maidl [1], Felipe Silveira Marques Lisboa [1], Laercio Pereira de Jesus [4], Renato de Arruda Penteado Neto [1], Leandro dos Santos Coelho [2,5,*] and Gideon Villar Leandro [2]

1. Mechanical Systems, Lactec, Comendador Franco Avenue, 1341-Jardim Botânico, Curitiba 80215-090, Brazil; paulo.piratelo@lactec.org.br (P.H.M.P.); rodrigonegria@gmail.com (R.N.d.A.); eduardo.yamao@lactec.org.br (E.M.Y.); gabriel.maidl@lactec.org.br (G.M.); felipe.lisboa@lactec.org.br (F.S.M.L.); renato@lactec.org.br (R.d.A.P.N.)
2. Department of Electrical Engineering-PPGEE, Federal University of Parana-UFPR, Coronel Francisco Heráclito dos Santos Avenue, 100, Curitiba 80060-000, Brazil; jose.filho@lactec.org.br (J.F.B.F.); gede@eletrica.ufpr.br (G.V.L.)
3. Power Systems, Lactec, Comendador Franco Avenue, 1341-Jardim Botânico, Curitiba 80215-090, Brazil
4. Department of Logistics and Supplies-DIS, Copel Distribuição S.A., Estrada da Graciosa Street, 730-Atuba, Curitiba 82840-360, Brazil; ldejesus@copel.com
5. Industrial and Systems Engineering-PPGEPS, Pontifical Catholic University of Parana-PUCPR, Imaculada Conceição Street, 1155-Prado Velho, Curitiba 80215-901, Brazil
* Correspondence: leandro.coelho@pucpr.br
† This paper is an extended version of our paper published in Piratelo, P.H.M.; de Azeredo, R.N.; Yamao, E.M.; Maidl, G.; de Jesus, L.P.; Neto, R.D.A.P.; Coelho, L.D.S.; Leandro, G.V. Convolutional neural network applied for object recognition in a warehouse of an electric company. In Proceedings of the 2021 14th IEEE International Conference on Industry Applications (INDUSCON), Paulo, Brazil, 15–18 Auguest 2021.

Citation: Piratelo, P.H.M.; de Azeredo, R.N.; Yamao, E.M.; Bianchi Filho, J.F.; Maidl, G.; Lisboa, F.S.M.; de Jesus, L.P.; Penteado Neto, R.d.A.; Coelho, L.d.S.; Leandro, G.V. Blending Colored and Depth CNN Pipelines in an Ensemble Learning Classification Approach for Warehouse Application Using Synthetic and Real Data. *Machines* **2022**, *10*, 28. https://doi.org/10.3390/machines10010028

Academic Editors: Marcos de Sales Guerra Tsuzuki, Marcosiris Amorim de Oliveira Pessoa and Alexandre Acassio

Received: 30 October 2021
Accepted: 22 December 2021
Published: 31 December 2021

Publisher's Note: MDPI stays neutral with regard to jurisdictional claims in published maps and institutional affiliations.

Copyright: © 2021 by the authors. Licensee MDPI, Basel, Switzerland. This article is an open access article distributed under the terms and conditions of the Creative Commons Attribution (CC BY) license (https://creativecommons.org/licenses/by/4.0/).

Abstract: Electric companies face flow control and inventory obstacles such as reliability, outlays, and time-consuming tasks. Convolutional Neural Networks (CNNs) combined with computational vision approaches can process image classification in warehouse management applications to tackle this problem. This study uses synthetic and real images applied to CNNs to deal with classification of inventory items. The results are compared to seek the neural networks that better suit this application. The methodology consists of fine-tuning several CNNs on Red–Green–Blue (RBG) and Red–Green–Blue-Depth (RGB-D) synthetic and real datasets, using the best architecture of each domain in a blended ensemble approach. The proposed blended ensemble approach was not yet explored in such an application, using RGB and RGB-D data, from synthetic and real domains. The use of a synthetic dataset improved accuracy, precision, recall and f1-score in comparison with models trained only on the real domain. Moreover, the use of a blend of DenseNet and Resnet pipelines for colored and depth images proved to outperform accuracy, precision and f1-score performance indicators over single CNNs, achieving an accuracy measurement of 95.23%. The classification task is a real logistics engineering problem handled by computer vision and artificial intelligence, making full use of RGB and RGB-D images of synthetic and real domains, applied in an approach of blended CNN pipelines.

Keywords: convolutional neural networks; warehouse management; image classification; ensemble learning; synthetic data; depth image; electrical maintenance

1. Introduction

Companies need to adapt continually. Organizations that identify and react with greater agility and intelligence have an advantage in the business environment [1]. This paper is an expanded version of a previous work [2] (© 2021 IEEE. Reprinted, with permission, from 14th IEEE International Conference on Industry Applications (INDUSCON)).

Advanced digitization within factories, aggregated with Internet technologies and smart devices, seems to change the fundamental paradigm in industrial production [3]. Some companies are capturing artificial intelligence (AI) application value at the corporate level and many others are increasing their revenues and reducing costs at least at the functional level [1,4,5]. The most-reported increase of revenues is for inventory and parts optimization, pricing, analysis of customer-services, sales and forecasting [1].

According to a study published by Massachusetts Institute of Technology (MIT) Sloan Management Review in collaboration with Boston Consulting Group [6], almost 60% of respondents on a global survey of more than 3000 managers assert that their companies are employing AI, and 70% know the business value proportioned with AI applications. Notwithstanding the evidence, only 1 in 10 companies achieve financial benefits with AI. The survey found that organizations apply basic AI concepts, and even with adequate data and technology, financial returns are minimal. Companies need to learn from AI and implement organizational learning, adapting their strategies over time. Hence, their chances of generating financial benefits increase to 73% [6].

A study compared the main trends in the digitization present in a "Factory of the Future" [7], showing that a digital factory has the goal to automate and digitize the intra-factory level, using virtual and augmented reality, and simulations, aiming to optimize production. The study reports that at the lowest layer of the supply chain, the Internet of Things plays an important role in the paradigms of real-time analysis, smart sensing, and perception. From the perspective of an inter-factory collaboration, the trends are related to cloud manufacturing and virtual factory. Big data is been used to support production, to help in internal business, and to guide new discoveries [8].

Technology implies accurate processes, better tools, and business innovation. It is compelling to blend automation, computer vision, and deep learning (DL) to improve warehouse management. Modular solutions, embedded intelligence, and data collection technologies are the key points to flexible automated warehouses [9]. Amazon invested heavily in drive units, picking robots, autonomous delivery vehicles, and programs to help workers learn software engineering and machine learning skills [10]. Researches conducted on literature show that in the last decade several studies on the theory of inventory management improvements were developed and new technologies could be applied in warehouse management systems [11]. In addition, over the last ten years, Artificial Intelligence played an important role in the supply chain management field, with customer demand predictions, order fulfillment, and picking goods. Nonetheless, it is reported a lack of study on the warehouse receiving stage [11]. According to an exploratory study, the implementation of warehouse management systems can bring benefits related to increasing inventory accuracy, turnaround time, throughput, workload management, productivity, besides reducing labor cost and paperwork [12]. In a proposed architecture for virtualization of data in warehouses, the authors replace conventional data with a non-subjective, consistent, and time-variant type of data, making a synthetic warehouse for analytical processing, aiming scalability and source dynamics [13]. Radiofrequency identification, short-term scheduling and online monitoring are largely used in warehouses and industry [14–16].

In this scenario, a Brazilian electrical company is facing obstacles in its logistics. The main problems that demand solutions are the outlays of inventory control, the time-consuming tasks, and the lack of reliability in maintaining a manual flow. There is a burgeoning need for automated processes in order to reduce costs. The identification of products in this company's warehouse is the flagship of a project to automate flow control and inventory. By this means, it is essential to build an intelligent application that can classify the products. In order to handle a classification task for this project, a real dataset was created. The dataset building represents a challenge given the variety of steps. Therefore, they were required hours of shooting, labeling, and filtering data process. Consequently, the project seeks to develop an intelligent system capable of assisting in the organization and management of the inventory of a company in the energy sector, an automated solution that checks the disposal of items in the warehouse and controls the flow of inputs and outputs, involving applications of computer vision, deep learning, optimization techniques, and autonomous robots. As part of the project, it is intended to build a tool to classify inventory items. This tool will be encapsulated in the automated system.

A blend of pipelines for colored and depth images is proposed, in a soft voting type of ensemble approach. Each pipeline corresponds to a classifier, and two CNN models are used for this task. The final classification of a scene is performed by this ensemble, not explored for this application yet. The decision is the average of the probabilities of each CNN multiplied by equal weights, meaning that the models have the same influence on the final result.

The remainder of this article is structured as follows. Section 2 presents the problem description, illustrating the electrical utility warehouse, devices, and sensors applied to capture the images. Section 3 depicts the dataset developed for this application, which is divided into synthetic and real data. Then, Section 4 brings the image processing, explaining the selection and the control of captured images during the data gathering step. Section 5 explains deep learning methods and their hyperparameter tuning to meet the requirements of the project. In Section 6 the ensemble learning approach is illustrated, followed by Section 7 illustrates the methodology to assess the different datasets. The results are presented and discussed in Section 8. Finally, Section 9 describes the conclusion and future works.

2. Problem Description

The electrical utility warehouse of this company is an 11,000 square meters building, containing more than 3000 types of objects used in the electrical maintenance field. The objects are distributed on shelves across the entire facility. There are a total of twenty-four shelves of 5 m tall, 3 m wide, and 36 m long across the entire building. The company has a constant flow and cannot allow absences of material. The stock needs to be up to date to deliver a reliable and fast service. Therefore, the company is facing management problems that need to be solved: miscounting, time-consuming processes of flow control, inventory check, as well as the costs of such processes. The project aims to reduce the impact of these issues, combining new technology and intelligent solutions, improving inventory management. Thus, an automated inventory check is proposed, using artificial intelligence, computer vision, and automation.

With the aim of keeping the inventory up to date, the project foresees two different procedures: flow control and periodic verification. The first procedure is designed to check every item that enters or leaves the warehouse using gateways and conveyor belts. Large objects will be handled by electric stackers, while small ones will be manually placed in the conveyor belts. To verify the handling products, gateways and conveyor belts will be equipped with cameras. The second procedure consists of an Automated Guided Vehicle (AGV) that checks all shelves inside the warehouse, counting the number of items.

The AGV is a Paletrans PR1770 electric stacker that will be fully automated in order to work remotely and to take pictures of the products. The AGV will receive a retractable robotic arm with 5 cameras, one in the front of the arm and the others in a straight-line arrangement, enabling a better field of view and capturing the full dimension of the shelves. Figure 1 (a) shows an overview of the AGV with the robotic arm. Figure 1 (b) represents the robotic arm fully extended and Light Detection and Ranging (LiDAR) positions. Finally, Figure 1 (c) illustrates the AGV capturing images of the shelves.

Since the gateways, conveyor belts, and the AGV are still in development, a mechanical device was built to manually capture the images inside the warehouse. The device was designed to emulate the AGV and it is manually placed on the shelves. Consequently, it was possible to build a dataset of products before the project conclusion. The data acquisition allows studying and developing techniques of image classification. Figure 2 (a) shows the project and (b) the constructed mechanical device.

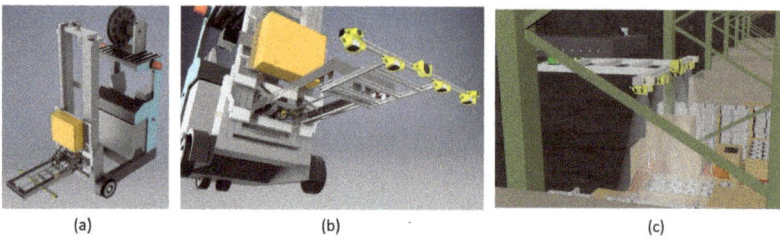

Figure 1. AGV with retractable robotic arm: (**a**) overview; (**b**) fully extended arm and (**c**) LiDARs capturing images.

Figure 2. Mechanical device: (**a**) project and (**b**) constructed.

The device is a mechanical structure equipped with the same technology that will be used in the AGV, gateways, and the conveyor belts: Intel RealSense L515 LiDAR cameras with laser scanning technology, a depth resolution of 1024 × 768 at 30 fps, and a Red–Green–Blue (RGB) resolution of 1920 × 1080 at 30 fps. They are arranged identically as in the AGV design shown in Figure 1. The technology embedded in the mechanical device for the data acquisition is described as follows: a portable computer with Ubuntu, five L515 cameras, a Python script with OpenCV, Intel Librealsense and Numpy libraries, an uninterruptible power supply (UPS) device for power autonomy, and a led strip for lighting control.

The warehouse harbors mainly materials for maintenance services for electrical distribution systems. Switches, contactors, utility pole clamps and brace bands, mechanical parts, screws, nuts, washers, insulators, distribution transformers, and wires are stored in the warehouse, among other products.

One of the essential parts of this project is to build a tool that classifies the products inside the warehouse using deep learning techniques. This tool analyzes the quality of captured images and classifies the objects placed on the shelves by RGB and Red–Green–Blue-Depth (RGB-D) data. The warehouse is considered an uncontrolled environment, increasing the difficulties for a computer vision application. It means that the shelves and pallets do not have a default background, like in some computer vision competitions. Moreover, at the time of the dataset building, there were no rules applied related to the arrangement of materials. There are issues that will be faced in order to accomplish this task, such as this random displacement of products. Moreover, the warehouse presents difficulties related to lighting distribution. Some places have poor lighting conditions. The LED strip installed on the AGV provides a better distribution of light in the shelves that has no sufficient conditions for data collection.

The depth information provided from the LiDAR can be used in order to extract more features of the scenes, where sometimes colored images do not hold these features. The arrangement of the five cameras has the role to avoid problems related to occlusion. The straight line of four cameras facing down is a setup that was brought to maximize the field of view of the scene. The robotic arm will slide from the beginning to the end of each pallet, allowing cameras to capture the most information possible for this arrangement. The frontal camera setup was also designed to give another perspective of the scene, taking pictures from the front of the pallets. Some objects are stored partially occluded. These objects are, for instance, insulators inside wooden boxes. This is a challenge that needs to be addressed as well.

3. Image Processing

The image processing operation for the captured images is divided into two parts. First, the quality of the images is checked, and then, two histogram analyses are performed. If the captured colored image is not satisfactory according to the procedure described in this section, the RGB and RGB-D images are discarded and another acquisition is made.

3.1. Image Quality Assessment

The image quality assessment (IQA) algorithms examine any image and generate a quality score on its output. The performance of these models is measured based on subjective human quality judgments, since the human being is the final receiver of the visual signal, as mentioned in [17]. Additionally, IQAs can be of three types, such as full reference (FR), reduced reference (RR), and no-reference (NR) [18]. The FR type is where an image without distortion (reference) is compared with its distorted image. This measure is generally applied in the evaluation of the quality of image compression algorithms. Another possibility is RR, without a reference image, but an image with some selective information about it. Finally, NR (or blind object) is the type where the only input from the algorithm is an image whose quality is to be checked [19].

In this research, an NR IQA algorithm called blind/referenceless image spatial quality evaluator (BRISQUE) [17] was applied, in order to evaluate the quality of the images at the time of mechanical device acquisition. The BRISQUE algorithm is a low complexity model since it uses only pixels to calculate features, with no need to perform image transformations. Based on natural scene statistics (NSS), which considers the normalized pixel intensity distribution, the algorithm identifies the unnatural or distorted images considering if they do not follow a Gaussian distribution (Bell Curve). Then, a comparison is made between the pixels and their neighbors by pairwise products. After a feature extraction step, the dataset feeds a learning algorithm that performs image quality score predictions. In this case, the model used was a support vector regressor (SVR) [17].

The objective is to have an image quality control at the time of stock monitoring. In this scenario, the main problem is the brightness of the environment. Therefore, in addition to the assessment value of the algorithm, an analysis of the distribution of the image histogram is performed. Thus, 25 images captured were evaluated, which cover the most diverse histogram distributions, seeking to define the limits of the mean of the histogram distribution and the quality limit value. Based on the subjective judgments of the project's developers, the threshold for the quality score of the images was set at 30, with the scale of the algorithm varying from 0 (best) to 100 (worst). Hence, images with an IQA score below 30 are considered acceptable. If the value is higher, the system considers the image unsatisfactory.

3.2. Image Adjustment

During this step, an analysis of image histogram distribution is performed to verify if there is a lack or excess of exposure. According to the evaluation of the images, the acceptable limits for the distribution of the histograms of the images were defined between 75 and 180, being a scale of 0 to 255 levels of gray. When the value is less than 75, it is a poor light environment. On the other hand, a result higher than 180 means light in excess.

An example is shown in Figure 3, where two images are presented with different IQA values and their respective average of gray levels for each histogram. Figure 3 (a) has the best quality according to the BRISQUE algorithm. Analyzing its histogram in (c), the mean value ($\bar{\mu}$) was 98.26. According to this assessment, the image is considered acceptable to provide information about the location. However, Figure 3b had an IQA value higher than the limit of 30, being unsatisfactory for the classification of the algorithm. In this case, the mean of the histogram is analyzed, where it is identified that the image problem is of low exposure since the mean values were less than 75, as shown in (d).

If the images are rejected by the assessment analysis, luminosity correction is carried out by adjusting the intensity of the light-emitting diodes (LEDs) present in the AGV structure. If after three consecutive attempts it is not possible to obtain an acceptable image in consonance with the defined parameters, the vehicle system registers that it was unable to collect data from that location and recommends the manual acquisition of images/information. Thus, an employee must go to the site and check the content of that pallet. The disadvantage of a manual check is that it goes against the main objectives of automated inventory control. Besides the time misspending by an employee, performing a manual verification, there is a time to register it in the system. During this period, the inventory will not be updated correctly.

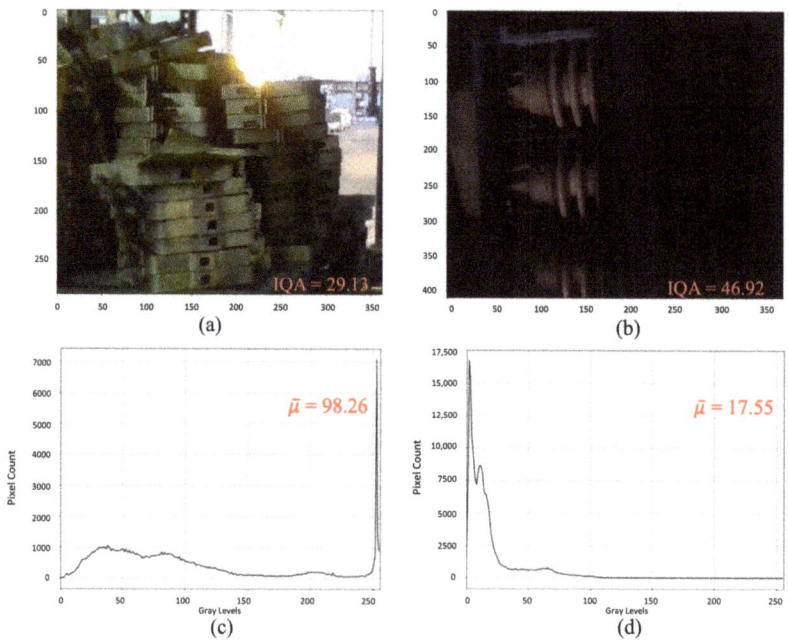

Figure 3. IQA score for (**a**) brace bands and (**b**) insulators, and histogram for brace bands (**c**) and (**d**) insulators.

Finally, if an image is considered valid using BRISQUE and histogram analysis criteria, then a new histogram analysis is performed, looking for the values that appear most frequently (peaks). Hence, if there are peaks close to the extremes, in a range of 20 levels on each side (0 represents black and 255 white color), an adaptive histogram equalization (AHE) [20,21] is performed in order to improve regions of the image that have problems with luminosity. Analyzing the histogram of the first image in Figure 3 (c), there is a peak close to the 255 level. Performing an AHE, the image will present a better distribution of the gray levels, providing an enhanced description of its characteristics to the neural network. The result is shown in Figure 4, where (a) indicates the original image and (b) the image after the AHE process.

Figure 4. Enhancing image features: (**a**) original image and (**b**) adaptive histogram equalization.

4. Synthetic and Real Data

Commonly, the use of deep learning techniques to deal with classification problems requires a great amount of data [22]. However, if there are not sufficient images on the dataset, the deep learning models might not be able to learn from the data. One way to solve this issue is to apply data augmentation processing, which is divided into two categories: classic and deep learning data augmentation. The classic or basic methods are flipping, rotation, shearing, cropping, translation, color space shifting, image filters, noise, and random erasing. Deep learning data augmentation techniques are Generative Adversarial Networks (GANs), Neural Style Transfer, and Meta Metric Learning [23]. Moreover, 3D Computer-aided design (CAD) software and renders are useful for generating synthetic images to train algorithms to perform object recognition [24]. Game engines are great tools to build datasets as well [22,25].

Another point is that manually labeling a dataset requires a great human effort and it has an expensive cost [25]. In contrast, on synthetic datasets, the process of labeling can be done automatically and it is easier to achieve more variations [26]. In addition, there are some situations where the creation of a dataset is critical or new samples acquisition are rare, and therefore, the synthetic approach can balance these problems, as presented in critical road situations [27], volcano deformation [28], radiographic X-ray images [29] and top view images of cars [30].

The use of generated images is not exclusively done by critical datasets. This method is used in applications of common activities and places as well. For instance, in a method that places products on synthetic backgrounds of shelves on a grocery object localization task [31], detection of pedestrian [25,27], cyclists [32], vehicles [26] and breast cancer [33], classification of birds and aerial vehicles [22], and synthetic Magnetic Resonance Imaging (MRI) [34]. However, generated images sometimes present a lack of realism, making models trained on synthetic data perform poorly on real images [24].

For this reason, there are some methods of using synthetic data during the training step of a deep learning model. The first one is training and validating only on the synthetic domain, and testing on real data. As reported by Ciampi et al. [25] when the model is trained only on synthetic, it shows a performance drop. In contrast, Saleh et al. [32] point out the ability to generalize their framework by training with synthetic data and testing on the real domain, increasing 21% the average precision in comparison with classical object localization methods. In Öztürk et al. [22], they demonstrated that models trained on synthetic images can be tested on real images in the task of classification.

The second method consists in mixing both domains when training the model, and testing on real images [27]. Reference Anantrasirichai et al. [28] presented that the training process with synthetic and real data improved the ability of the network to detect volcano deformation. The third method is the use of a transfer learning approach to mitigate the difference between synthetic and real domain shifts. The CNN models are trained on synthetic domain and fine-tuned on real dataset [26]. In an approach based on convolutional neural networks for MRI application, Moya-Sáez et al. [34] proved that fine-tuning a model trained with synthetic data improved performance, while a model trained only with an actual small dataset showed degradation. Reference Ciampi et al. [25] explored methods two and three to mitigate the domain shift problem, mixing synthetic and real data and training the CNN on a synthetic dataset, and fine-tuning on real images on training step. Both adaptations improved performance on specific real-world scenarios. Some techniques adapt CAD to real images, like the use of transfer learning to perform a domain adaptation loss, aligning both domains in feature space [24].

When working with 3D images, Talukdar et al. [35] evaluated different strategies to generate and improve synthetic data for detection of packed food, achieving an overall improvement of more than 40% on Mean Average Precision (mAP) object detection. The authors used the 3D rendering software Blender-Python. These strategies are a random packing of objects, data with distractor objects, scaling, rotation, and vertical stacking.

In this project, to generate the synthetic dataset, Blender [36] by Blender Foundation was selected for being an open-source software that provides a large python API [37] to create scenes for rendering. Moreover, Blender provides physically based rendering (PBR) shaders, getting the best out of its engines, Cycles, and Eevee, respectively a ray tracing and rasterization engine. A ray-tracing engine computes each ray of light that travels from a source and bounces throughout the scene and a rasterization engine is a computational approximation of how light interacts with the materials of the objects in the scene [38].

A nice synthetic dataset of rendered images must approximate real-world conditions, so the best choice is to use the Cycles engine for its ray tracing capability as mentioned by Denninger et al. [39] compared to rasterization engines because shading and light interaction is not consistent. Nonetheless, this method is computationally expensive considering the complexity of the scenes to render, number of samples, and resolution, making it difficult to generate large datasets. The Cycles approach is more accurate [38]. However, Eevee was chosen for rendering due to its Physically Based Rendering (PBR) capabilities differentiating from other rasterization engines, being the best choice when taking into account quality and speed.

The scene was created with Blender's graphic interface, setting the base to receive objects in a single blend file. Blender's python API [37] was used to create the pipeline, as shown in Figure 5, to configure intrinsic parameters of the real camera that was used to take real-world photos, to choose objects, to simulate physics to place then randomly, to create or set shades for materials and to prepare post-processing and generate depth images based on the rendered scene.

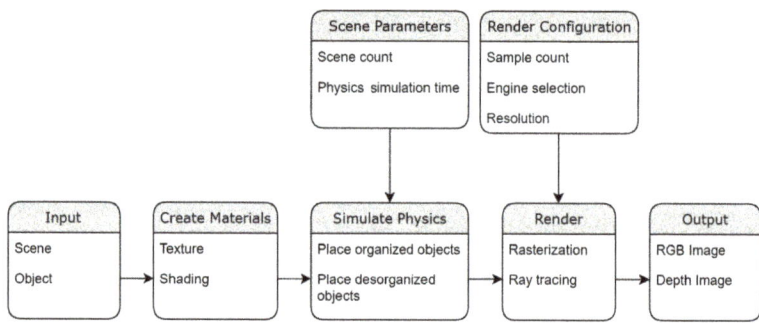

Figure 5. Pipeline diagram of rendering process.

This way the pipeline could be executed in a loop, considering the parameters scene count number, in an autonomous form to randomize the scenes and create a large dataset of synthetic RGB and depth images separated by objects classes to be used in training and test.

As a comparison, a test was conducted by rendering a scene with 5 cameras and 7 camera positions, giving a total of 35 rendered images for each engine, as shown in Figure 6. Cycles (a) took approximately 19.99 s to render one image and 759.2316 s to render all images. Eevee (b) took 2.97 s on one image and 140.3429 s to generate all images. Whileough with different light sets, object positions, and quantity due to randomization, the overall shading of both engines got the same look as the materials are processed the same way. However, Cycles can create better shadows and reflection as light rays interact with all objects, in exchange for render speed.

Figure 6. Rendered images: (**a**) Cycles and (**b**) Eevee.

To create the real dataset, an acquisition took place inside the warehouse using the mechanical device to manually capture the images. Two classes of materials were chosen to compose the dataset, making this a binary problem to be solved. The classes are utility pole insulators in Figure 7 (a) and brace bands in Figure 7 (b). These classes are the objects that appear more frequently in the warehouse, and due to their high transport flow, they are stored close to the first shelves to facilitate their flux.

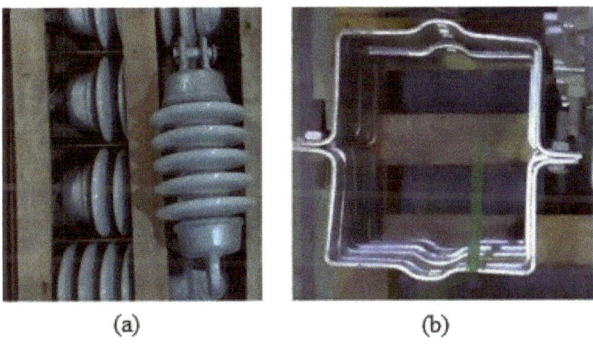

Figure 7. Classes: (**a**) utility pole insulators and (**b**) brace bands.

To build the dataset, the mechanical device was allocated on the pallets, performing the data collection, as described in Section 2. A scene (pallets with a displacement of objects) generated seven RGB and seven RGB-D pictures per camera from the straight line of 4 cameras, and the frontal camera took one picture for each domain. The result is a total of 58 pictures per scene.

The datasets used for these projects are from the synthetic and real domains. According to the literature review presented in this section, the use of synthetic images has improved training and the final accuracy for object classification tasks. However, synthetic data is not robust enough to be used solo in the application. This is the reason why the real data collection approach was also chosen to be included in the project. In this way, real and synthetic data can be applied as input to the deep learning models, to help train and achieve better accuracy in classifying the objects.

5. Convolutional Neural Networks

Deep Learning has been used in areas like computer vision, speech processing, natural language processing, and medical applications. To summarize, deep learning is many classifiers based on linear regression and activation functions [40]. It uses a high amount of hidden layers to learn and extract features of various levels of information [41]. Computer vision and convolutional neural networks accomplished what was considered impossible in the last centuries: recognition of faces, vehicles that drive without supervision, self-service supermarket, and intelligent medical solutions [42]. Computer vision is the ability of computers to understand, taking digital images or videos as input, aiming to represent an artificial system [40,43].

CNN is a type of neural network that delivered a promising performance on many competitions of computer vision and captivated the attention of industry and academia over the last years being a feedforward neural network that automatically extracts features using convolution structures [41–47]. CNN is a hot topic in image recognition [40]. Classification of images consists of allocating an image within a class category and CNN generally needs a large amount of data for this learning process.

Some of the advantages of a CNN are local connections, reducing parameters, and making it faster to converge. Moreover, they have a down-sampling dimensionality reduction, holding information while decreasing the amount of data. On the other hand, some challenges and disadvantages are: it may lack in interpretation and explanation; noise on input can cause a drop in performance; the training and validating step requires labeled data; a few changes in its hyperparameters can affect the overall performance; it cannot hold spatial information and they are not sensitive to slight changes in the images. Furthermore, the generalization ability is poor and they do not perform well in crowded scenes. Lastly, training a model requires time and computational cost and updating a trained model is not simple [40,42,43].

Optimization of deep networks is a non-stopped research area [48] and CNN improvements usually come from a restructuration of processing units and design of blocks related to depth and spatial exploitation [43]. The following CNN architectures presented great performance, state-of-the-art results, and innovation to this field of study.

5.1. AlexNet

The AlexNet architecture is a well-known CNN and caught the attention of the researchers when it won the ILSVRC-2012 competition by a large difference, achieving a top-5 error rate of 15.3% [49]. Moreover, one of the most important contributions of the paper was training the model with Graphics Processing Unit (GPU). The use of GPU allowed training deeper models with bigger datasets.

AlexNet is a five convolutional layer network, with three fully connected layers. The softmax is the final layer. The third and last layer is fully connected with a softmax function of 1000 neurons. The AlexNet's architecture is illustrated in Figure 8.

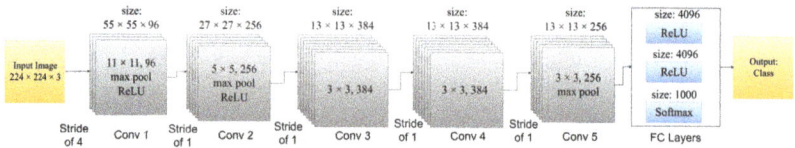

Figure 8. AlexNet architecture.

The AlexNet architecture [49] also proposed a local response normalization. To achieve a generalization to the normalization, a function was implemented in between the first three convolutional layers. The activity of a neuron when given a certain kernel i at the position x and y is measured by $a_{x,y}^i$. N is the total of kernels and n represents the number of adjacent kernel maps at the same spatial position. Three constants are used, being k, α, and β. The term $b_{x,y}^i$ measures the response normalized activity and follows the Equation (1)

$$b_{x,y}^i = a_{x,y}^i / \left(k + \alpha \sum_{j=max(0,i-n/2)}^{min(N-1,i+n/2)} (a_{x,y}^j)^2 \right)^\beta. \quad (1)$$

In the equation, $k = 2$, $n = 5$ and $\alpha = 10^4$ and β are hyperparameters determined on validation. Besides the use of GPU, another contribution of the authors in this paper is the use of dropout and data augmentation. The first technique was implemented in all neurons of FC layers 1 and 2. The outputs of the neurons were multiplied by a constant of 0.5.

5.2. VGGNet

The Visual Geometry Group (VGG) architecture is proposed by Simonyan and Zisserman [50]. VGG-11 is a network with 8 convolutional layers and 3 FC layers. The network performs a pre-processing on the image, subtracting the average values from the pixels in the training set. The input then goes to several convolutional layers of 3×3 filters. The VGG-11 achieved a top-5 test accuracy of 92.7% on ILSVRC.

The authors use five max pooling operations to achieve spatial pooling. This operation has a window of 2×2 pixels, using a stride of 2. The number of channels of the network starts with 64 on the first layer and increases the width by 2, ending with 512 channels on the last layer. Then, the last three layers are FC layers. The architecture of VGG-11 can be seen in Figure 9.

Figure 9. VGGNet architecture.

A novelty of VGG in comparison with Alexnet is the use of 3×3 convolutions instead of 11×11 ones. The authors showed that these new narrow convolutions performed better. For instance, if input and output of the mentioned 3×3 operation has C channels, the parametrization follows Equation (2), resulting in the number of weights

$$3(3^2 C^2) = 27 C^2. \quad (2)$$

On the other hand, with the use of just one 7×7 convolutional layer, this number of parameters is shown in Equation (3). As it can be seen, there is an increase of 81% of parameters

$$7^2 C^2 = 49 C^2. \quad (3)$$

VGG-11 has 133 million parameters. To bring more data to train the model, the authors performed a data augmentation on the dataset. This data augmentation helps to train the model, however, it delays the process and requires more computational power.

5.3. Inception

The neural networks proposed by Szegedy et al. [51] explored the other dimension of neural networks. Inception version 1 (Inception V1) came to increase the width of the architectures, not only the depth. Mainly, Inception V1 has three different sizes of filters, 1×1, 3×3, and 5×5. The second proposed architecture (Inception V2) reduces its dimensions, adding a one squared convolution just before the mentioned 3×3 and 5×5 filters from Inception V1. That operation focuses on reducing the parameters of the network, allowing a decrease of computational cost on training.

Inception version 3 [52] proposed a factorization, decreasing the size of convolutions. Two convolutions of 3 squared pixels replace the mentioned 5×5 convolutions of previous versions in a block called Block A. The authors implemented a factorization into asymmetric convolutions, changing the 3 × 3 convolution by a 1 × 3 and 3 × 1. Then, one of the 3 × 3 convolutions is replaced by a 3 × 1 and 1 × 3 pixel size, known as Block C. Following the same principle of factorization, Block B is composed of a 1 × 7 convolution, a 7 × 1 in parallel with two 7 × 1 and 1 × 7 convolutions. These blocks are represented in Figure 10, as well as their respective reduction blocks, where n is set with the value of 7.

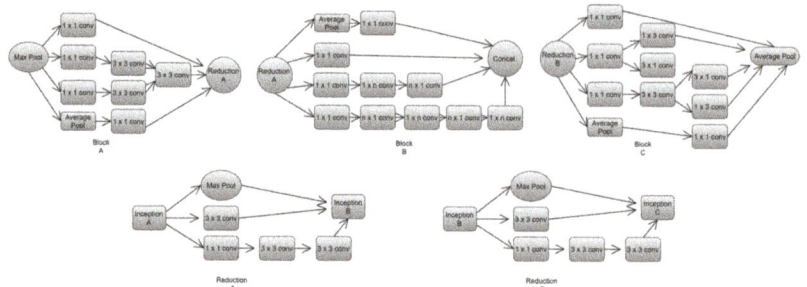

Figure 10. Inception blocks.

The Inception version 3 has 48 layers. Inception has an intermediate classifier. This classifier is a softmax layer that is used to help with varnish gradient problems by simply applying the loss in the second softmax function, in order to improve the results. Inception-v3 is shown in Figure 11.

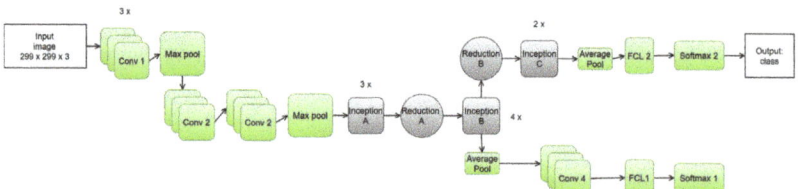

Figure 11. Inception V3 architecture.

The authors evaluated inception-v3 [52] on ILSVRC-2012 ImageNet validation set. The results were considered state-of-the-art, with a top-1 and top-5 error of 21.2% and 5.6% respectively.

5.4. ResNet

The Residual Networks (Resnets) are CNNs that learn residual functions referenced to inputs. Reference He et al. [53] proposed a framework for deep networks that prevent saturation and degradation of the accuracy by shortcut connections that add identity map outputs to the output of skipped stacked layers. ResNets showed that the idea of stacking layers to improve results in complex problems has a limit, and the capacity of learning is not progressive. Reference He et al. [53] introduced a residual block, enabling the creation of an identity map, bypassing the output from a previous layer into the next layer. Figure 12 shows the residual block of Resnet.

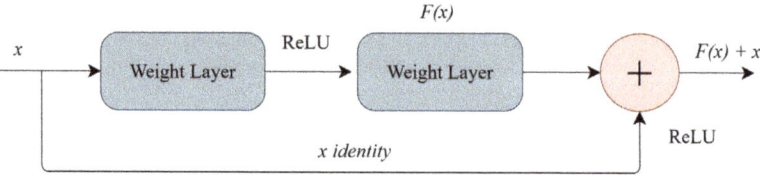

Figure 12. ResNet's residual block.

The x represents the input of a layer, and it is added together with the output $F(x)$. Since x and $F(x)$ can have different dimensions due to convolutional operations, a certain W weight function is used to adjust the parameters, allowing the combination of x and $F(x)$ by changing the channels to match with the residual dimension. This operation is illustrated in Equations (4) and (5), where x_i represents the input and $x_i + 1$ is the output of an i-th layer. F represents the mentioned residual function. The identity map $h(x_i)$ is equal to xi and f is a Rectified Linear Unit (ReLU) [54] function implemented on the block, which provides better generalization and enhances the speed of convergence. Equation (6) shows the ReLU function

$$y_i = h(x_i) + F(x_i, W_i); \tag{4}$$

$$x_{i+1} = f(y_i); \tag{5}$$

$$f(x_{relu}) = max(0, x_{relu}). \tag{6}$$

Moreover, the Resnets use stochastic gradient descent (SGD) in the opposite of adaptative learning techniques [55]. The authors introduced to the networks a pre-processing stage on the data, dividing the input in patches to introduce it to the network. This operation has the goal of improving the training performance of the network, which stacks residual blocks rather than layers. The activation of units is the combination of activation of the unit with a residual function. This tends to propagate the gradients through shallow units, improving the efficiency of training in Resnets [56].

One of the versions of Resnets, Resnet50 was originally trained on the ImageNet 2012 dataset for the classification task of 1000 different classes [57]. In total, Resnet50 runs 3.8×10^9 operations. Figure 13 illustrates the Resnet50 architecture.

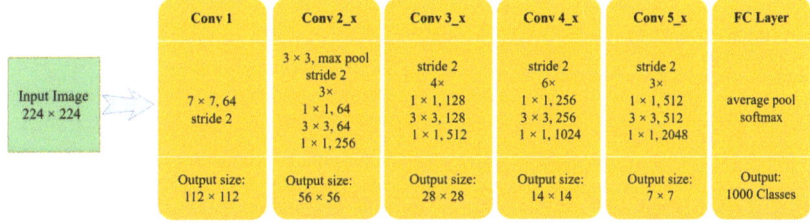

Figure 13. ResNet 50 architecture.

5.5. SqueezeNet

In their architecture, Iandola et al. [58] proposed a smaller CNN with 50 times fewer parameters than AlexNet, however, keeping the same accuracy, bringing faster training, making model updates easier to achieve, and a feasible FPGA and embedded deployment. This architecture is called SqueezeNet.

SqueezeNet uses three strategies to build its structure. The 3×3 filters are replaced by 1×1, in order to reduce the number of parameters. The input channel is also modified, limiting its number to 3×3 filters. SqueezeNet also uses a late downsample approach, creating larger activation maps. Iandola et al. [58] made use of fire modules, a squeeze convolution, and an expanded layer. Figure 14 shows the fire module. It can be tuned in three locations: the number of 1×1 filters in the squeeze layer, represented by S_{1x1}. The number of 1×1 filters on expanding layer (E_{1x1}) and finally, the number of 3×3 filters also on expanding layer, represented by E_{3x3}. The first two hyperparameters are responsible for the implementation of strategy 1. The fire blocks also implement strategy 2, limiting the number of input channels by following a rule: S_{1x1} has to be less than $E_{1x1} + E_{3x3}$.

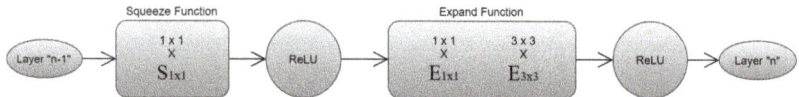

Figure 14. SqueezeNet's fire block.

SqueezeNet has three different architectures, resumed in Figure 15: SqueezeNet, SqueezeNet with skip connections, and SqueezeNet with complex skip connections, using 1 × 1 filters on the bypass. The fire modules are represented by the "fire" blocks, and the three numbers on them are the hyperparameters s1 × 1, e1 × 1, and e3 × 3 respectively. SqueezeNets are fully convolutional networks. The late downsample from strategy three is implemented by a max pooling, creating larger activation maps.

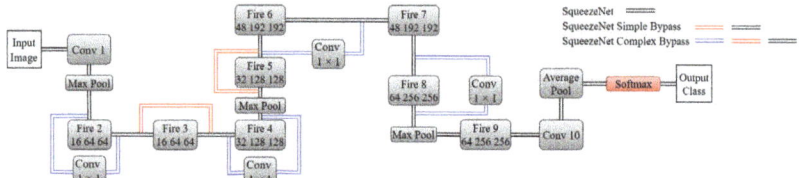

Figure 15. SqueezeNet architectures.

These bypasses are implemented to improve the accuracy and help train the model and alleviate the representational bottleneck. The authors trained the regular SqueezeNet on ImageNet ILSVRC-2012, achieving 60.4% top-1 and 82.5% top-5 accuracy with 4.8 megabytes as model size.

5.6. DenseNet

A CNN architecture that connects every layer to all layers was proposed by Huang et al. [59]. This architecture called DenseNet works with dense blocks. These blocks connect each of the feature maps from preceding layers to all subsequent layers. This approach reuse features from the feature maps, compacting the model in an implicit deep supervision way.

DenseNet has four variants, each one with a different number of feature maps in their four "dense blocks". The first variant, DenseNet-121 has the smallest depth on its blocks. It is represented in Figure 16. Transition layers are in between the dense blocks and are composed of convolution and max pool operations. The architecture ends with a softmax to perform the final classification. As the depth increases, accuracy also increases, reporting no sign of degradation or overfitting [59].

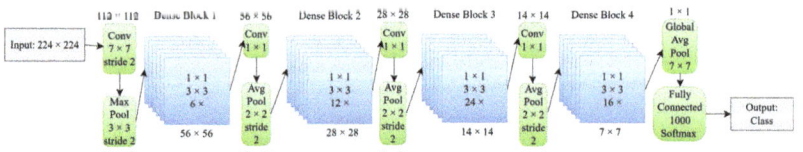

Figure 16. DenseNet architecture.

5.7. EfficientNet

EfficientNets [60] are architectures that have a compound scaling, balancing depth, width, and resolution to achieve better performance. The authors proposed a family of models that follow a new compound scale method, using a coefficient ϕ to scale their three dimensions. For depth: $d = \alpha^\phi$. Width: $w = \beta^\phi$. Finally, resolution: $r = \gamma^\phi$. So that, $\alpha \times \beta^2 \times \gamma^2 \approx 2$ and $\alpha \geq 1, \beta \geq 1, \gamma \geq 1$. The α, β, and γ are constants set by a grid search. The coefficient ϕ indicates how many computational resources are available.

The models are generated by a multi-objective neural architecture search, optimizing accuracy and Floating-point operations per second, or FLOPS. They have a baseline structure inspired by residual networks and a mobile inverted bottleneck MBConv. The base model is presented in Figure 17, known as EfficientNet-B0.

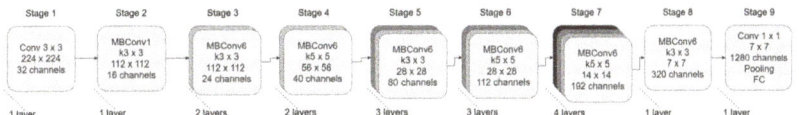

Figure 17. EfficientNet-b0 architecture.

As more computational units are available, the blocks can be scaled up, from EfficientNet-B1 to B7. The approach searches for these three values on a small baseline network, avoiding spending that amount of computation [60]. EfficientNet-B7 has 84,3% top-1 accuracy on ImageNet, a state-of-the-art result.

6. Ensemble Learning Approaches

Ensemble learning has gained attention on artificial intelligence, machine learning, and neural network. It builds classifiers and combines their output to reduce variances. By mixing the classifiers, ensemble learning improves the accuracy of the task, in comparison to only one classifier [61]. Improving predictive uncertainty evaluation is a difficult task and an ensemble of models can help in this challenge [62]. There are several ensembles approaches, suitable for specific tasks: Dynamic Selection, Sampling Methods, Cost-Sensitive Scheme, Patch-Ensemble Classification, Bagging, Boosting, Adaboost, Random Forest, Random Subspace, Gradient Boosting Machine, Rotation Forest, Deep Neural Decision Forests, Bounding Box Voting, Voting Methods, Mixture of Experts, and Basic Ensemble [61,63–68].

For deep learning applications, an ensemble classification model generates results from "weak classifiers" and integrates them into a function to achieve the final result [65,66]. This fusion can be processed in ways like hard voting, choosing the class that was most voted or a soft voted, averaging or weighing an average of probabilities [68]. Equation (7) illustrates the soft voting ensemble method. The vector $P(n)$ represents the probabilities of classes in an n-classifier. The value of $W(n)$ is the weight of the n-classifier, in the range from 0 to 1. The final classification is the sum of all vectors of probabilities used on the ensemble, multiplied with their respective weights. The sum of weights must follow the rule described in Equation (8)

$$E = \sum_{1}^{n} P(n)W(n); \tag{7}$$

$$\sum_{1}^{n} W(n) = 1. \tag{8}$$

The ensemble is a topic of discussion and it has been applied with deep learning and computer vision in tasks like insulator fault detection method using aerial images [67], to recognize transmission equipment on images taken by an unmanned aerial vehicle [68], and application of remote sensing image classification [69]. There are also studies that compare imbalanced ensemble classifiers [63].

7. Methodology

The authors defined a methodology, divided into two stages. The first one intends to check the performance of CNNs trained on a synthetic dataset and then, improve it by training the CNNs with real images. In the second stage, the experiment evaluates the use of a blend type ensemble approach using RGB and RGB-D images.

7.1. Stage I-Training the CNNs

The first stage consists of training different architectures on both synthetic domains and then training the best models with real data. Figure 18 shows the procedure of Stage I. Each CNN will have two models, one trained on RGB and the other one trained on RGB-D images. From the previous training on Imagenet, a fine-tuning will update the models. The datasets used for this training procedure are S-RGB and S-RGBD (step 1 in the Figure 18).

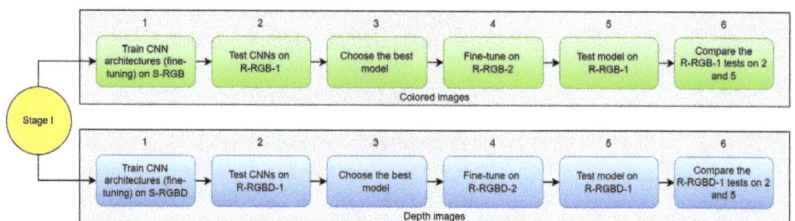

Figure 18. Diagram for Stage I - Training the CNNs.

After the training step, all models will be tested on real datasets corresponding to each domain, being R-RGB-1 and R-RGBD-1 (step 2). The CNN with the best overall performance in RGB will be selected (step 3). The chosen CNNs will be fine-tuned on a real dataset called R-RGB-2, in order to mitigate the domain shift (step 4). The same procedure will be done for the best model in RGB-D, which will be fine-tuned on R-RGBD-2, a real dataset.

Then, they will be tested again on R-RGB-1 and R-RGBD-1 (step 5). The intuition is to achieve improvements over the models trained only in synthetic images. Finally, a comparison of the tests conducted on steps 2 and 5 will show if the procedure outperformed the CNNs trained only on the synthetic domain (step 6).

7.2. Stage II-Blending Pipelines

In this stage, the chosen fine-tuned models will be blended in an ensemble approach. The goal here is to extract information from the scenes using a pipeline for RGB and another for RGB-D, seeking to increase classification performance over the single CNN approach on Stage I. This ensemble consists in blending the CNNs by applying soft voting to their outputs, averaging the probabilities with equal weights for the pipelines. However, the architectures were trained separately. The Equation (9) illustrates the result of the ensemble approach

$$E = P_c W_c + P_d W_d. \tag{9}$$

The P_c in Equation (10) is the vector of probabilities resulted from the RGB pipeline. The c_{ci} value is the probability of the class being an insulator whereas c_{cb} is the probability of it being a brace band

$$P_c = [c_{ci}, c_{cb}]. \tag{10}$$

The P_d in Equation (11) is the vector of probabilities resulted from the RGB-D pipeline. The c_{di} value is the probability of the class being a insulator whereas c_{db} is the probability of it being a brace band

$$P_d = [c_{di}, c_{db}]. \tag{11}$$

The vector of probabilities in the colored pipeline P_c is multiplied by its correspondent weight W_c. The same is true for the depth pipeline, where the vector P_d is multiplied by its W_d weight. Both pipeline weights receive the value of 0.5 in order to guarantee the same influence of the classifiers. The ensemble results in a vector of probabilities, which will be handled by the final decision step. The class with the higher probability will be chosen as the final decision for the scene.

The blend will be tested on the real scenes and compared with the results from the best CNNs on Stage I. Figure 19 shows the structure of the blended approach.

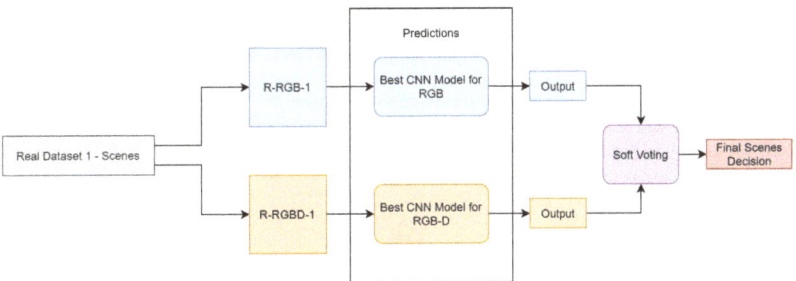

Figure 19. Blending CNN pipelines approach.

One advantage of the blended pipeline can be a more general classification, taking into account features extracted from both colored and depth images. The domains have features that can be different from each other, and using it in a blend may improve classification. The tool would be more sensitive to capture these features from both RGB and RGB-D images. Nevertheless, the pipelines must be tuned to have a similar and suitable accuracy, otherwise, the one with discrepancy would pull the average down.

7.3. Dataset Description

There are four domains in this experiment, being Synthetic RGB, Synthetic RGB-D, Real RGB, and Real RGB-D. Each scene has data in two domains (colored and depth), and for instance, scene 01 will have a correspondent image in both RGB and RGB-D. This is valid for scenes in synthetic and real datasets. Figure 20 shows all domains of the experiment.

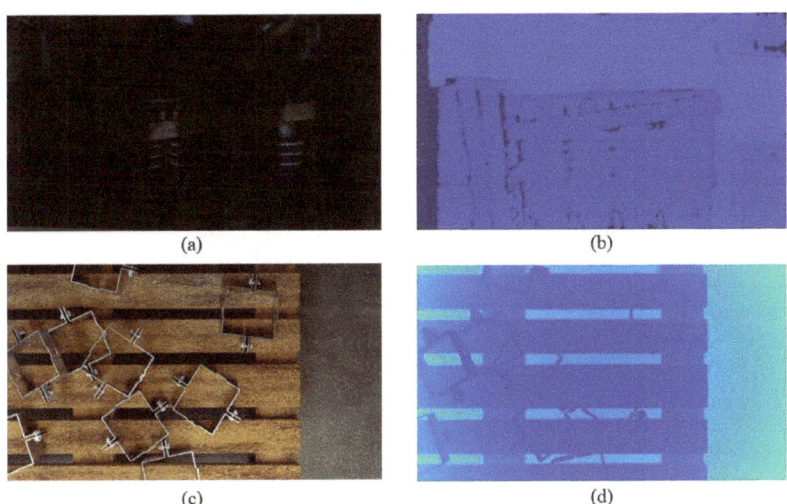

Figure 20. The four domains of the experiment: (**a**) real RGB; (**b**) real RGB-D; (**c**) synthetic RGB and (**d**) synthetic RGB-D.

In Figure 20 it can be seen two scenes with colored and depth images. Figure 20 (a,b) show a good example of a blend to classify a scene as Figure 20 (a) has poor light conditions and (b) can be used to capture features that are difficult for (a) to deliver.

For the experiments, six datasets were separated using synthetic and real images. Two synthetic datasets, S-RGB and S-RGBD were used to train the CNNs. Two real RGB datasets, R-RGB-1 and R-RGB-2 were used to test the CNNs and to fine-tune them, respectively. The same is true for real RGB-D, two sets were created, being R-RGBD-1 and R-RGBD-2, to test and fine-tune the CNNs. Table 1 illustrates these datasets.

Table 1. Datasets of the experiment.

Datasets	R-RGB-1	R-RGBD-1	R-RGB-2	R-RGBD-2	S-RGB	S-RGBD
Insulator	96	96	153	153	720	720
Brace band	156	156	250	250	720	720
Division	test set	test set	80% train 20% valid	80% train 20% valid	75.6% train 24.4% valid	75.6% train 24.4% valid

8. Results

The extensive tests conducted on this paper lead to the results shown in this section. As mentioned in Section 7, they are separated into subtopics, where the first one is used to assess the CNNs performance applying synthetic data. The best models were fine-tuned on a set of real images. The second one evaluates the ensemble method using color and depth images.

8.1. Stage I-Training the CNNs

In this stage, the proposed architectures were trained with a transfer learning technique. The models were pre-trained on Imagenet and fine-tuned on the synthetic datasets. The experiment was conducted as follows: 2 classes; batch-size of 16; 30 epochs; 5-fold cross-validation; adaptive moment estimation or Adam as optimizer (except for ResNet and AlexNet due to incompatibilities, it was used Stochastic Gradient Descent or SGD), and learning rate of 0.001 [55,70–73].

At the end of each training and validation, the CNNs received the test dataset R-RGB-1 and R-RGBD-1. This test set was not used for tuning the hyperparameters and weights of the net. It is a separated set of data, containing real images, to verify the ability of the models trained on synthetic to perform on a real domain. The metrics used to evaluate the models are accuracy, precision, recall, and f1-score. Figure 21 shows the evaluation of the models trained on synthetic RGB and tested on R-RGB-1.

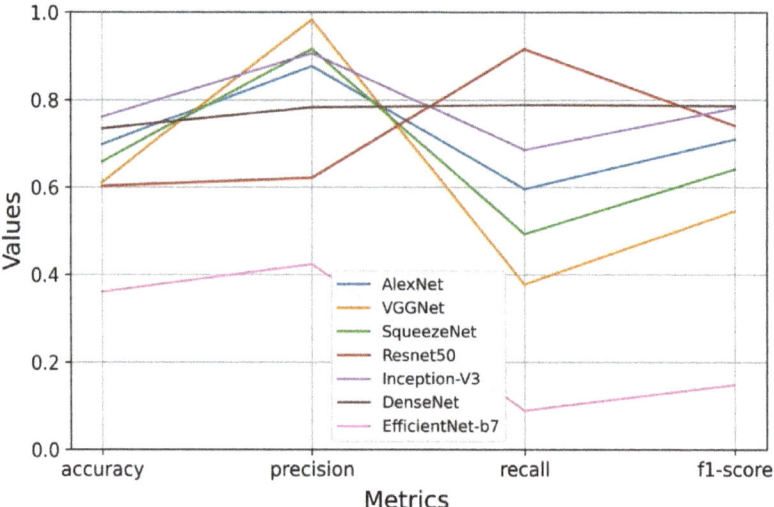

Figure 21. Evaluation of models on R-RGB-1.

DenseNet achieved the second-highest accuracy and recall, as well as the best f1-score. Since f1-score takes into account precision and recall, DenseNet was more stable than Inception-V3 in these metrics. Inception-V3 achieved the best result in accuracy. However, DenseNet was chosen as the best overall performance mainly due to its stability and also a marginally higher evaluation of f1-score. EfficientNet and VGGNet suffered from the domain shift and did not perform well on the test, with an f1-score lower than 50%.

For the depth domain, the results are shown in Figure 22. The models were trained on synthetic S-RGB-D and tested on R-RGBD-1.

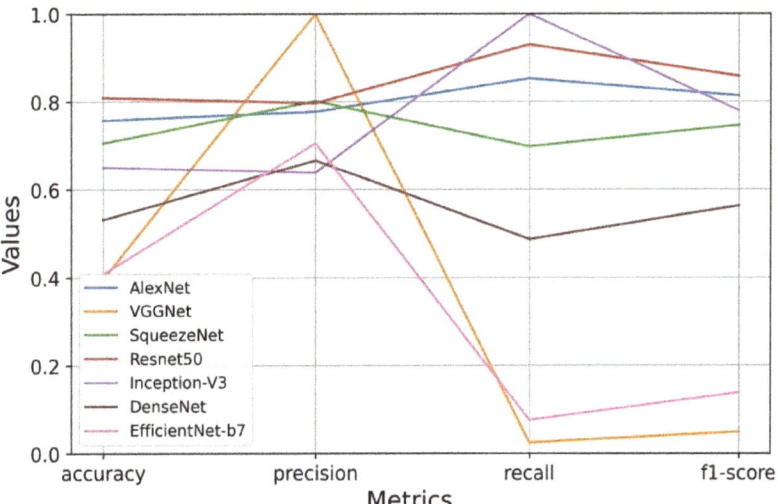

Figure 22. Evaluation of models on R-RGBD-1.

Resnet50 performed better in comparison with other models. It achieved the best accuracy and f1-score on the test. VGGNet and EfficientNet had the lowest values on accuracy, recall, and f1-score.

The best model for each domain was DenseNet (RGB images) and Resnet (RGB-D images). Therefore, these architectures were selected to be fine-tuned using sets of real images, R-RGB-2 and R-RGBD-2. The models trained on synthetic were fine-tuned on the real sets to attack the domain shift problem.

DenseNet was fine-tuned on R-RGD-2, with 8 as batch size, 30 epochs, 5-fold, Adam optimizer, and learning rate of 0.001. The average accuracy and loss on training were 91.31% and 0.236. The model was tested on R-RGB-1. To verify if training with synthetic and then with real images is the best approach, Densenet was also straight fine-tuned on R-RGB-2. The confusion matrixes for these two tests are shown in Figure 23.

Fine-tuning the model on S-RGB and then on R-RGB-2 (a) outperformed the model straight trained on R-RGB-2 (b). The use of a pre-trained model on synthetic domain missed only 5 insulators and 8 brace bands with an accuracy of 94.84%. While (b) missed only 1 insulator, it also missed 20 brace bands, achieving an accuracy of 91.67%.

ResNet was fine-tuned on the real set, with a batch size of 8, 30 epochs, and 5-fold, Adam optimizer, and a learning rate of 0.001. The average accuracy and loss on training were 95.78% and 0.097, respectively. The model was tested on R-RGBD-1. To verify the proposed approach, ResNet was straight fine-tuned on a real dataset as well, R-RGBD-2. The confusion matrixes for these two tests are shown in Figure 24.

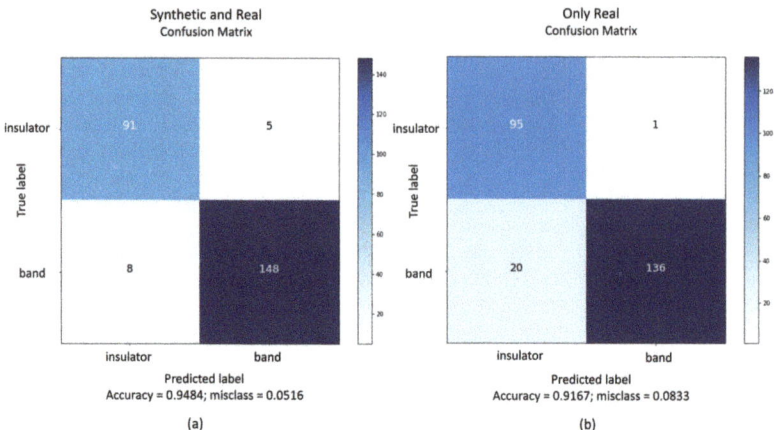

Figure 23. Confusion Matrixes of R-RGB-1 tested on: (**a**) DenseNet trained on S-RGB and R-RGB-2 datasets and (**b**) DenseNet trained on R-RGB-2.

The approach of fine-tuning the model in synthetic S-RGB dataset and then in a real set R-RGB-2 Figure 24 (a) outperformed the model trained straight on the real set R-RGB-2 Figure 24 (b). The use of a pre-trained model on the synthetic domain (a) missed only 4 insulators and 19 brace bands, with an accuracy of 90.87% in contrast to 7 insulators and 22 brace bands and accuracy of 88.49% (b).

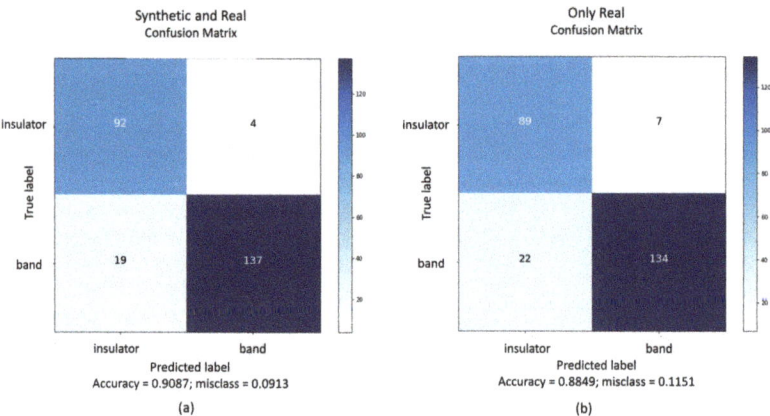

Figure 24. Confusion Matrixes of R-RGBD-1 tested on: (**a**) Resnet trained on synthetic and real datasets and (**b**) Resnet trained on real dataset.

8.2. Stage II-Blending Pipelines

DenseNet and Resnet fine-tuned on synthetic and real domains were blended in an ensemble approach. The pipeline for RGB (DenseNet) and for RGB-D (ResNet) had their probabilities of a class inference summed and averaged in a soft voting operation, with equal weights, meaning that the pipelines had the same influence on the final result. The test set used in the blended pipelines is the same used for testing the CNNs on Stage I, the scenes from R-RGB-1 and R-RGBD-1. The confusion matrix for this test is shown in Figure 25.

As it can be seen in Figure 25, the blended approach did not miss insulators and missed only 12 brace bands. The accuracy for the test was 95.24%.

The blended approach is then compared with the performance of each CNN individually. This comparison is shown in Table 2. The Blend outperformed Densenet and ResNet in accuracy, precision, and f1-score. The Blend also achieved better results on recall in comparison with Resnet, although it did not perform better than DenseNet on this particular metric. Since DenseNet misclassed only eight brace bands, it performed better in the recall. However, DenseNet classified wrongly five insulators.

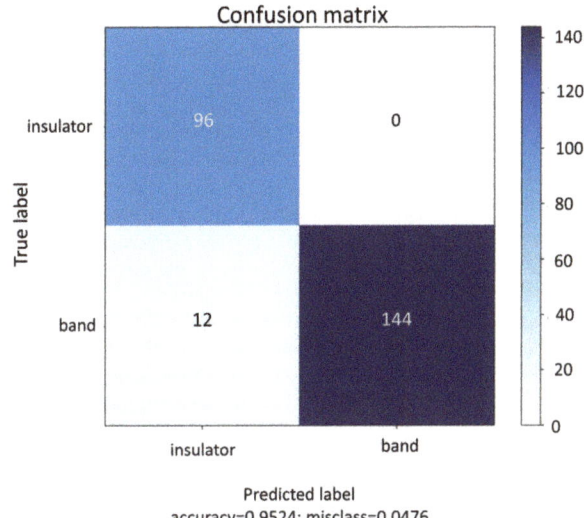

Figure 25. Confusion matrix of blended pipelines.

Table 2. Comparison of single CNN and Blend results.

Metrics	Accuracy	Precision	Recall	F1-score	Misclassed Insulator	Brace band	Total
DenseNet	0.9484	0.9673	**0.9487**	0.9579	5	8	13
Resnet	0.9087	0.9716	0.8782	0.9225	4	19	23
Blended CNNs	**0.9523**	**1**	0.9230	**0.9600**	**0**	12	**12**

Table 3 shows the difference in the percentage of all metrics of the single CNNs in comparison with the blend. The proposed mixed pipelines achieved an improvement of 0.39% in accuracy, 2.84% in precision, and 0.21% in f1-score over the best result of each single CNN. It also had a drop of 2.57% in recall in comparison with DenseNet.

Table 3. Difference in percentage of Blended CNN and single CNNs.

	Accuracy	Precision	Recall	F1-Score
DenseNet	0.39	3.27	−2.57	0.21
Resnet	4.36	2.84	4.48	3.75

9. Discussion and Conclusions

To create the synthetic dataset, Blender was used with Eevee as the render engine due to speed. Eevee was almost seven times faster than Cycles engine.

For the real dataset, the mechanical device performed the data gathering. It was built prior to the automated stacker being constructed. This enabled to testing of the CNNs with the real dataset. The Brisque IQA algorithm and the histogram analysis provided quality control for the real images.

From the seven CNNs fine-tuned on the synthetic dataset and tested on the real domain, the best performance in RGB and RGB-D was respectively DenseNet and Resnet-50. The training procedure on synthetic datasets and testing on real samples showed a domain shift problem, a common issue discussed in recent studies. To contour this problem, a set of real images was separated and used to fine-tune DenseNet and Resnet. Fine-tuning the models with synthetic and then with real images outperformed classification in comparison with a straight fine-tuning on real images. The use of synthetic images generated by Blender and rendered by Eevee proved to help the classification performance.

The proposed blended RGB and RGB-D pipelines were used to improve the recognition of insulators and brace bands. Since the scenes have colored and depth information, they were applied in an ensemble approach. Each pipeline contributed equally to the output prediction of the scenes, in a soft voting decision. The final classification is the average of probabilities of colored and depth pipelines. The blended approach outperformed the best results of the single CNNs, with the only exception being the metric recall in the DenseNet colored test. However, the blended pipelines misclassed fewer items in comparison with DenseNet. Blending colored and depth CNN pipelines achieved better accuracy in comparison with the previous study, where a Resnet-50 was used to classify the insulators and brace bands through RGB images [2]. This present approach resulted in an accuracy of 95.23% in contrast to 92.87% seen in Piratelo et al. [2].

This paper presented a blend of convolutional neural network pipelines that classifies products from an electrical company's warehouse, using colored and depth images from real and synthetic domains. The research also compared the results of training the architectures only with real and then synthetic and real data. The stage I consisted in training several CNNs on a synthetic dataset and testing them in the real domain. The architectures that performed better in RGB and RGB-D images were DenseNet and Resnet-50, although, they all suffered from the domain shift. A procedure to overcome this issue was done by fine-tuning the CNNs on a real set of data. The procedure improved the accuracy, precision, recall, and f1-score of the models in comparison to only training with real data, proving that synthetic images helped to train the models. It was time-consuming and it required a computational effort to train all CNNs on synthetic and real images, yet it showed to be a valid method to improve accuracy.

On stage II, the DenseNet model trained on RGB images was used as the first pipeline, and Resnet trained on RGB-D images composed the ensemble as the second pipeline. Each one contributed equally to the final classification, using the average of their probabilities in a soft voting method. A poorly performance of one CNN could drag down final accuracy for the ensemble. Both pipelines must be tuned with suitable accuracy. Moreover, this requires more computational resources in comparison with inferences from single CNNs on Stage I, since it is working with two models. However, the ensemble lead to a more robust classification, taking advantage of colored and depth information provided by the images, extracting features from both domains. The blended pipelines outperformed accuracy, precision, and f1-score in comparison with the single CNNs.

This approach intends to be encapsulated and used to keep the inventory of the warehouse up to date. The classification task is handled by computer vision and artificial intelligence, making full use of RGB and RGB-D images of synthetic and real domains, applied in an approach of blended CNN pipelines. The quality of captured images is also examined by an IQA called BRISQUE, equalizing the images and enhancing their features.

This study classifies two types of objects: insulators and brace bands. A transfer learning method called fine-tuning took advantage of pre-trained models, utilizing its feature detection capacity and adapting it to the problem. This transfer learning technique was used twice, first with synthetic and then with real images. The extensive tests conducted on this paper evaluated the performance of different CNNs individually and blended in an ensemble approach. In the end, the blend was able to classify the objects with an average accuracy of 95.23%. The tool will be included in the prototype of an AGV that travels the entire warehouse capturing images of the shelves, combining the application of automation, deep learning, and computer vision in a real engineering problem. This is the first step to automate and digitize the warehouse, seeking to achieve the paradigms of real-time data analysis, smart sensing, and perception of the company.

As future work, it is intended to test weights for the pipelines and seek their best combination. It is compelling to find different alternatives to ensemble learning techniques. A study of the combination between synthetic and real domain and their influence on training the models is also intended to be accomplished.

Author Contributions: Conceptualization, P.H.M.P.; methodology, P.H.M.P.; software, P.H.M.P., R.N.d.A. and F.S.M.L.; validation, E.M.Y., L.d.S.C. and G.V.L.; formal analysis, J.F.B.F.; writing—original draft preparation, P.H.M.P. and R.N.d.A.; writing—review and editing, P.H.M.P.; R.N.d.A.; E.M.Y.; J.F.B.F.; G.M.; F.S.M.L.; L.P.d.J.; R.d.A.P.N.; L.d.S.C.; G.V.L.; supervision, E.M.Y.; project administration, E.M.Y.; funding acquisition, E.M.Y. All authors have read and agreed to the published version of the manuscript.

Funding: This research was supported by COPEL (power utility from the state of Paraná, Brazil), under the Brazilian National Electricity Agency (ANEEL) Research and Development program PD-02866-0011/2019. The APC was funded by LACTEC.

Institutional Review Board Statement: Not applicable.

Informed Consent Statement: Not applicable.

Acknowledgments: The authors thank LACTEC for the assistance.

Conflicts of Interest: The funders had no role in the design of the study; in the collection, analyses, or interpretation of data; in the writing of the manuscript, or in the decision to publish the results".

References

1. McKinsey & Company. The state of AI in 2020. Available online: https://www.mckinsey.com/business-functions/mckinsey-analytics/our-insights/global-survey-the-state-of-ai-in-2020 (accessed on 5 February 2021).
2. Piratelo, P.H.M.; de Azeredo, R.N.; Yamao, E.M.; Maidl, G.; de Jesus, L.P.; Neto, R.D.A.P.; Coelho, L.D.S.; Leandro, G.V. Convolutional neural network applied for object recognition in a warehouse of an electric company. In Proceedings of the 14th IEEE International Conference on Industry Applications, São Paulo, Brazil, 15–18 August 2021; pp. 293–299.
3. Lasi, H.; Fettke, P.; Kemper, H.; Feld, T.; Hoffmann, M. Industry 4.0. *Bus. Inf. Syst. Eng.* **2014**, *4*, 239–242. [CrossRef]
4. Rubio, J.D.J.; Pan, Y.; Pieper, J.; Chen, M.Y.; Sossa, J.H.A. Advances in Robots Trajectories Learning via Fast Neural Networks. *Front. Neurorobot.* **2021**, *15*, 29. [CrossRef]
5. López-González, A.; Campaña, J.M.; Martínez, E.H.; Contro, P.P. Multi robot distance based formation using Parallel Genetic Algorithm. *Appl. Soft Comput.* **2020**, *86*, 105929. [CrossRef]
6. MIT Sloan Management Review and Boston Consulting Group. Available online: https://sloanreview.mit.edu/projects/expanding-ais-impact-with-organizational-learning (accessed on 3 February 2021).
7. Salierno, G.; Leonardi, L.; Cabri, G. The Future of Factories: Different Trends. *Appl. Sci.* **2021**, *11*, 9980. [CrossRef]
8. Davenport, T. How strategists use "big data" to support internal business decisions, discovery and production. *Strategy Leadersh.* **2014**, *42*, 45–50. [CrossRef]
9. Custodio, L.; Machado, R. Flexible automated warehouse: a literature review and an innovative framework. *Int. Adv. Manuf. Technol.* **2019**, *106*, 533–0558. [CrossRef]
10. Laber, J.; Thamma, R.; Kirby, E.D. The Impact of Warehouse Automation in Amazon's Success. *Int. J. Innov. Sci. Eng. Technol.* **2020**, *7*, 63–70.
11. Yang, J.X.; Li, L.D.; Rasul, M.G. Warehouse management models using artificial intelligence technology with application at receiving stage-a review. *Int. J. Mach. Learn. Comput.* **2021**, *11*, 242–249. [CrossRef]
12. Hokey, M. The applications of warehouse management systems: An exploratory study. *Int. J. Logist. Res. Appl.* **2006**, *9*, 111–126.
13. Nasir, J.A.; Shahzad, M.K. Architecture for virtualization in data warehouse. In *Innovations and Advanced Techniques in Computer and Information Sciences and Engineering*; Springer: Dordrecht, The Netherlands, 2007; pp. 243–248.
14. Ferdous, S.; Fegaras, L.; Makedon, F. Applying data warehousing technique in pervasive assistive environment. In Proceedings of the 3rd International Conference on Pervasive Technologies Related to Assistive Environments, Samos, Greece, 23–25 June 2010; pp. 1–7.

15. Chryssolouris, G.; Papakostas, N.; Mourtzis, D. Refinery short-term scheduling with tank farm, inventory and distillation management: An integrated simulation-based approach. *Eur. J. Oper. Res.* **2005**, *166*, 812–827. [CrossRef]
16. Stavropoulos, P.; Papacharalampopoulos, A.; Souflas, T. Indirect online tool wear monitoring and model-based identification of process-related signal. *Adv. Mech. Eng.* **2020**, *12*, 1687814020919209. [CrossRef]
17. Mittal, A.; Moorthy, A.K.; Bovik, A.C. No-reference image quality assessment in the spatial domain. *IEEE Trans. Image Process.* **2012**, *21*, 4695–4708. [CrossRef]
18. Virtanen, T.; Nuutinen, M.; Vaahteranoksa, M.; Oittinen, P.; Hakkinen, J. Cid2013: A database for evaluating no-reference image quality assessment algorithms. *IEEE Trans. Image Process.* **2015**, *24*, 390–402. [CrossRef] [PubMed]
19. Wang, Z.; Bovik, A.C. Reduced-and no-reference image quality assessment. *IEEE Signal Process. Mag.* **2011**, *28*, 29–40. [CrossRef]
20. Stark, J. Adaptive image contrast enhancement using generalizations of histogram equalization. *IEEE Trans. Image Process.* **2000**, *9*, 889–896. [CrossRef] [PubMed]
21. Pizer, S.M.; Amburn, E.P.; Austin, J.D.; Cromartie, R.; Geselowitz, A.; Greer, T.; ter Haar Romeny, B.; Zimmerman, J.B.; Zuiderveld, K. Adaptive histogram equalization and its variations. *Comput. Vision, Graph. Image Process.* **1987**, *38*, 99. [CrossRef]
22. Öztürk, A.E.; Erçelebi, E. Real UAV-Bird Image Classification Using CNN with a Synthetic Dataset. *Appl. Sci.* **2021**, *11*, 38–63. [CrossRef]
23. Khalifa, N.E.; Loey, M.; Mirjalili, S. A comprehensive survey of recent trends in deep learning for digital images augmentation. *Artif. Intell. Rev.* **2021**, 1–27. [CrossRef]
24. Peng, X.; Saenko, K. Synthetic to real adaptation with generative correlation alignment networks. In Proceedings of the 2018 IEEE Winter Conference on Applications of Computer Vision, Lake Tahoe, NV, USA, 12–15 March 2018; pp. 1982–1991.
25. Ciampi, L.; Messina, N.; Falchi, F.; Gennaro, C.; Amato, G. Virtual to real adaptation of pedestrian detectors. *Sensors* **2020**, *20*, 5250. [CrossRef]
26. Wang, Y.; Deng, W.; Liu, Z.; Wang, J. Deep learning-based vehicle detection with synthetic image data. *IET Intell. Transp. Syst.* **2019**, *13*, 1097–1105. [CrossRef]
27. Poibrenski, A.; Sprenger, J.; Müller, C. Toward a methodology for training with synthetic data on the example of pedestrian detection in a frame-by-frame semantic segmentation task. In Proceedings of the 2018 IEEE/ACM 1st International Workshop on Software Engineering for AI in Autonomous Systems, Gothenburg, Sweden, 28 May 2018; pp. 31–34.
28. Anantrasirichai, N.; Biggs, J.; Albino, F.; Bull, D. A deep learning approach to detecting volcano deformation from satellite imagery using synthetic datasets. *Remote Sens. Environ.* **2019**, *230*, 111179. [CrossRef]
29. dharani Parasuraman, S.; Wilde, J. Training Convolutional Neural Networks (CNN) for Counterfeit IC Detection by the Use of Simulated X-Ray Images. In Proceedings of the 2021 22nd International Conference on Thermal, Mechanical and Multi-Physics Simulation and Experiments in Microelectronics and Microsystems, St. Julian, Malta, 19–21 April 2021; pp. 1–7.
30. Narayanan, P.; Borel-Donohue, C.; Lee, H.; Kwon, H.; Rao, R. A real-time object detection framework for aerial imagery using deep neural networks and synthetic training images. In *Signal Processing, Sensor/Information Fusion, and Target Recognition XXVII*; International Society for Optics and Photonics: Bellingham, DC, USA, 2018; Volume 10646, p. 1064614.
31. Varadarajan, S.; Srivastava, M.M. Weakly Supervised Object Localization on grocery shelves using simple FCN and Synthetic Dataset. In Proceedings of the 11th Indian Conference on Computer Vision, Graphics and Image Processing, Hyderabad, India, 18–22 December 2018; pp. 1–7.
32. Saleh, K.; Hossny, M.; Hossny, A.; Nahavandi, S. Cyclist detection in lidar scans using faster r-cnn and synthetic depth images. In Proceedings of the 2017 IEEE 20th International Conference on Intelligent Transportation Systems, Yokohama, Japan, 16–19 October 2017; pp. 1–6.
33. Das, A.; Mohanty, M.N.; Mallick, P.K.; Tiwari, P.; Muhammad, K.; Zhu, H. Breast cancer detection using an ensemble deep learning method. *Biomed. Signal Process. Control* **2021**, *70*, 103009. [CrossRef]
34. Moya-Sáez, E.; Peña-Nogales, Ó.; de Luis-García, R.; Alberola-López, C. A deep learning approach for synthetic MRI based on two routine sequences and training with synthetic data. *Comput. Methods Programs Biomed.* **2021**, *210*, 106371. [CrossRef] [PubMed]
35. Talukdar, J.; Gupta, S.; Rajpura, P.S.; Hegde, R.S. Transfer learning for object detection using state-of-the-art deep neural networks. In Proceedings of the 2018 5th International Conference on Signal Processing and Integrated Networks, Noida, India, 22–23 Febuary 2018; pp. 78–83.
36. Blender Foundation. Blender Foundation. Available online: http://www.blender.org (accessed on 28 July 2021).
37. Blender 3.0.0. Blender 3.0.0 Python API Documentation. Available online: https://docs.blender.org/api/current/index.html (accessed on 28 July 2021).
38. Blender 2.93 Manual. Getting Started. Available online: https://docs.blender.org/manual/en/2.93/getting_started/index.html (accessed on 28 July 2021).
39. Denninger, M.; Sundermeyer, M.; Winkelbauer, D.; Zidan, Y.; Olefir, D.; Elbadrawy, M.; Lodhi, A.; Katam, H. BlenderProc. *arXiv* **2019**, arXiv:1911.01911.
40. Dong, S.; Wang, P.; Abbas, K. A survey on deep learning and its applications. *Comput. Sci. Rev.* **2021**, *40*, 100379. [CrossRef]
41. Mishra, M.; Nayak, J.; Naik, B.; Abraham, A. Deep learning in electrical utility industry: A comprehensive review of a decade of research. *Eng. Appl. Artif. Intell.* **2020**, *96*, 104000. [CrossRef]

42. Li, Z.; Liu, F.; Yang, W.; Peng, S.; Zhou, J. A survey of convolutional neural networks: Analysis, applications, and prospects. *IEEE Trans. Neural Netw. Learn. Syst.* **2021**, doi:10.1109/TNNLS.2021.3084827. [CrossRef]
43. Khan, A.; Sohail, A.; Zahoora, U.; Qureshi, A.S. A survey of the recent architectures of deep convolutional neural networks. *Artif. Intell. Rev.* **2020**, *53*, 5455–5516. [CrossRef]
44. de Rubio, J.J. Stability analysis of the modified Levenberg-Marquardt algorithm for the artificial neural network training. *IEEE Trans. Neural Netw. Learn. Syst.* **2020**, *32*, 3510–3524. [CrossRef]
45. de Jesús R.J.; Lughofer, E.; Pieper, J.; Cruz, P.; Martinez, D.I.; Ochoa, G.; Islas, M.A.; Garcia, E. Adapting H-infinity controller for the desired reference tracking of the sphere position in the maglev process. *Inf. Sci.* **2021**, *569*, 669–686.
46. Chiang, H.S.; Chen, M.Y.; Huang, Y.J. Wavelet-based EEG processing for epilepsy detection using fuzzy entropy and associative petri net. *IEEE Access* **2019**, *7*, 103255–103262. [CrossRef]
47. Vargas, D.M. Superpixels extraction by an Intuitionistic fuzzy clustering algorithm. *J. Appl. Res. Technol.* **2021**, *19*, 140–152. [CrossRef]
48. Ghods, A.; Cook, D.J. A survey of deep network techniques all classifiers can adopt. *Data Min. Knowl. Discov.* **2020**, *35*, 46–87. [CrossRef]
49. Krizhevsky, A.; Sutskever, I.; Hinton, G.E. Imagenet classification with deep convolutional neural networks. *Adv. Neural Inf. Process. Syst.* **2012**, *25*, 1097–1105. [CrossRef]
50. Simonyan, K.; Zisserman A. Very deep convolutional networks for large-scale image recognition. *arXiv* **2014**, arXiv:1409.1556.
51. Szegedy, C.; Liu, W.; Jia, Y.; Sermanet, P.; Reed, S.; Anguelov, D.; Erhan, D.; Vanhoucke, V.; Rabinovich, A. Going deeper with convolutions. In Proceedings of the IEEE Conference on Computer Vision and Pattern Recognition, Boston, MA, USA, 7–12 June 2015; pp. 1–9.
52. Szegedy, C.; Vanhoucke, V.; Ioffe, S.; Shlens, J.; Wojna, Z. Rethinking the inception architecture for computer vision. In Proceedings of the IEEE Conference on Computer Vision and Pattern Recognition, Las Vegas, NV, USA, 27–30 June 2016; pp. 2818–2826.
53. He, K.; Zhang, X.; Ren, S.; Sun, J. Deep residual learning for image recognition. In Proceedings of the IEEE Conference on Computer Vision and Pattern Recognition, Las Vegas, NV, USA, 27–30 June 2016; pp. 770–778.
54. Nair, V.; Hinton, G.E. Rectified linear units improve restricted boltzmann machines. In Proceedings of the 27th international conference on machine learning (ICML-10), Haifa, Israel, 21–24 June 2010; pp. 807–814.
55. Keskar, N.S.; Socher, R. Improving generalization performance by switching from adam to sgd. *arXiv* **2017**, arXiv:1712.07628.
56. Gu, J.; Wang, Z.; Kuen, J.; Ma, L.; Shahroudy, A.; Shuai, B.; Liu, T.; Wang, X.; Wang, G.; Cai, J.; et al. Recent advances in convolutional neural networks. *Pattern Recognit.* **2018**, *77*, 354–377. [CrossRef]
57. Deng, J.; Dong, W.; Socher, R.; Li, L.-J.; Li, K.; Fei-Fei, L. Imagenet: A large-scale hierarchical image database. In Proceedings of the 2009 IEEEConference on Computer Vision and Pattern Recognition, Miami, FL, USA, 20–25 June 2009.
58. Iandola, F.N.; Han, S.; Moskewicz, M.W.; Ashraf, K.; Dally, W.J.; Keutzer, K. SqueezeNet: AlexNet-level accuracy with 50x fewer parameters and <0.5 MB model size. *arXiv* **2016**, arXiv:1602.07360.
59. Huang, G.; Liu, Z.; Van Der Maaten, L.; Weinberger, K.Q. Densely connected convolutional networks. In Proceedings of the IEEE Conference on Computer Vision and Pattern Recognition, Honolulu, HI, USA, 21–26 July 2017; pp. 4700–4708.
60. Tan, M.; Le, Q. Efficientnet: Rethinking model scaling for convolutional neural networks. In Proceedings of the International Conference on Machine Learning, Long Beach, CA, USA, 10–15 June 2019; pp. 6105–6114.
61. Rincy, T.N.; Gupta, R. Ensemble learning techniques and its efficiency in machine learning: A survey. In Proceedings of the 2nd International Conference on Data, Engineering and Applications, Bhopal, India, 28–29 Febuary 2020; pp. 1–6.
62. Abdar, M.; Pourpanah, F.; Hussain, S.; Rezazadegan, D.; Liu, L.; Ghavamzadeh, M.; Fieguth, P.; Cao, X.; Khosravi, A.; Acharya, U.R.; et al. A review of uncertainty quantification in deep learning: Techniques, applications and challenges. *Inf. Fusion* **2021**, *76*, 243–297. [CrossRef]
63. Zhao, D.; Wang, X.; Mu, Y.; Wang, L. Experimental study and comparison of imbalance ensemble classifiers with dynamic selection strategy. *Entropy* **2021**, *23*, 822. [CrossRef]
64. González, S.; García, S.; Del Ser, J.; Rokach, L.; Herrera, F. A practical tutorial on bagging and boosting based ensembles for machine learning: Algorithms, software tools, performance study, practical perspectives and opportunities. *Inf. Fusion* **2020**, *64*, 205–237. [CrossRef]
65. Dong, X.; Yu, Z.; Cao, W.; Shi, Y.; Ma, Q. A survey on ensemble learning. *Front. Comput. Sci.* **2020**, *14*, 241–258. [CrossRef]
66. Sagi, O.; Rokach, L. Ensemble learning: A survey. *Wiley Interdiscip. Rev. Data Min. Knowl. Discov.* **2018**, *8*, 1249. [CrossRef]
67. Jiang, H.; Qiu, X.; Chen, J.; Liu, X.; Miao, X.; Zhuang, S. Insulator fault detection in aerial images based on ensemble learning with multi-level perception. *IEEE Access* **2019**, *7*, 61797–61810. [CrossRef]
68. Huang, S.; Zhang, Z.; Yanxin, L.; Hui, L.; Wang, Q.; Zhang, Z.; Zhao, Y. Transmission equipment image recognition based on Ensemble Learning. In Proceedings of the 2021 6th Asia Conference on Power and Electrical Engineering, Chongqing, China, 8–11 April 2021; pp. 295–299.
69. Han, M.; Liu, B. Ensemble of extreme learning machine for remote sensing image classification. *Neurocomputing* **2021**, *149*, 65–70. [CrossRef]
70. Jafar, A.; Myungho, L. Hyperparameter optimization for deep residual learning in image classification. In Proceedings of the IEEE International Conference on Autonomic Computing and Self-Organizing Systems, Washington, DC, USA, 17–21 August 2020; pp. 24–29.

71. Habibzadeh, M.; Jannesari, M.; Rezaei, Z.; Baharvand, H.; Totonchi, M. Automatic white blood cell classification using pre-trained deep learning models: Resnet and inception. *Tenth Int. Conf. Mach. Vis.* **2017**, *10696*, 1069612.
72. Radiuk, P.M. Impact of training set batch size on the performance of convolutional neural networks for diverse datasets. *Inf. Technol. Manag. Sci.* **2017**, *20*, 20–24. [CrossRef]
73. Reyes, A.K.; Caicedo, J.C.; Camargo, J.E. Fine-tuning deep convolutional networks for plant recognition. *CLEF (Work. Notes)* **2015**, *1391*, 467–475.

Article

Speed Control with Indirect Field Orientation for Low Power Three-Phase Induction Machine with Squirrel Cage Rotor

Robert R. Gomes *, Luiz F. Pugliese, Waner W. A. G. Silva, Clodualdo V. Sousa, Guilherme M. Rezende and Fadul F. Rodor

Institute of Technological Sciences, Federal University of Itajubá, Itabira 35903-087, Brazil; pugliese@unifei.edu.br (L.F.P.); waner@unifei.edu.br (W.W.A.G.S.); clodualdosousa@unifei.edu.br (C.V.S.); guilhermemre@unifei.edu.br (G.M.R.); fadulrodor@unifei.edu.br (F.F.R.)
* Correspondence: robertribeiro@unifei.edu.br; Tel.: +55-(31)-98815-1828

Citation: Gomes, R.R.; Pugliese, L.F.; Silva, W.W.A.G.; Sousa, C.V.; Rezende, G.M.; Rodor, F.F. Speed Control with Indirect Field Orientation for Low Power Three-Phase Induction Machine with Squirrel Cage Rotor. *Machines* **2021**, *9*, 320. https://doi.org/10.3390/machines9120320

Academic Editors: Marcos de Sales Guerra Tsuzuki, Marcosiris Amorim de Oliveira Pessoa and Alexandre Acássio

Received: 27 October 2021
Accepted: 23 November 2021
Published: 27 November 2021

Publisher's Note: MDPI stays neutral with regard to jurisdictional claims in published maps and institutional affiliations.

Copyright: © 2021 by the authors. Licensee MDPI, Basel, Switzerland. This article is an open access article distributed under the terms and conditions of the Creative Commons Attribution (CC BY) license (https://creativecommons.org/licenses/by/4.0/).

Abstract: Induction machines are widely used in the industry due to their many advantages compared to other industrial machines. This article presents the study and implementation of speed control applied to a Three-phase Induction Machine (MIT) of the squirrel cage type. The induction motor was modeled using the rotor flux in the synchronous reference to design Proportional-Integral (PI) type controllers for the current and velocity control loops. It is the objective of the article also to present in detail the development of converter hardware that comprises the stages of power, acquisition, and conditioning of engine signals. The system was simulated using computational tools and validated using a prototype designed, constructed, and commissioned.

Keywords: induction machines; electrical machines; vector control; SVPWM modulation; frequency inverter

1. Introduction

Electrical machines play an essential role in the industry and the development of society, they make up an efficient means of electromechanical conversion. The industrial modernization process increasingly requires the application of electric motors in drive systems, which often requires control of torque, acceleration, speed or position.

Variable speed drives are mainly used in applications, such as electrical vehicles, pumps, elevators, fans, ventilation, heating, robotics, ship propulsion, and air conditioning [1–3]. The benefits of squirrel-cage induction motors—high robustness and low maintenance—make it widely used through various industrial modern processes, with growing economical and performing demands [1,4].

Previously, the use of motors in industrial applications that required good performance in position and speed control were limited to DC machines. Such use was due to the inherent decoupling characteristics of the machine: one electrical circuit to impose magnetic flux on the machine (field) and another to impose torque (power), which simplifies the control of torque, position, and speed of the machine [5].

The advancement of power electronics, control theory, growth of studies in the field of microelectronics and the advancement of digital processor technology, allowed the efficient use of the electric motor in the activation of the most varied industrial loads, enabling the use of alternating current machines in applications of high dynamic performance [6,7].

Thus, the alternating current motor has become widely used in the industry since it presents characteristics such as robustness, low manufacturing cost, absence of sliding contacts and possibility of operating a wide range of torque and speed [8].

The scalar control technique is simple to implement. Although the constant voltage/frequency control method is the simplest, the performance of this method is not good enough [9–11]. Vector-based control methods enable the control of voltage and frequency amplitude unlike the scalar method. They also provide the instantaneous position of the current, voltage and flux vectors [12].

The dynamic behavior of the IM drives is also improved significantly using the vector-based control method. However, the existence of the coupling between the electromagnetic torque and the flux increases the complexity of the control. To deal with this inherent disadvantage, several methods have been proposed for the decoupling of the torque and flux.

One difficulty of its use for position or speed control is its complex mathematical modeling, requiring greater computational effort for its implementation [5]. In addition, in induction machines there is no unique physical circuit for the field, and this makes its control more complex [13,14].

In Induction Machines (IM), the power phases of the machine each contribute to maintaining the concatenated magnetic flux of the machine and are thus coupled to each other through mutual inductances. To solve the problem of induction machine control, axis transformation techniques are used, which make it possible to transform the electrical and mechanical model of an induction machine so that the flux and torque of the machine can be controlled independently, known as field-oriented control or vector control [15].

The field-oriented control technique is widely used in high-performance motion control applications of induction motors. However, in real applications, accurate decoupling of torque and flux cannot be achieved with accuracy due to the uncertainties of the machine parameters present in the model. These uncertainties are associated with external disturbances, unpredictable parameter variations, and unmodeled nonlinear plant dynamics. Consequently, this deteriorates the dynamic performance of flux and speed significantly. In general, the performance of this technique is dependent on the accuracy of the mathematical model of the induction motor [16].

Field oriented control consists of concentrating the rotor flux on the direct axis of the synchronous reference system, thus achieving decoupling of the flux and torque loops, like DC machines. It is a highly computational technique that involves many mathematical transformations and requires powerful microcontrollers such as Digital Signal Processors (DSPs) and Digital Signal Controllers (DSC) [17].

In general, the field-oriented control performance is sensitive to the deviation of motor parameters, particularly the rotor time constant [7,18,19]. To deal with this problem, there are many fluxes measurement and estimation mechanisms in the published literature [18–21].

Speed information is required for the operation of vector-controlled IM drive. The rotor speed can measure through a mechanical sensor or can be estimated using voltage, current signals, and machine parameters [22,23]. The use of speed sensors is associated with some drawbacks, such as noise present in the measurement, requirement of shaft extension, reduction of mechanical robustness of the motor drive, not suitable for hostile environments and more expensive [24]. On the other hand, using the sensorless strategy reduces hardware complexity, reduces costs, elimination of the sensor cable, better noise immunity, higher reliability, and less maintenance requirements [22].

In this context, the work presents the design and commissioning steps of a three-phase frequency inverter for the activation of a small three-phase induction machine. In addition, the simulation and control of the machine will be addressed through the Indirect Oriented Field Control strategy (IFOC). In this work, the main components that make up the power circuit, current, voltage and speed measurement circuits will be presented. It is also presented the models used for the design of the current and speed controllers of the vector control and, later, the validation in the simulation environment and the experimental bench. In addition, in this work an encoder attached to the machine shaft is used to measure the position and speed of the rotor to prevent parametric variations and speed estimation errors from affecting the quality of the indirect field control. Also, the tests performed in this work are for lower speeds and without a load attached to the machine shaft.

The main contribution of this paper is to propose the design and commissioning of a low-cost converter to drive an induction machine using the indirect field control strategy with a sensor attached to the shaft.

The work is organized as follows: Section 2 will deal with the description of all the hardware used in the construction of the prototype. Section 3 will describe the dynamic modeling of the field oriented directly to the three-phase induction motor. Section 4 will describe the concepts related to Space Vector Pulse Width Modulation (SVPWM). Section 5 will display the tuning of the controllers for the current and speed loops. Section 6 is responsible for presenting the results obtained in the computational simulation as well as the practical tests performed in the prototype developed. Finally, Section 7 presents the conclusions of the work.

2. Hardware Description

The designed converter is based on the FNA41560 IC (Integrated Circuit) produced by Fairchild Semiconductor's. This Insulated Gate Bipolar Transistor (IGBT) module is designed for low-power AC (Alternating Current) motor drive applications, cooler and air conditioning. The device responsible for the data acquisition, processing and execution of the vector control algorithm is the C2000 Delfino MCU TMS320F28379D LaunchPad microprocessor from Texas Instruments.

Figure 1 presents the system built to perform the experimental tests.

Figure 1. Experimental bench built.

The experimental bench is composed of:

1. Variable auto transformer;
2. Auxiliary source power supply circuit breaker;
3. Circuit breaker of the capacitor bank pre-charge circuit;
4. AC power input;
5. Auxiliary voltage source;
6. Rectifier circuit;
7. DC link;
8. Three-phase inverter;
9. Current acquisition board;
10. Voltage acquisition board;
11. Encoder signal conditioning board;
12. TMS320F28379D DSP (Digital Signal Processor);
13. Digital multimeter;
14. Incremental encoder;
15. Three-phase induction motor.

There are two basic types of inverters, Voltage Source Inverter (VSI), powered by a voltage source, and Current Source Inverters (CSI) powered by a current source. VSI topology is more common and synthesizes a well-defined voltage in the machine terminals while CSI provides a current signal in the machine terminals [5].

Figure 2 presents the simplified schematic diagram of a VSI. This setting uses a rectifier to transform the AC input signal into DC. After rectification, the signal is filtered by the DC link to obtain a signal without oscillations. Next, the DC signal is applied to the three-phase IGBT bridge with the sole purpose of creating an alternating signal at the output by switching the semiconductor frequently defined by the modulation technique employed.

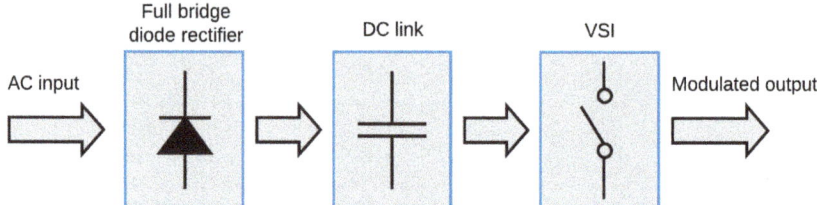

Figure 2. Diagram of a voltage inverter.

2.1. Digital Signal Processor

The TMS320F28379D microcontroller is designed for application in control systems, drive drivers, signal detection and processing. The DSP has two Central Processing Units (CPU), two Control Law Accelerator (CLA) and a maximum adjustable clock of 200 MHz [25,26].

LaunchPad features 32-bit architecture, 1 MB of flash memory, 204 kB of RAM, 14 analog-to-digital conversion channels (ADC) with 16/12-bit resolution and 14 channels designated for PWM function [25,27,28].

2.2. Rectifier

Rectifiers can be classified according to the number of phases of the input AC voltage source, i.e., single-phase or three-phase. Depending on the semiconductors used, they can be classified as uncontrolled, semi-controlled, or fully controlled. Also, concerning topology, they can be classified as half-wave or full-wave rectifiers [29]. The three-phase uncontrolled full-wave rectifier topology is common in converter drive applications because it uses diodes as the rectifying element, connected in a full-bridge arrangement.

In this configuration, the entire cycle of the alternating voltage of the power supply is rectified, providing in the voltage output a constant average value and with smaller oscillations [29]. For this work, the full-wave bridge rectifier configuration encapsulated in the KBPC3510 IC is used. This rectifier allows a maximum current of 10 A, enough to supply the consumption of the three-phase induction motor used.

2.3. DC Link

The design of the DC link capacitors considers the power of the converter load, the maximum permissible voltage variation, and the hold-up time of the load, which is defined as the time that the output voltage should remain constant in the event of a momentary fault in the capacitor bank's input voltage. The Equation (1) is used to design the capacitor bank, given by

$$C_o = \frac{2P_{rated}t_H}{0.19V_{rated}^2} \quad (1)$$

where P_{rated}, t_H, V_{rated} represent the rated power, hold-up time, and rated feed voltage of the machine, respectively. The variation of the output voltage was defined at 10% and the hold-up time of 8.33 ms, that is, a half cycle of the nominal frequency of the load. Thus, the capacitance value of the projected DC link is 4.4 mF, in which a total of eight EPCOS

B43845 capacitors of 2200 µF and voltage of 200 V_{dc} were used. The designed DC link can work with voltages up to 400 V_{dc}, enabling AC voltage connection at the rectifier input of 127 V_{rms} or 220 V_{rms}.

The DC link also contains pre-charge and bank discharge resistors. The pre-charge resistor is responsible for preventing a high in-rush current from flowing through the converter elements and damaging the components when the rectifier's AC supply circuit breaker is tripped. The single-phase variable autotransformer at the input of the rectifier circuit was also employed for this purpose, allowing the input AC voltage to rise gradually to the value of 220 V_{rms}.

The discharge resistor has the function of draining the energy stored in the capacitors when the converter is switched off, ensuring that the voltage reduces to zero slowly, also avoiding accidents. Another function of the discharge resistor is to work as a braking resistor when the machine starts to work as a generator due to the type of load it drives. In this context, the DC link voltage tends to rise due to the reverse power flow. Figure 3 illustrates the simplified DC link circuit with the pre-charge (R_{pc}) and discharge (R_d) resistors.

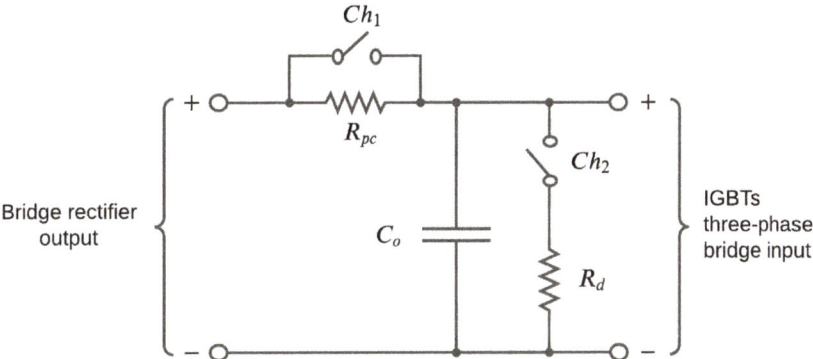

Figure 3. DC link with pre-charge and discharge resistors [30].

The AC voltage at the input of the rectifier circuit is 220 V_{rms}. Thus, the DC voltage at the output of the rectifier corresponds to the maximum value of the input voltage, i.e., 311 V_{dc}. The pre-charge resistor was chosen so that the link voltage reaches the maximum value (311 V_{dc}) in 5 s, while the discharge resistor was chosen so that the bank voltage is zero in 132 s. The Ch_1 switch is open only at the beginning of bank charging, and after the bank voltage stabilizes the switch is closed since the pre-charge resistor is no longer needed. Similarly, the switch (Ch_2) is closed when the converter is switched off, allowing the stored energy in the bank to be drained, and when in the normal operation of the converter, Ch_2 is open.

2.4. Power Module

The choice of Fairchild Semiconductor's FNA41560 driver is a consequence of the need for a compact and high-performance solution for the frequency inverter. The chip features optimized circuit protection and a combined IGBT drive to reduce losses. The module is also equipped with overcurrent protection, under-voltage interlocks, temperature monitoring output, bootstrap diodes and features gate drive circuits and internal dead-time.

The FNA41560 must be powered with a voltage of 15 V_{dc}, supports 600 V_{dc} (drain-source voltage) in the IGBT while dissipating 41 W of power in each semiconductor. The module operates with a maximum switching frequency of 20 kHz, a maximum current of 15 A at 25 °C, an insulation rating of 2000 V_{rms}/min and terminals for individual current monitoring of each phase [31].

The power module has six PWM outputs to drive the IGBTs, which can be controlled by the DSP via optocoupler ICs. The PWM modules of the DSP have complementary outputs and internal dead-time production. However, a gate drive circuit was designed to receive the command signals from the upper switches of the three-phase IGBT bridge and, from these, the command signals of the lower switches and the dead-time between the switches of the same arm are produced via hardware. In this way, configuration errors of the PWM channels and dead-time of the DSP that could damage the FNA41560 are avoided. Figure 4 illustrates the dead-time generation circuit for the U-arm of the three-phase IGBT bridge.

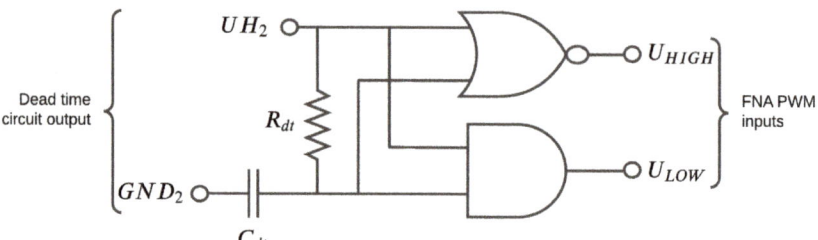

Figure 4. Circuit to generate the complementary PWM and dead-time signals [30].

From Figure 4 it is noted that the circuit to generate the complementary signals has a resistor (R_{dt}) and a capacitor (C_{dt}) to generate the dead-time between the switches of the same arm along with AND and NOR logic gates. The switching frequency chosen is 10 kHz, within the operating range of the FNA41560. Thus, the resistor and capacitor were set to guarantee a dead-time of 1 µs equivalent to 1% of the switching period.

The designed PCB of the IGBT module contains three dead-time generation circuits and complementary signals. Looking at Figure 4, when applying a high logic level signal to the UH_2 input of the first arm, the U_{HIGH} output also assumes high logic level. However, the U_{LOW} output, which drives the lower IGBT, has a low logic level with the addition of the time delay, preventing simultaneous driving of the switches. The same principle is extended to the switches of the V and W arms of the three-phase IGBT bridge.

In addition, to drive the module's switches, optical isolation circuits or optocoupler circuits were used. These are circuits used with the purpose of electrical decoupling, the PWM channels of the DSP do not have a direct electrical connection with the inputs of the FNA41560. Its working principle is based on a Light Emitting Diode (LED) that, when energized, emits a light that puts a phototransistor in a conduction state [32].

The necessity for electrical isolation comes from the fact that there are voltages and currents in the power stage that could cause damage to the control and data acquisition circuits. For this reason, the reference of the power module is different from the reference of the DSP and the data acquisition boards, ensuring the isolation of the control and measurement circuits from the power circuits.

For this project, it was chosen to use optocouplers of the HCPL2630 type. These ICs work with Transistor-Transistor Logic (TTL) logic level, considerable electrical isolation, 12 ns delay time and have two independent channels in the same encapsulation [33].

Figure 5 illustrates the circuit developed to isolate PWM signals from the DSP to the FNA41560 driver.

The PWM signals coming from the DSP have as reference GND_1, while at the output of the optocouplers (power stage), the reference is GND_2. At the output of the isolation circuit, the PWM signals are injected into the dead-time circuits and there is the generation of the complementary PWM to drive the IGBTs of the power module.

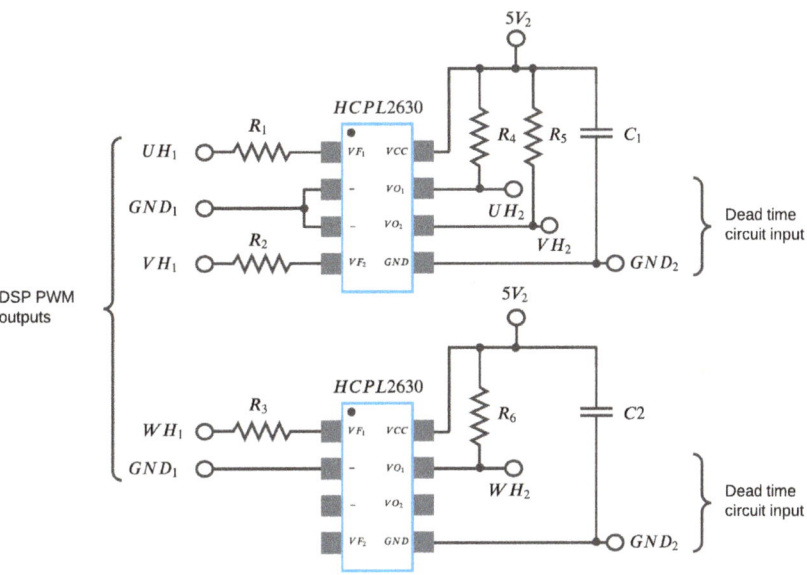

Figure 5. PWM signal isolation circuit [30].

To monitor the temperature of the power module, a voltage comparator circuit with a variable resistor was designed to adjust the maximum temperature limit of the FNA41560. The IGBT module has an internal thermistor that changes its resistance according to the temperature change. If the module temperature exceeds the set limit, the circuit containing the operational amplifier LM2904 changes the state of the output from logic level low to high. In this context, the signal at a logic high level triggers an LED as a visual alert, in addition to sending a pulse to the DSP to interrupt the PWM signals from the IGBTs.

Figure 6 presents the circuit used for under-voltage blocking recommended by the FNA41560 manufacturer. Also, a 4N25 optocoupler and a voltage divider circuit were used to transmit the signal to the DSP.

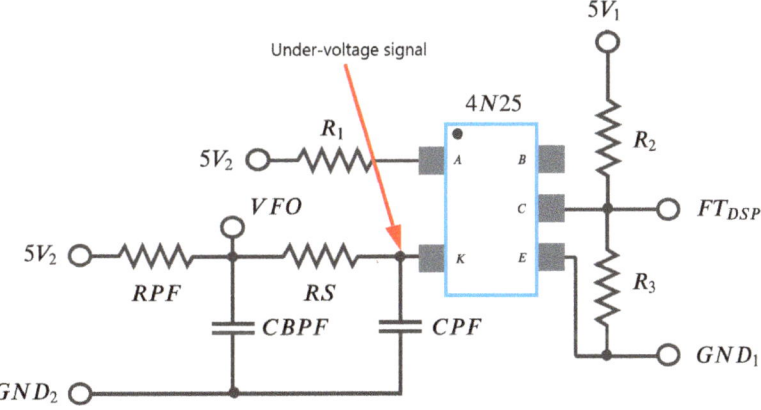

Figure 6. FNA41560 under-voltage detection circuit [30].

The resistors RPF, RS and the capacitors CPF, CBPF are defined according to the module datasheet. The resistor R_1 is used to limit the current of the optocoupler LED, the resistors R_2 and R_3 are the voltage divider resistors. The values of 200 Ω, 10 kΩ and 20 kΩ respectively have been set. The terminals "A" and "K" of the 4N24 IC indicate the Anode and Cathode terminals of the LED. The terminals "C", "B" and "E" represents the collector, base and emitter terminals of the phototransistor, respectively.

Short-circuit current detection of the FNA41560 is provided by a shunt resistor located at the source terminals of the lower IGBTs of the three-phase bridge. If the voltage across the resistor exceeds the short-circuit threshold voltage trip level (0.5 V_{dc}) at the C_{sc} input of the module, a fault signal is assigned and the lower arm IGBTs are turned off [34].

The protection circuit contains a low-pass filter, formed by resistors R_f and capacitor C_{sc}. The module manufacturer suggests values for these components in such a way that the time constant is between 1 µs and 2 µs. A 15 Ω resistor and 100 nF capacitor were used, i.e., filter time constant of 1.5 µs. A shunt resistor with a value of 0.1 Ω was sized. Thus, for the protection to operate, a current greater than or equal to 5 A circulating through the module is required.

2.5. Current Measurement

The current measurement of the converter was made using the LA 55-P Hall-effect sensor from the LEM manufacturer. Table 1 presents the characteristics of the transducer.

Table 1. LA 55-P sensor characteristics [35].

Electrical Data	Value
Primary nominal R.M.S current	50 A
Secondary nominal R.M.S current	50 mA
Conversion ratio	1:1000
Supply voltage	±12 or 15 V

When an electrical current circulates through the primary winding (I_p) of the current sensor, a current also circulates in the secondary (I_s) proportional to the conversion ratio of the transducer. The current in the secondary in turn produces a voltage (V_{in}) on the shunt resistor R_m and thus this voltage signal is fed into the non-inverting adder circuit. In the design of the current acquisition circuit, it was chosen to use the TL084 operational amplifier and the R_m resistor was set to a value of 100 Ω. The current conditioning circuit is shown in Figure 7.

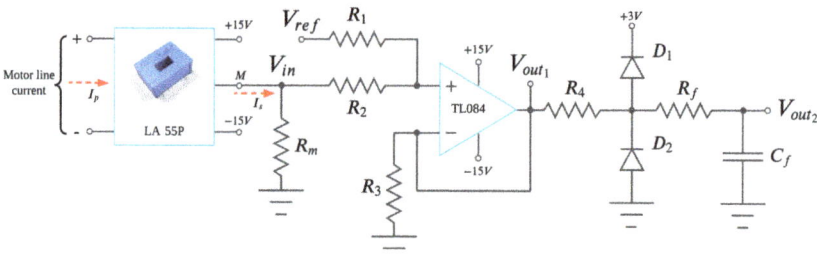

Figure 7. Current signal conditioning circuit.

The voltage signal V_{ref} is responsible for adding an offset voltage to the signal. Since the analog-to-digital conversion channels of the DSP operate with input voltages between 0 and 3 V_{dc}, an offset voltage must be added to the input signal such that this voltage represents the null measurements of the sensor, while voltages above the offset value represent positive measurements of the sensor and voltages below the offset voltage represent negative measurements. The Equation (2) presents the relationship between output voltage V_{out1} with inputs V_{ref} and V_{in}, given by

$$V_{out1} = \frac{V_{in}}{2} + \frac{V_{ref}}{2} \qquad (2)$$

Therefore, the output voltage is given by half of the reference voltage and half of the input voltage. The output signal of the non-inverting adder circuit passes through a clipper circuit, composed of two diodes and a resistor, whose function is to limit the output voltage of the operational amplifier between 0 and 3 V_{dc}.

A simulation was developed to evaluate the operation of the current conditioning circuit. Considering a current in the secondary of the transducer of 10 mA and reference voltage of 3 V_{dc}, i.e., offset voltage of 1.65 V_{dc} to ensure that the post-processing signal remains in the middle of the reading range. In addition, a signal with a frequency of 10 kHz and amplitude of 0.05 V was added to the input signal to represent the noise present in the current measurement as a function of power module switching.

Figure 8 shows the result of the current conditioning circuit simulation.

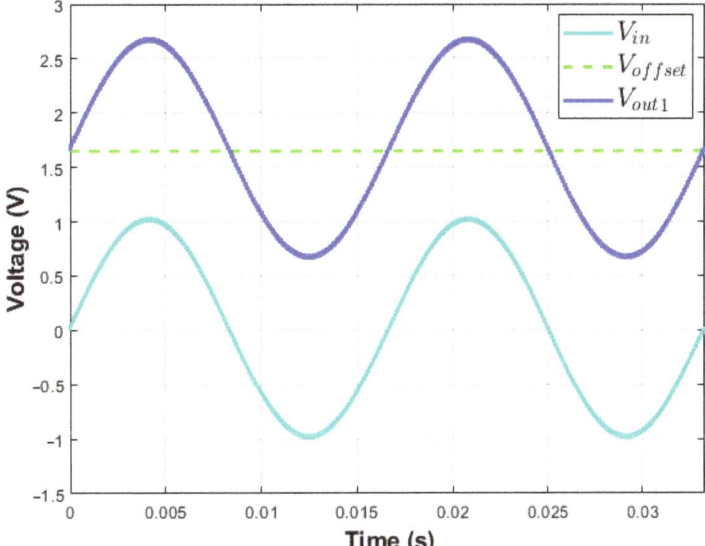

Figure 8. Simulation of the conditioning circuit of the current sensor.

By inspecting Figure 8, it is possible note that the V_{out1} voltage signal oscillates over the offset voltage value and remains between 0 and 3 V_{dc}. However, the voltage signal representing the current is not exclusively sinusoidal, it contains a high-frequency component, coming from the switching of the power module. Thus, it is desirable to remove the ripple caused by switching. The RC loop consisting of R_f and C_f, operating as a passive anti-aliasing low-pass filter was inserted into the output of the conditioning circuit, as presented in Figure 7. The Equation (3) is used to calculate the filter cutoff frequency, given by

$$f_c = \frac{1}{2 \cdot \pi \cdot R_f \cdot C_f} \qquad (3)$$

A cutoff frequency of 2.4 kHz was adopted for the filter and, thus, the value of the resistor used was 2 kΩ and the capacitor of 33 nF. Figure 9 shows the comparison between the input voltage signal and the voltage signal after the low-pass filter.

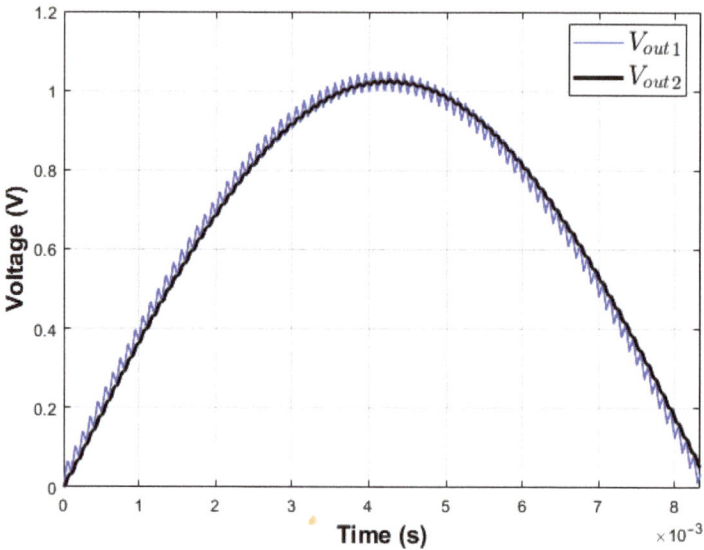

Figure 9. Simulation of the RC anti-aliasing filter.

From Figure 9 it is noted that the ripple caused by switching has been considerably attenuated by the filter used. It can also be seen that a delay has been incorporated into the signal, but not very significantly.

2.6. Voltage Measurement

The LV 20-P voltage transducer from the manufacturer LEM was used for voltage measurement. Table 2 shows the characteristics of the sensor.

Table 2. LV 20-P sensor characteristics [36].

Electrical Data	Value
Measuring nominal voltages	10–500 V
Primary nominal R.M.S current	10 mA
Secondary nominal R.M.S current	25 mA
Conversion ratio	2500:1000
Supply voltage	±12 or 15 V

When applying a voltage to the sensor measuring terminals, a current (I_p) is produced in the primary winding. Thus, in the secondary, the current (I_s) is produced due to the transducer conversion ratio. In turn, the current in the secondary produces a voltage on the shunt resistor (R_m) which reflects the voltage applied to the primary of the sensor.

The voltage conditioning circuit is like the current conditioning circuit, shown in Figure 10.

From Figure 10 it is noted the presence of resistors on the primary side of the voltage sensor. The function of this resistor is to limit the transducer primary current according to the information presented in Table 2. Thus, it was adopted for the R_1 resistor a value of 75 kΩ, enabling the measurement of voltages up to 500 V. For the shunt resistor R_m a value of 100 Ω was adopted.

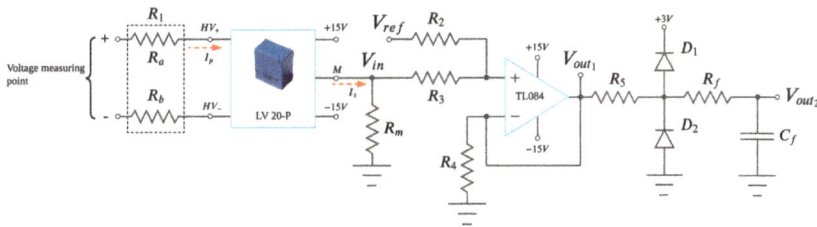

Figure 10. Voltage signal conditioning circuit.

2.7. Measurement of Rotor Position and Speed

The measurement of the machine's speed, position and direction of rotation can be accomplished through the encoder. The encoder used, called incremental with quadrature output, is a device capable of converting shaft rotation information into pulse train signals. State-of-the-art encoders mostly operate on potentiometric, capacitive, magnetic, or optical principles [37].

The incremental encoder is composed of a disk with a series of slots, an infrared light source, and a photoelectric sensor that produce the electrical pulses of channel A. The direction of rotation is obtained by the second strip of slots, channel B, positioned such that they produce an electrical 90° lag concerning channel A. The frequency of the pulse train of channels A and B are directly proportional to the rotational speed of the machine's rotor. The third channel (index) outputs a high logic level signal with each complete revolution of the encoder disk. Figure 11 shows the encoder conditioning circuit and the pulse train of each output channel.

The output voltage of the channels is 12 V_{dc}, as a function of the encoder supply which is 12 V_{dc}. To connect these signals to the DSP's eQEP (Enhanced Quadrature Encoder Pulse) inputs, voltage divider circuits were used to reduce the voltage of the three output channels. In addition, operational amplifiers were used in the voltage buffer configuration, isolating the input signal from the DSP.

The encoder model used for the development of this work is the E30S4 and has an output with a resolution of 360 pulses per revolution.

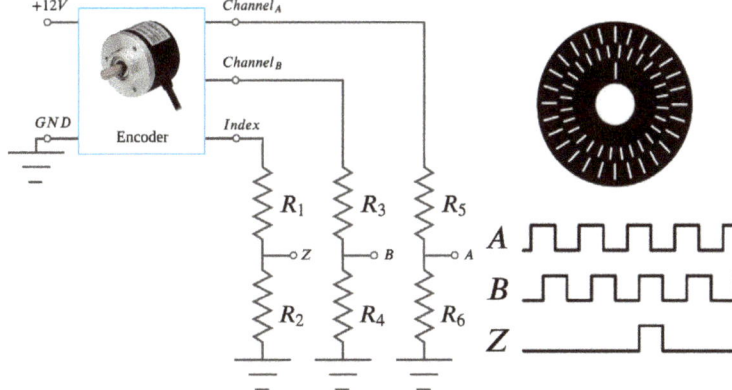

Figure 11. Encoder signal conditioning circuit.

2.8. Auxiliary Voltage Source

The auxiliary voltage source has multiple outputs: ±15 V_{dc}, ±12 V_{dc}, 5 V_{dc} and 3.3 V_{dc}. These voltages are required to power the integrated circuits and sensors of the current measurement boards, voltage measurement, encoder, and complementary power driver circuits. It is also important to mention the need to use voltage sources isolated concerning the sources that power the power driver, to prevent short circuits in the power step damage the inverter measurement and control circuits.

2.9. Induction Motor Parameters

For the experiments, an induction motor from the manufacturer Voges was used. Table 3 shows the electrical and mechanical parameters of the machine.

Table 3. Parameters of the motor used.

Board Data			
Potency	1 HP	Rotation	1730 RPM
Voltage (Δ)	220 V	I(Δ)	4.0 A
Frequency	60 Hz	Ip/In	5.4
Test Data			
R_s	7.8650 Ω	R_r	4.9925 Ω
L_s	0.1340 H	L_r	0.1340 H
L_m	0.1249 H	σ	0.1312
J	0.0012 kgm²	B	0.000397 Nms

Some parameters were collected directly from the motor board data and others were obtained through the blocked-rotor and no-load tests.

3. Dynamic Modeling of the Indirect Field Oriented for the MIT

The field-oriented control strategy is classified according to the method of acquisition of the rotor flux angle. In the direct method (Direct Field Oriented Control—DFOC), sensors installed in the machine's air gap are used to measure the flux. In the indirect method (IFOC), there is no flux measurement, the position and slip are used to obtain the position of the rotor flux angle [5].

The IFOC makes use of the fact that satisfying the relationship between slipping and stator current is a necessary and sufficient condition to produce field orientation [5].

In Field Oriented Control (FOC), one can make a direct analogy with the field-independent DC machine control [38]. In DC machines, the magnetic flux established by the armature and field currents is orthogonal to each other, regardless of the rotor's position and mechanical load. Thus, for a constant field current level, the armature current is responsible for the machine's torque production. In asynchronous machine FOC, the same principle applies, where the direct axis current is responsible to establish the machine's magnetic flux and quadrature axis current is responsible for the torque level.

Figure 12 presents the block diagram of the proposed speed control system using the indirect oriented field control strategy for the three-phase induction machine [39]. The system consists of an induction motor with a three-phase inverter, SVPWM modulation block, orientation block in which the rotor flux angle is calculated, block for the referential transformations (ABC to dq0) through Clarke and Park transformations, current control loop, speed control loop and position control loop. In this paper, only the design and tuning of the current and speed control loops will be addressed.

The equation of the dynamic model of the induction machine with indirect orientation presented in this paper is supported on the models and simplifications described in [5,39,40].

In an ideal field oriented of a three-phase induction machine, there is decoupling between the direct and quadrature axes, and the rotor flux is aligned to the direct axis (Figure 13). Thus, the flux and its derivative on the quadrature axis are null.

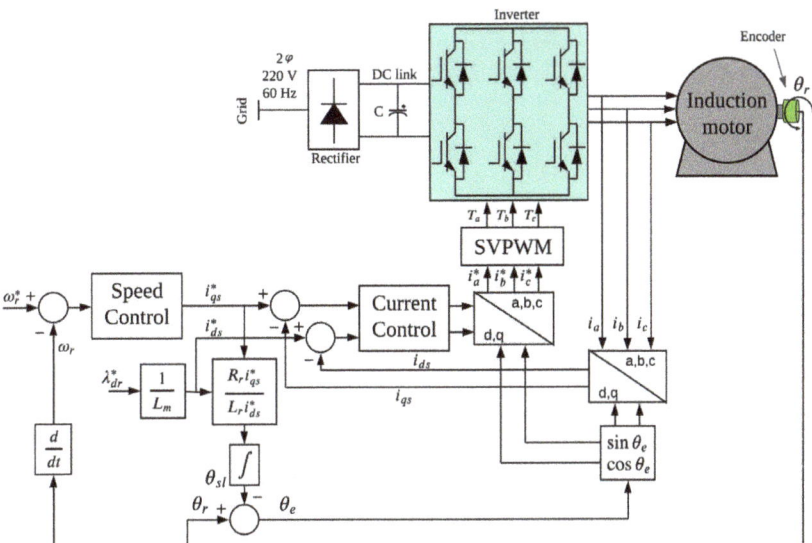

Figure 12. Indirect field oriented control scheme of induction machine.

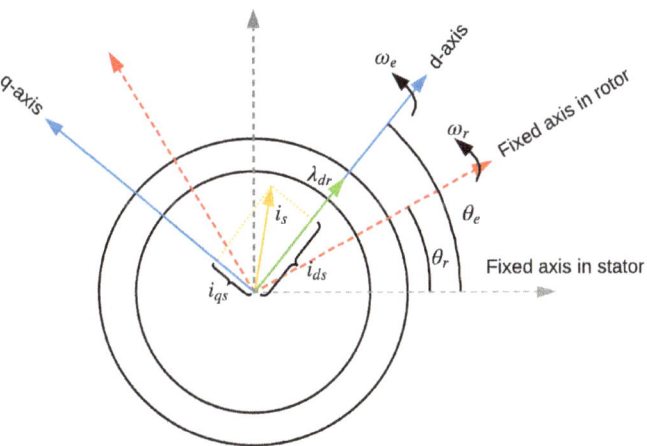

Figure 13. Analysis of the induction motor in the synchronous rotating referential.

Thus, the Equation (4) is used to calculate the rotor flux, given by

$$\lambda_{dr} = \frac{L_m i_{ds}}{1 + p\frac{L_r}{R_r}} \quad (4)$$

where L_m, L_r, R_r and i_{ds} represent mutual inductance, rotor inductance, rotor resistance, direct axis current, and p represents the derivative in time (d/dt), respectively.

The equation of the electromagnetic torque (Equation (5)) can be found considering that the electrical time constant of the system is negligible concerning the mechanical constant in the Equation (4), thus, it is obtained

$$T_e = \frac{3}{2}\frac{P}{2}\frac{L_m}{L_r} i_{qs} \lambda_{dr} \quad (5)$$

where T_e is the electromagnetic torque, P are the number of poles, λ_{dr} the direct axis flux and i_{qs} the quadrature axis current that denotes the torque command of the machine. The Equation (6) shows that the direct axis current is directly related to the magnetizing current of the machine, given by

$$i_{ds} = \frac{\lambda_{dr}}{L_m} \qquad (6)$$

In the indirect field-oriented method, the frequency needs to be calculated in dq0 coordinates. Thus, the Equation (7), allows obtaining the slip frequency, given by

$$\omega_{sl} = \frac{L_m R_r i_{qs}^*}{L_r \lambda_{dr}} = \frac{R_r i_{qs}^*}{L_r i_{ds}^*} \qquad (7)$$

The Equation (8) relates the torque, rotor velocity and angular position, given by

$$\theta_r = p\omega_r = \frac{1/J}{p + B/J}(T_e(p) - T_c(p)) \qquad (8)$$

where θ_r, J, B and T_c denote the rotor position, moment of inertia, viscous coefficient of friction and load torque, respectively.

4. Space Vector Pulse Width Modulation—SVPWM

The SVPWM is a modulation technique that presents reduced switching number and allows better utilization of DC link voltage, compared to the SPWM (Sinusoidal Pulse Width Modutalion), widely used in scalar V/F AC drives. In addition, the SVPWM presents reduced current harmonic distortion and relatively simple digital implementation [3,41–43].

Figure 14 presents the waveforms of the eight possible states for a two-level three-phase inverter. Furthermore, it is considered that at the instant when the upper switch of one arm of the converter is closed, the lower one is open and reciprocal.

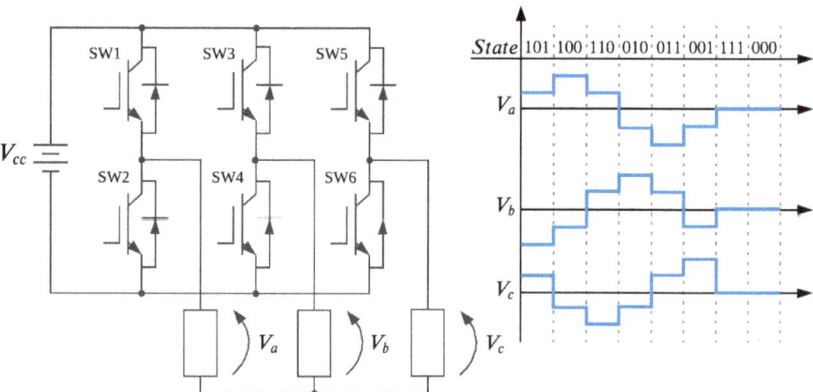

Figure 14. Single-phase voltage combinations for a two-level three-phase converter.

Each of the six nonzero states gives rise to a vector in the complex plane $\alpha\beta$ (Figure 15). The six active vectors have magnitude $2V_{cc}/3$ and lagged from each other by an angle of $60°$. The null vectors are represented at the origin of the complex plane.

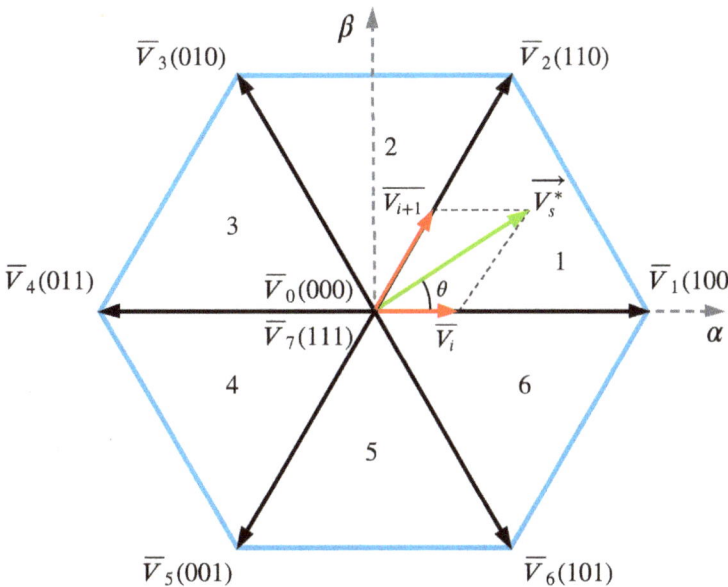

Figure 15. Hexagon of the converter output voltage spatial vectors.

To synthesize a reference voltage level ($\overline{V_s^*}$) during a sampling time interval (t_s) it is necessary to use adjacent voltage vectors and null vectors. The Equation (9) is used to calculate the reference voltage, given by

$$\overline{V_s^*} \cdot t_s = \overline{V_i} \cdot t_a + \overline{V_{i+1}} \cdot t_b + \overline{V_0} \cdot t_0 + \overline{V_7} \cdot t_7 \tag{9}$$

where $\overline{V_i}$ and $\overline{V_{i+1}}$ are adjacent vectors, $\overline{V_0}$ and $\overline{V_7}$ null vectors, t_a, t_b correspond to the length of time adjacent vectors are used, while t_0 and t_7 represent the duration of null vectors. The Equations (10)–(12) allow you to calculate the duration times of adjacent and null vectors, given by

$$t_a = \frac{\sqrt{3V_s^*}}{V_{cc}} \cdot t_s \cdot sen(\frac{\pi}{3} - \theta) \tag{10}$$

$$t_b = \frac{\sqrt{3V_s^*}}{V_{cc}} \cdot t_s \cdot sen(\theta) \tag{11}$$

$$t_0 = \frac{t_a + t_b - t_s}{2} \tag{12}$$

To avoid over-modulation in SVPWM modulation, the amplitude of the reference vector cannot be larger than the magnitude of the largest circumscribed radius of the hexagon. In SPWM the use of the DC link voltage is restricted to $V_{cc}/\sqrt{3}$. In, SVPWM the maximum magnitude of the voltage reference is $V_{cc}/\sqrt{2}$, i.e., 15% increase in the utilization of the voltage available on the DC link [44,45].

5. Controller's Tuning

Numerous types of speed controllers are available for induction motors. The Proportional Integral Derivative (PID) controller, which is widely utilized in industrial applications because of its simple design and structure [3].

To perform the tuning of the current and speed control loops, simplified linear representations of the system were used, which are based on the analysis and simplifications described in [5,46]. The values of the gains of the PI controllers were initially determined by the root locus method in conformity with performance specifications and subsequently adjusted in the real system by an iterative approach to meet the required performance.

5.1. Current Control

In the design of the current loop controller the switching is considered perfect, so the reference voltage produced by the controller is exactly equal to the voltage applied to the machine terminals. The Equations (13) and (14) relate the reference voltages on the d and q axes of the controller with the currents that circulate through the windings of the machine, given by

$$v_{ds} = \underbrace{R_s i_{ds} + \sigma L_s \frac{d}{dt} i_{ds}}_{v'_{ds}} + \underbrace{\frac{L_m}{L_r} \frac{d}{dt} \lambda_{dr} - \omega_e \sigma L_s i_{qs}}_{v_{ds,ffd}} \quad (13)$$

$$v_{qs} = \underbrace{R_s i_{qs} + \sigma L_s \frac{d}{dt} i_{qs}}_{v'_{qs}} + \underbrace{\omega_e \frac{L_m}{L_r} \lambda_{dr} + \omega_e \sigma L_s i_{ds}}_{v_{qs,ffq}} \quad (14)$$

where R_s, L_s, σ, i_{ds}, i_{qs}, ω_e, $v_{ds,ffd}$, $v_{qs,ffq}$ represent stator resistance, stator inductance, dispersion coefficient, direct axis current, squaring axis current, synchronous rotary reference system speed, direct axis feedforward compensation, and feedforward compensation of the squaring axis, respectively. Figure 16 shows the current control loop.

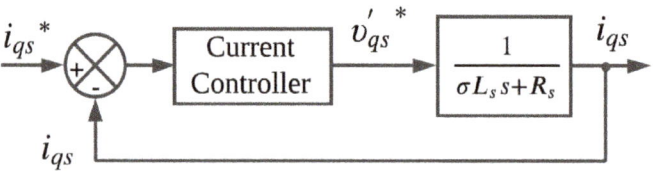

Figure 16. Current control loop diagram.

The voltage equations have coupling terms between them, in which the direct axis voltage depends on the quadrature axis current, and the quadrature axis voltage depends on the direct axis current. For control purposes, the compensation terms are considered disturbances and will therefore be disregarded for current controller design purposes. The reduced equations of the direct and quadrature axis voltages are given by Equations (15) and (16), assuming that the controller will be dominant enough to reject this error.

$$v'_{ds} = R_s i_{ds} + \sigma L_s \frac{d}{dt} i_{ds} \quad (15)$$

$$v'_{qs} = R_s i_{qs} + \sigma L_s \frac{d}{dt} i_{qs} \quad (16)$$

For the current loop controller design, the criteria presented in Table 4 were defined.

Table 4. Performance criteria for the current loops controllers (i_d and i_q).

Criteria	Value
Settling time (t_s)	0.008 s
Damping coefficient (ζ)	1.0
Overshoot (M_p)	0.0%

With the current loop model and performance criteria, the projected gains of the PI controller used in simulation environments and experimental tests are $K_p = 9.7158$ and $K_i = 4395$.

5.2. Speed Control

The mechanical model of the induction machine is linked to the electrical torque (T_e), load torque (T_c), the moment of inertia (J), mechanical velocity (ω_r) and viscous coefficient of friction, as shown in Equation (17)

$$\frac{d\omega_r}{dt} = \frac{1}{J}(T_e - B\omega_r - T_c) \tag{17}$$

Applying the Laplace transform in the Equation (17) and considering the load torque as an external disturbance, Equation (18) is obtained

$$p\Omega_r = \frac{1}{J}(T_e(p) - B\Omega_r(p)) \tag{18}$$

The quadrature axis current reference is provided from the PI speed controller, considering that the current control is perfect, and the reference current is reproduced in the machine windings, the speed controller model is presented in Figure 17, in which the relationship between the quadrature axis current and the electric torque is given by the machine's torque equation.

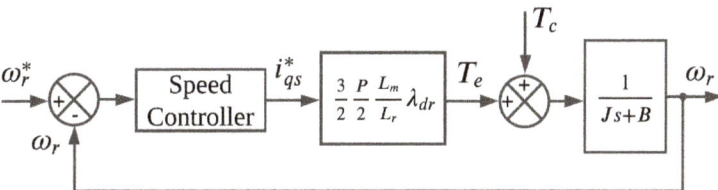

Figure 17. Speed control loop diagram.

For the speed loop controller design, the criteria presented in Table 5 were defined.

Table 5. Performance criteria for the speed loop controller.

Criteria	Value
Settling time (t_s)	0.4 s
Damping coefficient (ζ)	0.7
Overshoot (M_p)	20.0%

The projected gains of the speed loop controller used in simulation environments and experimental tests were $K_p = 0.0676$ and $K_i = 0.6939$.

6. Results

To analyze the behavior of control loops and vector control strategy, simulations were performed in the MATLAB/Simulink® software. Subsequently, the same tests were performed on the commissioned converter prototype. In this way, it is possible to examine the behavior of the converter both with regard to hardware operation and the practical implementation of the field-oriented control strategy on a real machine.

Initially, reference profiles were applied only to the direct axis current loop to evaluate the magnetization of the machine. Figure 18 presents the behavior of the direct axis loop for references of $i_d^* = 0.2$ A and $i_d^* = 0.4$ A. In addition, the bottom part shows the rotor speed of the machine.

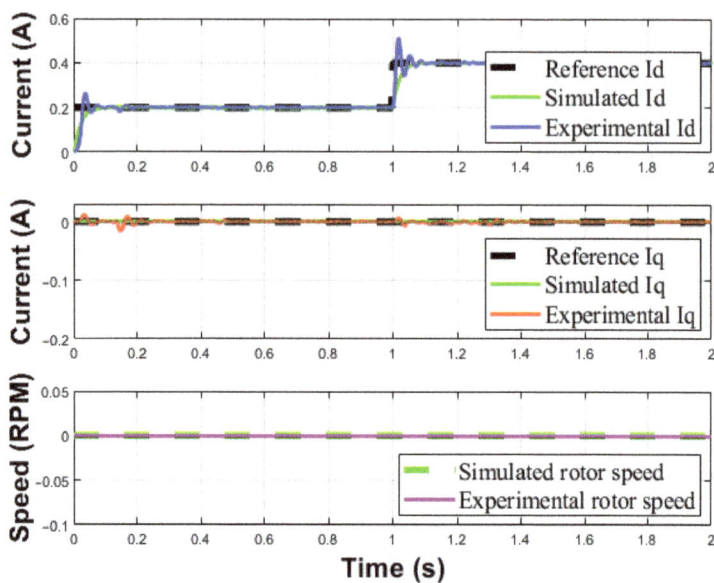

Figure 18. Current control loop behavior for $i_d^* = 0.2$ A and $i_d^* = 0.4$ A.

It is verified in Figure 18 that the controller can keep the currents at the reference values. Furthermore, it is possible to analyze the behavior of the rotor speed, which due to the presence of only magnetizing current, the rotor stands still. Table 6 presents the performance results of the direct axis current controller.

Table 6. Direct axis current controller performance.

Criteria	Simulated Value	Experimental Value
Settling time (t_s)	0.06 s	0.07 s
Overshoot (M_p)	0.0%	27%

The overshoot present in the experimental i_d signal is related to simplifications adopted for the current controller design. The closed-loop transfer function of the current plant presents a zero. This zero was disregarded to calculate the PI controller gains. Thus, in the real system, this zero will cause a proportional overshoot for fast responses, but the settling time will be preserved.

To verify the behavior of the quadrature axis current controller, reference profiles were applied in positive step of 1.0 A and negative step of -1.0 A, as shown in Figures 19 and 20, respectively.

Analyzing the Figures 19 and 20, the controller can maintain the imposed references. Still, it is possible to highlight that for positive current values in the quadrature axis, the machine rotor rotates clockwise. However, for negative values of current in the quadrature axis, the machine rotation is reversed, characterized by negative speed values.

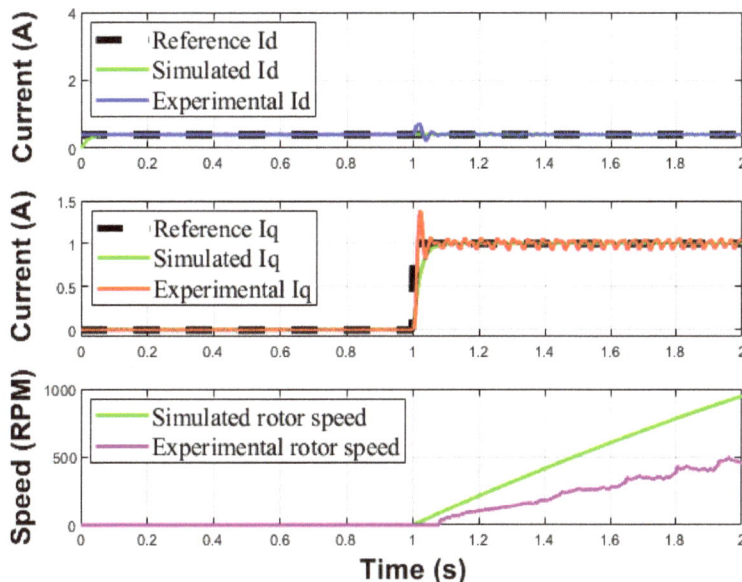

Figure 19. Current control loop behavior for $i_q^* = 1.0$ A.

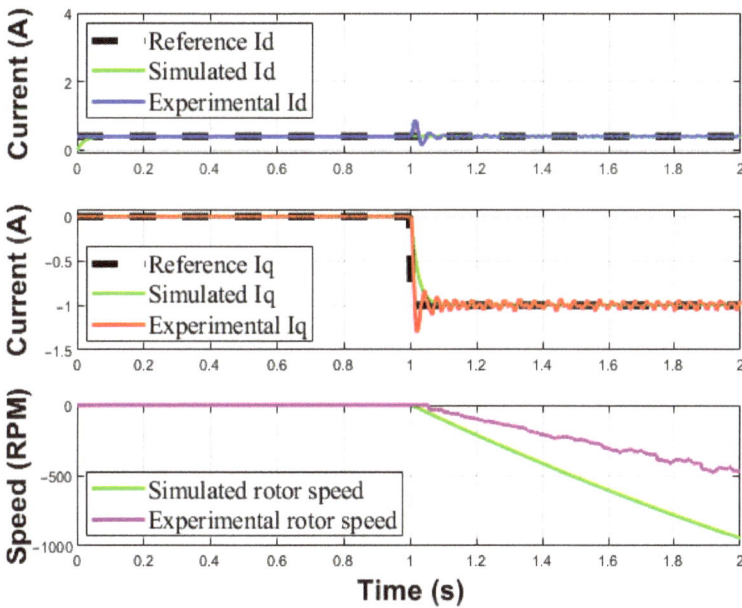

Figure 20. Current control loop behavior $i_q^* = -1.0$ A.

For both situations, the speed of the machine increases wildly, because there is no speed control loop. It is also noted that the speed of the machine in simulation assumes higher values than those collected in the experiments. The explanation for this is since in simulation many electrical and mechanical elements are considered ideal or even not present in the machine model. Thus, they do not contribute to the speed behavior of the machine. In the same way as clarified for the current i_d, the overshoot of the current i_q control loop is justified.

Table 7 presents the performance parameters of the quadrature axis current controller.

Table 7. Controller performance for reference $i_q^* = 1.0$ A and $i_q^* = -1.0$ A.

Reference	Criteria	Simulated Value	Experimental Value
$i_q^* = 1.0$ A	Settling time (t_s)	0.05 s	0.05 s
	Overshoot (M_p)	0.0%	40%
$i_q^* = -1.0$ A	Settling time (t_s)	0.05 s	0.05 s
	Overshoot (M_p)	0.0%	40%

After validating the correct operation of the current control loop, the speed control loop controller was simulated and implemented in the digital signal processor. The direct axis current reference was set at a value of 0.4 A, which represents the magnetization of the machine, and experimentally was the value that presented the best speed and torque results. Tests were performed with step, trapezoidal, and sinusoidal references. For a positive step reference, initially with a reference of 300 RPM and then increasing to 500 RPM and returning to 300 RPM, as shown in Figure 21.

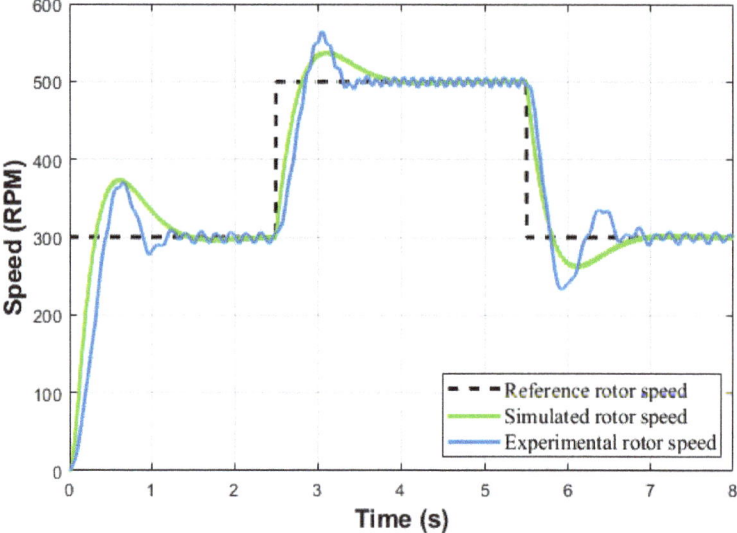

Figure 21. Behavior of the speed control loop with step reference.

Figure 22 shows the machine speed behavior for a trapezoidal reference, with a lower reference of 300 RPM and higher than 400 RPM.

Figure 23 shows the behavior of the machine speed for a sine reference with a frequency of 0.1 Hz and amplitude of 200 RPM.

Analyzing the results for the different speed references, it is evident the correct functioning of the implemented speed control loop. It is important to highlight that due to the accuracy of the encoder available during the tests, the speed signal has low amplitude oscillations. Table 8 presents the speed controller performance results for the three reference profiles.

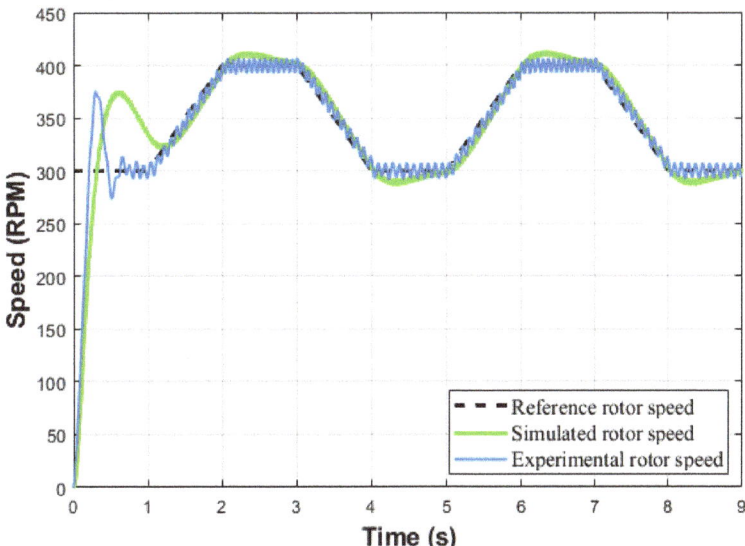

Figure 22. Behavior of the speed control loop with trapezoidal reference.

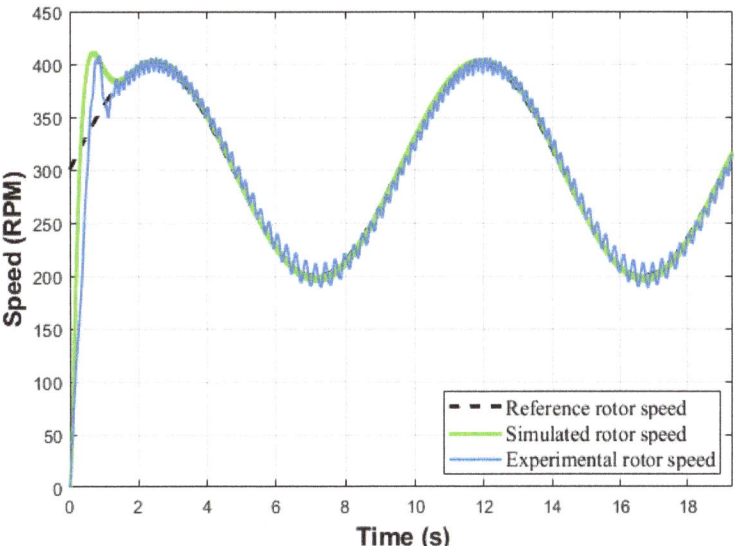

Figure 23. Behavior of the speed control loop with 0.1 Hz sinusoidal reference.

Table 8. Speed controller performance.

Reference	Criteria	Simulated Value	Experimental Value
Step	Settling time (t_s)	1.5 s	1.2 s
	Overshoot (M_p)	23%	33%
Trapezoidal	Settling time (t_s)	1.3 s	0.7 s
	Overshoot (M_p)	25%	26%
Sinusoidal	Settling time (t_s)	1.45 s	1.3 s
	Overshoot (M_p)	36%	35%

The overshoot signal for the speed loop with different speed profiles shows little difference from the performance specifications. The closed loop of the speed controller has a zero that was disregarded for the calculation of the controller gains. In the experimental environment, this zero contributes to the divergence of the peak speed from the simulated speed signal.

Due to the accuracy of the electrical parameters obtained through routine tests to make up the machine model, the dynamic performance of the current and speed controllers both in simulation and in the experimental tests did not rigidly follow the design specifications. In addition, simplifications in the current loop were adopted concerning the coupling terms that influence the actual behavior of the machine. However, the divergences present in the compliance with the performance criteria did not compromise the operation of the indirect field-oriented control strategy. The results presented validate the correct operation of the hardware, control loops and the implemented strategy.

7. Conclusions

The built AC drive presented results within the expected range. Thus, the methodology adopted for the commissioning of the power circuit and acquisition system met the intended goals. The use of the FNA41560 power module made the construction of the converter more versatile and economically viable, and it can be used for different drive applications for small induction machines.

The results show that both current and speed controllers presented similar results to the results obtained during the simulation step. To solve the divergences in performance, settling time and overshoot, verified in the tests, it is proposed to implement feedforward compensation in the current control loop. Furthermore, as the machine used is of small size, it is difficult to obtain electrical parameters with good accuracy only with conventional blocked-rotor and no-load tests.

Vector control allows you to control the torque and flux of the machine independently. However, this technique is dependent on the angle of the rotor flux and the fidelity of the machine's electrical and mechanical parameters. Thus, lines of research focus on approaches that use flux estimators or the use of controllers that are robust to parametric variations or adaptive or predictive type controllers.

SVPMW modulation proved to be an effective technique for driving the semiconductor switches of the IGBT bridge and relatively simplicity of implementation in the DSP. In addition, the use of this technique allows greater use of the voltage available on the DC link, reduced switching numbers and lower harmonic distortion in the machine line currents.

The developed test bench allows work focusing on other control strategies, activation of induction machines and PWM modulation techniques to be studied and validated.

Author Contributions: Conceptualization, R.R.G. and L.F.P.; methodology, R.R.G.; validation, R.R.G. and W.W.A.G.S.; writing—original draft preparation, R.R.G.; writing—review and editing, R.R.G., L.F.P. and W.W.A.G.S.; visualization, C.V.S., G.M.R. and F.F.R.; supervision, C.V.S., G.M.R. and F.F.R.; project administration, L.F.P. All authors have read and agreed to the published version of the manuscript.

Funding: This work has financial support from Federal University of Itajubá.

Institutional Review Board Statement: Not applicable.

Informed Consent Statement: Not applicable.

Data Availability Statement: The data presented in this study are available on request from the corresponding author.

Conflicts of Interest: The authors declare no conflict of interest.

References

1. Ustun, S.V.; Demirtas, M. Modeling and control of V/f controlled induction motor using genetic-ANFIS algorithm. *Energy Convers. Manag.* **2009**, *50*, 786–791. [CrossRef]
2. Reza, C.M.F.S.; Islam, M.D.; Mekhilef, S. A review of reliable and energy efficient direct torque controlled induction motor drives. *Renew. Sustain. Energy Rev.* **2014**, *37*, 919–932. [CrossRef]
3. Abd Ali, J.; Hannan, M.A.; Mohamed, A. A Novel Quantum-Behaved Lightning Search Algorithm Approach to Improve the Fuzzy Logic Speed Controller for an Induction Motor Drive. *Energies* **2015**, *8*, 13112–13136. [CrossRef]
4. Agamloh, E.; von Jouanne, A.; Yokochi, A. An Overview of Electric Machine Trends in Modern Electric Vehicles. *Machines* **2020**, *8*, 20. [CrossRef]
5. Novotny, D.W.; Lipo, T.A. *Vector Control and Dynamics of AC Drives*; Clarendon Press: Oxford, UK, 1996; ISBN 978-01-9856-439-3.
6. Morcos, M.M.; Lakshmikanth, A. DSP-based solutions for AC motor drives. *IEEE Power Eng. Rev.* **1999**, *19*, 57–59. [CrossRef]
7. Wai, R.-J.; Lin, K.-M. Robust decoupled control of direct field-oriented induction motor drive. *IEEE Trans. Ind. Electron.* **2005**, *52*, 837–854. [CrossRef]
8. Bim, E. *Máquinas Elétricas e Acionamento*, 4th ed.; Elsevier Editora Ltda: Rio de Janeiro, Brazil, 2018; ISBN 978-85-352-9067-7.
9. Zidani, F.; Nait-Said, M.-S.; Abdessemed, R.; Benoudjit, A. Induction machine performances in scalar and field oriented control. In Proceedings of the POWERCON'98. 1998 International Conference on Power System Technology, Beijing, China, 18–21 August 1998; Proceedings (Cat. No.98EX151); Volume 1, pp. 595–599.
10. Krein, P.T.; Disilvestro, F.; Kanellakopoulos, I.; Locker, J. Comparative analysis of scalar and vector control methods for induction motors. In Proceedings of the IEEE Power Electronics Specialist Conference—PESC '93, Seattle, WA, USA, 20–24 June 1993; pp. 1139–1145.
11. Finch, J.W. Scalar and vector: A simplified treatment of induction motor control performance. In Proceedings of the IEE Colloquium on Vector Control Revisited (Digest No. 1998/199), London, UK, 23–23 February 1998; pp. 1–4.
12. Liu, H.; Zhou, Y.; Jiang, Y.; Liu, L.; Wang, T.; Zhong, B. Induction motor drive based on vector control for electric vehicles. In Proceedings of the 2005 International Conference on Electrical Machines and Systems, Nanjing, China, 27–29 September 2005; Volume 1, pp. 861–865.
13. de Santana, E.S.; Bim, E.; do Amaral, W.C. A Predictive Algorithm for Controlling Speed and Rotor Flux of Induction Motor. *IEEE Trans. Ind. Electron.* **2008**, *55*, 4398–4407. [CrossRef]
14. Chang, G.-W.; Espinosa-Perez, G.; Mendes, E.; Ortega, R. Tuning rules for the PI gains of field-oriented controllers of induction motors. *IEEE Trans. Ind. Electron.* **2000**, *47*, 592–602. [CrossRef]
15. Lai, C.-K.; Lai, H.-Y.; Tang, Y.-X.; Chang, E.-S. Field Programmable Gate Array-Based Linear Shaft Motor Drive System Design in Terms of the Trapezoidal Velocity Profile Consideration. *Machines* **2019**, *7*, 59. [CrossRef]
16. Cabrera, R.S.; de Leon Morales, J. Some results about the control and observation of induction motors. In Proceedings of the 1995 American Control Conference—ACC'95, Seattle, WA, USA, 21–23 June 1995, Volume 3, pp. 1633–1637.
17. Shah, V. Implementation of Sensorless Field Oriented Current Control of Induction Machine. In Proceedings of the 2017 IEEE International Conference on Computational Intelligence and Computing Research (ICCIC), Coimbatore, India, 14–16 December 2017; pp. 1–5.
18. Vukosavic, S.N.; Stojic, M.R. On-line tuning of the rotor time constant for vector-controlled induction motor in position control applications. *IEEE Trans. Ind. Electron.* **1993**, *40*, 130–138. [CrossRef]
19. Moreira, J.C.; Lipo, T.A. A new method for rotor time constant tuning in indirect field oriented control. *IEEE Trans. Power Electron.* **1993**, *8*, 626–631. [CrossRef]
20. Jeon, S.H.; Oh, K.K.; Choi, J.Y. Flux observer with online tuning of stator and rotor resistances for induction motors. *IEEE Trans. Ind. Electron.* **2002**, *49*, 653–664. [CrossRef]
21. Jansen, P.L.; Lorenz, R.D. A physically insightful approach to the design and accuracy assessment of flux observers for field oriented induction machine drives. *IEEE Trans. Ind. Appl.* **1994**, *30*, 101–110. [CrossRef]
22. Pinheiro, D.D.; Carati, E.G.; Del Sant, F.S.; da Costa, J.P.; Cardoso, R.; de Stein, C.M.P. Improved Sliding Mode and PLL Speed Estimators for Sensorless Vector Control of Induction Motors. In Proceedings of the 2018 13th IEEE International Conference on Industry Applications (INDUSCON), Sao Paulo, Brazil, 12–14 November 2018; pp. 1030–1037.
23. Correa, M.B.R.; Jacobina, C.B.; dos Santos, P.M.; dos Santos, E.C.; Lima, A.M.N. Sensorless IFOC for single-phase induction motor drive system. In Proceedings of the IEEE International Conference on Electric Machines and Drives, San Antonio, TX, USA, 15 May 2005; pp. 162–166.
24. Kumar, A.; Ramesh, T. MRAS speed estimator for speed sensorless IFOC of an induction motor drive using fuzzy logic controller. In Proceedings of the 2015 International Conference on Energy, Power and Environment: Towards Sustainable Growth (ICEPE), Shillong, India, 12–13 June 2015; pp. 1–6.
25. Texas Instruments. TMS320F2837xD Dual-Core Delfino Microcontrollers—Technical Reference Manual. Available online: https://www.ti.com/lit/ug/spruhm8h/spruhm8h.pdf (acessed on 5 February 2021).
26. Mutlag, A.H.; Mohamed, A.; Shareef, H. A Nature-Inspired Optimization-Based Optimum Fuzzy Logic Photovoltaic Inverter Controller Utilizing an eZdsp F28335 Board. *Energies* **2016**, *9*, 120. [CrossRef]
27. Joy, P. Development of Software Using TMS320F28379D DSP. *Int. J. Res. Eng. Sci. Manag.* **2020**, *3*, 383–388.

28. Aravena, J.; Carrasco, D.; Diaz, M.; Uriarte, M.; Rojas, F.; Cardenas, R.; Travieso, J.C. Design and Implementation of a Low-Cost Real-Time Control Platform for Power Electronics Applications. *Energies* **2020**, *13*, 1527. [CrossRef]
29. Hart, D.W.; Júnior, A.P.; Abdo, R. *Power Electronics*, 1st ed.; The McGraw-Hill Companies: New York, NY, USA, 2011; ISBN 978-00-7338-067-4.
30. Gomes, R.R.; Pugliese, L.F.; Silva, W.W.A.G. Projeto de Dispositivo de Hardware para Acionamento de Máquinas de Baixa Potência. In Proceedings of the 2021 14th IEEE International Conference on Industry Applications (INDUSCON), São Paulo, Brazil, 15–18 August 2021; pp. 598–605.
31. Fairchild Semiconductor. FNA41560/FNA41560B2 Motion SPM 45 Series. Available online: https://br.mouser.com/datasheet/2/308/1/FNA41560B2_D-2313479.pdf (accessed on 6 June 2021).
32. de Oliveira, A.S.; de Andrade, F.S. *Sistemas Embarcados: Hardware e Firmware na Prática*, 2nd ed.; Editora Érica: São Paulo, Brazil, 2009; ISBN 978-85-3650-105-5.
33. Fairchild Semiconductor. High Speed-10 MBit/s Logic Gate Optocouplers. Available online: https://br.mouser.com/datasheet/2/308/1/HCPL2631_D-2314281.pdf (accessed on 28 May 2021).
34. Fairchild Semiconductor. AN-9070 Smart Power Module Motion SPM—Products in SPM45H Packages. Available online: https://au.mouser.com/pdfdocs/AN9070.pdf (accessed on 25 April 2021).
35. LEM. Current Transducer LA 55-P. Available online: https://www.lem.com/sites/default/files/products_datasheets/la_55-p_e.pdf (accessed on 10 May 2021).
36. LEM. Voltage Transducer LV 20-P. Available online: https://media.digikey.com/pdf/data%20sheets/lem%20usa%20pdfs/lv%2020-p.pdf (accessed on 08 May 2021).
37. Seybold, J.; Bülau, A.; Fritz, K.-P.; Frank, A.; Scherjon, C.; Burghartz, J.; Zimmermann, A. Miniaturized Optical Encoder with Micro Structured Encoder Disc. *Appl. Sci.* **2019**, *9*, 452. [CrossRef]
38. Bose, B.K. *Modern Power Electronics and AC Drives*; Prentice Hall: Upper Saddle River, NJ, USA, 2001; ISBN 978-01-3016-743-9.
39. Liaw, C.M.; Lin, F.J.; Kung, Y.S. Design and implementation of a high performance induction motor servo drive. *IEE Proc. B Electr. Power Appl.* **1993**, *140*, 241–248. [CrossRef]
40. Bose, B.K. *Power Electronics and Ac Drives*, 1st ed.; Prentice-Hall: Englewood Cliffs, NJ, USA, 1986; ISBN 978-01-3686-882-8.
41. Fang, H.; Feng, X.; Song, W.; Ge, X.; Ding, R. Relationship between two-level space-vector pulse-width modulation and carrier-based pulse-width modulation in the over-modulation region. *IET Power Electron.* **2014**, *7*, 189–199. [CrossRef]
42. Gopalakrishnan, K.S.; Narayanan, G. Space vector based modulation scheme for reducing capacitor RMS current in three-level diode-clamped inverter. *Electr. Power Syst. Res.* **2014**, *117*, 1–13. [CrossRef]
43. Zhang, Y.; Zhao, Z.; Zhu, J. A Hybrid PWM Applied to High-Power Three-Level Inverter-Fed Induction-Motor Drives. *IEEE Trans. Ind. Electron.* **2011**, *58*, 3409–3420. [CrossRef]
44. Tiwari, S.; Sahu, S.K. Space vector pulse width modulation based two level inverter. *Res. J. Eng. Sci.* **2017**, *6*, 8–12.
45. Kumar, N.; Jahangir, A.; Hoque, I. Modeling of Stand-Alone Three-phase Inverter for Various Loads using Space Vector Pulse Width Modulation. In Proceedings of the 2017 14th IEEE India Council International Conference (INDICON), Roorkee, India, 15–17 December 2017; pp. 1–6.
46. Mohan, N.; Undeland, T.M.; Robbins, W.P. *Power Electronics: Converters, Applications, and Design*, 3rd ed.; John Wiley & Sons: Hoboken, NJ, USA, 2003; ISBN 978-81-2651-090-0.

Article

A Digital Twin Architecture Model Applied with MLOps Techniques to Improve Short-Term Energy Consumption Prediction †

Tiago Yukio Fujii [1], Victor Takashi Hayashi [1,*], Reginaldo Arakaki [1], Wilson Vicente Ruggiero [1], Romeo Bulla, Jr. [1], Fabio Hirotsugu Hayashi [2] and Khalil Ahmad Khalil [1]

[1] Polytechnic School (EPUSP), University of São Paulo, São Paulo 05508-010, Brazil; tiago.fujii@usp.br (T.Y.F.); reginaldo.arakaki@poli.usp.br (R.A.); wilson@larc.usp.br (W.V.R.); romeo@larc.usp.br (R.B.J.); kha@usp.br (K.A.K.)
[2] Dry Educational Laboratories, Federal University of ABC (UFABC), Santo André 09210-580, Brazil; fabio.hayashi@ufabc.edu.br
* Correspondence: victor.hayashi@usp.br
† This paper is an extended version of our paper published in Fujii, T.Y.; Ruggiero, W.V.; do Amaral, H.L.; Hayashi, V.T.; Arakaki, R.; Khalil, K.A. Desafios para Aplicação de MLOps na Previsão do Consumo Energético. In Proceedings of the 2021 14th IEEE International Conference on Industry Applications (INDUSCON), Saint Paul, Brazil, 16–18 August 2021; IEEE: Piscataway, NJ, USA, 2021; pp. 455–462.

Abstract: Using extensive databases and known algorithms to predict short-term energy consumption comprises most computational solutions based on artificial intelligence today. State-of-the-art approaches validate their prediction models in offline environments that disregard automation, quality monitoring, and retraining challenges present in online scenarios. The existing demand response initiatives lack personalization, thus not engaging consumers. Obtaining specific and valuable recommendations is difficult for most digital platforms due to their solution pattern: extensive database, specialized algorithms, and using profiles with similar aspects. The challenges and present personalization tactics have been researched by adopting a digital twin model. This study creates a different approach by adding structural topology to build a new category of recommendation platform using the digital twin model with real-time data collected by IoT sensors to improve machine learning methods. A residential study case with 31 IoT smart meter and smart plug devices with 19-month data (measurements performed each second) validated Digital Twin MLOps architecture for personalized demand response suggestions based on online short-term energy consumption prediction.

Keywords: MLOps; digital twin; IoT; machine learning; prediction

Citation: Fujii, T.Y.; Hayashi, V.T.; Arakaki, R.; Ruggiero, W.V.; Bulla, R., Jr.; Hayashi, F.H.; Khalil, K.A. A Digital Twin Architecture Model Applied with MLOps Techniques to Improve Short-Term Energy Consumption Prediction. *Machines* **2022**, *10*, 23. https://doi.org/10.3390/machines10010023

Academic Editor: Xiang Li

Received: 30 October 2021
Accepted: 24 December 2021
Published: 28 December 2021

Publisher's Note: MDPI stays neutral with regard to jurisdictional claims in published maps and institutional affiliations.

Copyright: © 2021 by the authors. Licensee MDPI, Basel, Switzerland. This article is an open access article distributed under the terms and conditions of the Creative Commons Attribution (CC BY) license (https://creativecommons.org/licenses/by/4.0/).

1. Introduction

A smart grid enables bidirectional communication between utilities and consumers, which may be used to optimize energy usage by demand side management (i.e., demand-response). As increasing energy demand and peak of energy consumption are concerns for utilities, the demand side management enables an effective method to reduce costs of electricity, which in turn restrict the need for more investments in transmission and distribution infrastructure [1,2].

One example of demand side management is employing dynamic hourly energy prices to make consuming energy in peak hours more expensive. Even though demand response has the potential to reduce energy costs and foster more sustainable communities, investigating methods of change consumer behavior towards energy consumption management is an ongoing effort [2].

Future energy facilities for residential and industrial sectors should compose a consumption chain where the behavior of real-time energy usage will be enabled by digital platforms. These aggregated data will allow analyzing consumption, seasonality, costs and planning in terms of generation, and transmission and distribution capacity [3,4]. With this, the scenario of digital data, ready to be processed by algorithms and artificial intelligence platforms, is quite consistent with product innovations and services in this area.

Smart meters adoption for energy consumption monitoring enables analysing usage habits of home appliances. Added to the direct feedback received, user-customized services such as prediction and classification of energy consumption increase their user's energy awareness and help them reduce their electricity bills [5].

Machine learning (ML) techniques for forecasting and classification of energy consumption are broadly used both academically and in the industry [6,7]. However, academic research focuses on static or offline environments, without analyzing the degradation of accuracy over time due to unexpected changes in the behavior of the time series (concept drift) [5], the sensitivity of the configuration manual of hyper parameters, and training times and prediction of the models.

Residential energy consumption has a large dependence on time of year and temperature [6,8], resulting in concept drift that is not analyzed in experiments in static environments. It is possible to use outdoor temperature data and WiFi thermostat data to improve energy consumption prediction [9], and internal building temperatures can be predicted as well [10]. In addition, although the literature presents standardized metrics for measuring the accuracy of models, there is no consensus on the use of such metrics to measure the aptitude of machine learning systems as to its operation in online environments, rendering comparisons between solutions difficult.

The convergence of digital twin and machine learning is said to improve productivity and quality in smart manufacturing scenarios [11]. Physical appliances could adapt to operational changes in real time and forecast events based on historical data by using a digital twin. However, one of the relevant challenges to build and implement digital twins is the question of how to integrate different engineering models and foster cross-domain collaboration.

This paper has addressed the following research question in order to face the challenge of modeling real-time energy consumption data: Are there computational mechanisms that enable specialized insights from customers employing prediction models? This fundamental question generates the other questions listed below:

Research Question 1. *How do we obtain intelligent real-time database containing information from each user instead of using conventional database structures with raw data?*

The first research question demands that not only raw collected data by IoT are stored and managed in the proposed solution, but its metadata must also be included to allow energy consumption forecast customization.

Research Question 2. *How do we configure Machine Learning Prediction Services for each user that would consider the challenges of real world deployment?*

This second research question shows the need for the proposed solution to consider the constraints of a real world deployment: missing data, multiple time granularity, and diverse metrics.

This paper presents a different approach adding structural topology to build a new category of recommendation platform using the digital twin model fed with real-time data collected by IoT Sensors to improve the existing machine learning approach. Residential study cases with 31 IoT smart meter and smart plug devices with data of 19 months (measurements performed each second) were used to validate Digital Twin MLOps architecture for personalized demand response suggestions based on online short-term energy consumption prediction.

Our main contributions are related to closing the gap between machine learning models used for predicting residential energy consumption and real world deployment by presenting a solution that includes household metadata so that other systems make better use of prediction results. The results contribute to the state of the art with an approach robust to missing data with multiple time granularity.

This article is an extended version of a conference paper [12], which focused solely on MLOps tests. The text is organized as follows: Section 2 presents the related work, corresponding research gaps found in the literature, and the concepts used in our solution. The method is described in Section 3. MLOps and Digital Twin modeling results are described in Section 4, and the results analysis, comparison with related work, known limitations, and development considerations are presented in Section 5. The article is concluded in Section 6 with final thoughts and suggestions for future work.

2. Research Methodology
2.1. Research Context

Personalized recommendations concerning energy saving may be supported by specialized recommender systems. A proposal found in the literature is based on user profiling and micro-moment recommendations with a mobile user interface to foster energy saving behavior change [13,14]. The solution uses appliance-level energy consumption data collected by sensors deployed in the household to recognize micro-moments for timed recommendations. However, one shortcoming of employing user profiling with collaborative filtering is that the recommendations are not fully personalized, as they are aimed at a cluster of users and not at a specific user.

The gamified management platform application found in the literature exemplifies how gamification could be used to foster demand change based on device-level monitoring [15]. The approach was validated with four households within four months, achieving up to 30% peak period consumption. Even though it is based solely on an user dashboard (i.e., passive instead of the active method a chatbot might interact with users), it organized the platform by individual and group tasks, badges, and informative pages regarding benefits, such as CO_2 emission reductions, grid operation, and electricity bill savings.

Another work uses outside temperature prediction and smart home activity recognition models to propose a controller that concurrently considers both energy savings and comfort requirements at the same time [16]. The proposal was evaluated in four apartments, and it could achieve 5.14% Heating, Ventilation, and Air Conditioning (HVAC) energy consumption reduction over the on/off controller, while simultaneously maintaining the comfort level (i.e., maximum indoor temperature difference of 0.06 °F).

A proposal found in the literature used a digital twin to model energy providers and residences [17]. It employed a reinforcement learning algorithm to optimize smart home appliances scheduling to flatten total household energy consumption to avoid peak demands and reduce the energy bill. They used the the digital twin as a sandbox to test the optimization algorithm before enforcing it to physical devices. The solution presented 17.7% energy cost reduction for a real-life dataset.

One example application of the Digital Twin architecture is energy consumption prediction. Appliance level consumption is heterogeneous, requiring time granularity selection due to complex seasonality [18] of different house appliances. Choosing the wrong granularity might induce information loss [19] due to generalization or erroneous assumptions concerning trends and correlations with features [20].

Just as household data can be used to forecast district level consumption [6], appliance data could be used to forecast residential consumption, helping not only consumers but also utility companies. Most experiments are focused on forecasting only the total house consumption, with few studies on how to analyze and optimize appliances' energy consumption. The authors of [21] used major appliances' consumption data to increase entire house consumption forecasting accuracy. Other exogenous variables are also used as input features, such as weather [6], calendar [22], and socioeconomic and building conditions [8].

In [23], the time granularity for consumption forecasting was chosen by using the Mean Average Percentage Error (MAPE), while [8] used the Normalized Root Mean

Squared Error (NRMSE). In both cases, the normalized errors tended to favor low frequencies of granularity (hourly or daily), while resulting in greater errors for high frequencies (minutely). Conversely, non-normalized errors such as Mean Absolute Error (MAE) or Mean Squared Error (MSE) favor high frequencies to the detriment of lower ones. Thus, there is currently a research gap due to the inadequacy of using error metrics to choose an adequate time frequency, as the result might be biased according to the metric chosen.

To the best of the author's knowledge, [24] is the only study to consider real world deployment challenges on consumption forecasting by using hierarchical models when data such as weather forecasts are missing or unavailable during prediction. Despite not being focused on real-world deployment, [21] used the time taken to train and predict using the forecasting model as a metric to evaluate the trade-off between accuracy and computational resources.

One of the difficulties in comparing results between different studies found in the literature is related to the different metrics used, as observed in Table 1, as well as the various datasets considered, which all use different time granularities and experimental periods, and refer to different countries such as Australia [8], Canada [21], Germany [23], Ireland [6], Portugal [24,25], and the United States [23].

Table 1. Related studies regarding residential consumption forecasting.

Paper	# of Houses	Appliance Data	Models Used	Metrics	Time Granularity	Forecast Horizon	Experiment Period
[6]	25	no	SVR, ANN	NRMSE	30 min	1–24 h	18 months
[8]	27	no	Gradient boosting	RMSE	30 min	1–24 h	35 months
[23]	7	yes	Multiple	MAPE	15–60 min	15–1440 min	9 months
[21]	1	yes	LSTM	MAPE	30 min	30 min	24 months
[24]	20	no	Hierarchical	NMAE, NQS	hourly	24 h	12 months
[25]	93	no	ANN	R2, MAPE, SDE	hourly	1–24 h	17 months

2.2. Short Term Energy Consumption Prediction

Machine learning (ML) techniques for prediction [6] and classification [7] of energy consumption are widely used in both academia and the industry, applying different learning models, such as neural networks [26], support vector machines [8], and gradient boosting [22].

Residential energy consumption forecasting can be used to assist residents in decision making and conscious spending planning [27] and utilities in medium-scale and large-scale prediction and detection of customer consumption anomalies [28]. They can also facilitate energy transactions between prosumers in peer-to-peer (P2P) energy markets [29,30], promoting the efficient use of the power grid. Home Energy Management Systems (HEMS) can use consumption prediction as an input for predictive control models [31], assisting in planning the usage of controllable applications, such as washing machines, air-conditioning systems, and electric vehicles, in order to optimize the use of energy co-generation and financial savings for users in variable energy tariff schemes.

Unlike medium-scale and large-scale energy consumption, individual hourly consumption is more volatile, with daily consumption peaks occurring at different times. Due to this characteristic, traditional metrics for measuring forecasts, such as the mean absolute error (MAE), end up measuring only point-to-point accuracy and do not analyze temporal or shape errors.

Figure 1 shows an example of a constant forecast (F1), which does not introduce any significant value to its user and has a smaller point-to-point error than a forecast with behavior closer to the real one but displaced in time (F3). While the F1 forecast has an MAE of 0.82, the F3 forecast has an MAE of 0.99.

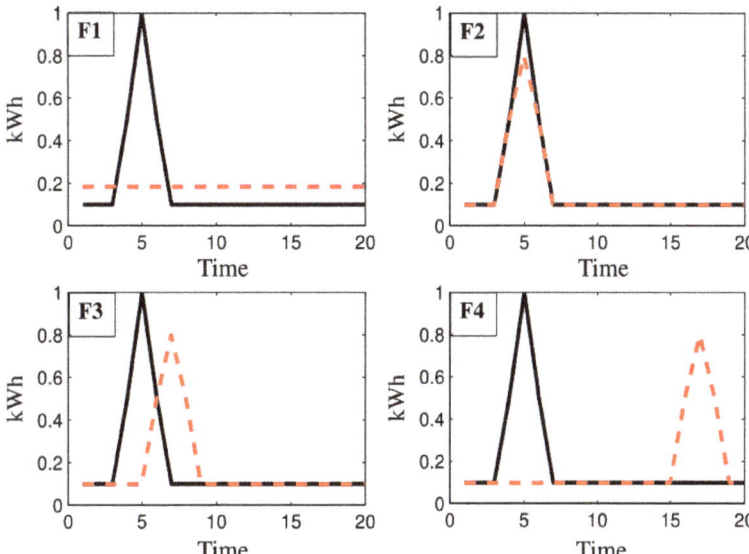

Figure 1. Four different predictions F1, F2, F3, and F4 (dotted lines) compared to the actual value (solid lines). Source: Reprinted with permission from [32].

For a satisfactory analysis of prediction models, it is necessary to use metrics that consider shape and temporal errors such as Dynamic Time Warping, Move-Split-Merge [33], DILATE [34], or the adjusted error [32].

Good forecasts are not measured only by their accuracy. Not only can different metrics can produce different results, but it is also important to consider other types of goodness, such as correspondence to human specialists judgment (consistency), similarity between forecast and previous observations (quality), and insight generation to their users (value) [35].

Additionally, the main features used to improve the accuracy of consumption forecasts in the short term (next hours or next days horizon) include weather data, such as temperature, precipitation, or wind speed, and calendar data, such as time of day, day of the week, or occurrence of holidays [6,8,24].

In order to create value for residential consumers, it is important to capture the multiple seasonalities and trends in their energy consumption. Energy Consumption has complex seasonality [18], with hourly, daily, weekly, and yearly components. In order to better analyze them, the Auto Correlation Function (ACF) plot can be used to compare similarity between time series and its lagged versions.

Most experiments performed, however, are performed in offline environments, not providing due importance to the treatment of erroneous or incomplete data in addition to the degradation of accuracy over time [24]. The use of MLOps has been deployed to address these challenges in other applications of machine learning [36,37].

2.3. MLops

In order to integrate the stages of software development and operations of information technology systems, DevOps culture uses test automation, monitoring and integration, and infrastructure management as code, among other techniques, thus allowing continuous delivery and deployment of the system [38].

The application of DevOps culture in Machine Larning (ML) systems, known as MLOps [39], seeks to adapt DevOps techniques to the area, distinguish itself from practices used in traditional software systems due to its dependence on data quality through correct extraction and processing its exploratory nature during development by testing different configurations, model architectures, and feature generation and its error monitoring derived not only from erroneous system programming but also caused by obsolete or biased models and training data.

Thus, testing systems before introducing them into production environments and monitoring their performance is considered good practice in the development and operation of software systems. However, due to their predictive nature, such practices are difficult to define and implement in ML systems [40].

Google Research uses 28 metrics to measure the readiness of ML systems in production [36]. These metrics involve tests related to 4 categories, which are input data, the model used, the infrastructure, and system monitoring. Each category has seven tests, such as ensuring privacy control for data, tuning hyper parameters, testing integration throughout the pipeline, and monitoring code dependencies.

In addition to these metrics, another good practice in ML projects is the separation of its steps into pipelines [37] to facilitate the integration of the different steps, the scalability of the system, and the reproducibility of the results.

One of the differentials of online systems is the need of continually training their models to avoid concept drift. In [41], a strategy is defined for simulating and evaluating the effects of periodic retraining in time series, finding the seasonality of the input data and updating the model at each seasonal cycle by using training and validation data that reflect the most recent cycle.

2.4. Digital Twin

One of the most critical aspects of creating a higher engagement level of human user and digital service interaction involves advanced personalization techniques. In this context, real-time data obtained from IoT devices (Technical IoT) and from humans (Human IoT) could be combined to represent digital users in both dimensions: structural/static and dynamic/behavior. The digital twin-based model might bring more engagement elements by offering helpful information with request–response interactions [42]. Instead of the Human IoT concept presented in the literature [43], which aims to develop IoT solutions focused on usability guidelines, the Human IoT is used to refer to cooperation between humans and machines, considering that IoT may enable machine–machine [44] and machine–human cooperation.

Digital modeling of a naval building, an oceanic petroleum platform, civil construction, and health care are examples of digital twin techniques for improving operational efficiency. Real-time data collected from IoT devices are mapped directly to the corresponding element digitally created in these cases.

With the digital twin model, each part of a physical structure is linked to precise data, and each behavior is recognized and registered to help in such operational procedures. Moreover, applying prediction models allow efficiency in terms of cost reduction or risk mitigation in some use cases [45–47].

How could all this be performed in the smart-home demand-side management scenario? Energy-consuming profiles can be collected and analyzed in a real-time fashion and specifically to each customer. A digital twin model organizes structural and behavioral data, which means precision and prediction information. This prediction and meta-data information may orient customers with customized suggestions to help people reduce their energy bill.

One of the alternatives to model a smart home digital twin that could be useful to our approach is by using ontologies. These semantics-related knowledge representations are understandable by humans and readable by machines. As found in the literature, it is possible to use ontologies to model a smart-home digital twin [48,49]. For example, a digital twin based on the Web of Things (W3C) description [50] is compatible with JSON format and supports SPARQL queries [48]. Other authors designed modular and independent ontologies with the Protégé editor tool to model a home automation system digital twin with the environment, equipment, resources, and their possible relations [49].

3. Digital Twin Mlops Method

3.1. Digital Twin Architecture Requirements

This project considered some requirements to prioritize efforts related to scientific and industrial lines of using a digital twin (DT) model, as described below:

1. How may it improve human–computer interaction (HCI) by applying personalization techniques considering the customer as an energy consumer, home environment, home places, appliances, and specific and distributed IoT devices to measure power consumption?
2. Considering that HCI may use natural language to implement natural language interactions, it is crucial to consider the memory aspect to create continuous and evolutive engagement levels. In this context, investigating how using a digital twin supports natural language interactions is a must.
3. How does it integrate digital twin and machine learning models to map seasonal behavior of energy consumption and to execute prediction functions and to help with energy awareness personalized suggestions?

The project requirements can be met with conventional engineering mechanisms to build human–computer Interaction (HCI). However, in this research project, the decision was to apply Digital Twin technology to facilitate integration with other emergent technologies, including IoT and Machine Learning tools, to obtain a more effective HCI. Table 2 shows some characteristics to compare digital twin to a conventional implementation. It also summarizes conclusions regarding architecture decisions in both implementation alternatives: conventional and enhanced by digital twin.

In conventional modeling, the database structure is centralized, and register fields are sufficient for adding such attributes to static data and events collected in real-time integration. The personalization configuration to HCI, Machine Learning for Seasonal and Prediction, and Natural Language Processing uses a set of user profile parameters. Note that personalization considers a set of similar profiles to deal with the trade off between volume and performance. It is an impracticable process individually for each user-customer: low performance with a substantial impact on usability. It is a crucial highlight—the centralized database is enormous in volume and not prepared for individual access and processing—that the processing balance is performed by grouping users with the same profiles to process a set of registers. For customized interactions, seasonal profile analysis, and energy-consumption prediction, the same rationale is valid. Therefore, personalization is limited to similar profiles parameterization.

The digital twin applied to this research project is different in crucial aspects, and the results are more effective in terms of personalization in general with a positive impact on all requirements listed. The first difference appears in modeling. Each user and his/her home, places, appliances, and IoT devices correspond to a digital twin that is different from conventional implementation, whereby the centralized database includes registers for both static and temporal series of events.

Software objects with data and functions organize and implement each user energy-consuming database; that is, software objects are connected with abstraction: User-customer is connected to home; it connects to place-spaces of home; it connects to each family member; and it connects to each appliance and to each IoT device. The organization forms a structural ontology that supports all personalized natural language interactions and all prediction functions in each user database.

Table 2. Architectural aspects comparison using the Digital Twin.

Aspect	Conventional	Using Digital Twin
Database entities (energy consumption data)	Energy-consumption registers Database entities: 1. House; 2. User (energy consumer); 3. Family (people and energy consumer); 4. Convenience (places of the house); 5. Electrical appliances; 6. IoT devices.	Digital space implementing real world elements as software objects from Digital Twin mechanisms includes the following: 1. A digital object of the house; 2. A digital object of user-consumer; 3. A digital object of each family member (energy consumers); 4. A digital object of each convenience (each one with energy consumer devices); 5. A digital object of electrical appliances; 6. A digital object of each IoT device.
Data titular and controller (According to the Brazilian's LGPD law)	All data stored as registers of a centralized database, and operated by the digital platform. Each data titular can go along only with his/her registers processing.	Each digital twin stores the corresponding data collected. According to law requirements, the user owns data and acts as titular and controller, in cooperation with the platform that acts as a data operator.
Real-time data (IoT)	Registers on database (events): 7. Timestamp; 8. Measurements; 9. IoT devices; 10. Relationship (House, User-consumer, Convenience).	Digital software objects register their own collected events: Convenience(4), Electrical appliance(5), and IoT device (6): 7. TimeStamp (4) (5) (6); 8. Measurements (4) (5) (6).
Natural language for information obtained from the user	The centralized database registers provides information in the right column table.	Each digital software objects register data provided by the user, connecting directly to digital twin implementation (house, user-consumer, convenience, electrical appliance, and IoT device)
Machine Learning (e.g., Chatbot)	Historical data include talking with limited and centralized memory. The conversational interaction is almost repetitive and focused on a set of users profiles.	With the digital twin, all interaction and memorization connect the correct user. In this implementation, the more historical data, the more maturity accomplished.
Machine Learning (Prediction)	Machine Learning implemented for: 11. Learning and showing seasonal information, using events data; 12. Predicting energy-consuming data, using events data. All data parametrization refers to sets of the same profile user.	Machine Learning parameters mapping directly for each digital twin software objects; that is, all data and all objects relations of real-world elements (house, user, family, conveniences, appliances, and IoT): 9. Learn and show seasonal data with precision and helpful information; 10. Predict energy-consumption data with precision according to all parameters related to his software objects.
Data organization	Huge centralized database, where sets of user-profile foundation to process intelligent services.	A federation of databases. Each database corresponds to one group of user-consumer implementing digital twin of real-world energy consumers.

IoT devices collect each event, and each energy-consumption datum corresponds to one user-consumer software object. Conversational interactions may be supported by the memory associated with this user-consumer software object. This implementation uses a NoSQL implementation tool and organizes one database for each energy-consumption user as detailed, with a digital twin implemented as a set of software objects. With this architectural decision, personalization is superior to the conventional approach.

A machine learning parametrization procedure is superior when compared to the conventional approach because all data applied are corrected to their own control: user-customer structure data (home, family, convenience, appliances, IoT devices) and time series. With this, seasonal modeling is accurate and valuable according to the data provided with respect to energy-consumption prediction, considering that the parameters show more precision than the conventional implementation.

Digital Twin adoption brings research challenges, especially for industrial applications. One relevant aspect is data organization. As described in Table 2, the database federation is the foundation of data organization, whereby each user-consumer is the owner—there is no centralized and colossal database.

In this context, other opportunities arise about data usage: the European GDPR [51] (General Data Protection Regulation) and Brazil's LGPD [52] (Lei Geral de Proteção de Dados) are laws regarding data privacy. This database federation creates the condition to enable user empowerment as a data principal and controller. In this case, the platform acts as a data operator service, providing user autonomy and coverage.

3.2. Smart Home Testbed

The smart home testbed was based on data collection architecture of energy consumption presented in [3], implemented in the early 2020 in four Brazilian households.

Specifically, the digital twin Proof of Concept for this work is built upon a household with four inhabitants and 31 energy consumption time series collected with smart meter and smart plug IoT devices.

The smart meters have a data collection system tolerant to connection failures, ensuring the integrity of data during network outages through the connection with an intermediary for data temporary storage [53].

The hourly consumption and internal temperature data of the residence are sent to a remote database, which are used by the proposed solution to forecast energy consumption and train machine learning models.

Currently, the database has information on the period from January 2020 to December 2021, with a gap from January 2021 to April 2021 due to modifications made to the smart meters used, resulting in a total of 19 months of data. Figure 2 shows load profiles of minute granularity in a weekday from different appliances monitored, such as television, refrigerator, computer, air conditioner, and living room light bulbs.

3.3. Mlops

Each stage of the pipeline has multiple steps, as shown in Figure 3. In the offline environment, tests were performed for prototyping models and experimenting new functionalities.

The online pipeline, on the other hand, albeit similar to the offline environment, has differences regarding the degree of automation, runtime constraints, and error handling. In the first stage of the pipeline, which only occurs in online environments, the automated search for data is performed, either in internal databases or through external interfaces, requiring the correct handling of exceptions due to unavailability or transfer errors. The next pre-processing step includes feature cleaning and engineering, in addition to the treatment of anomalies and missing values performed manually in offline environments.

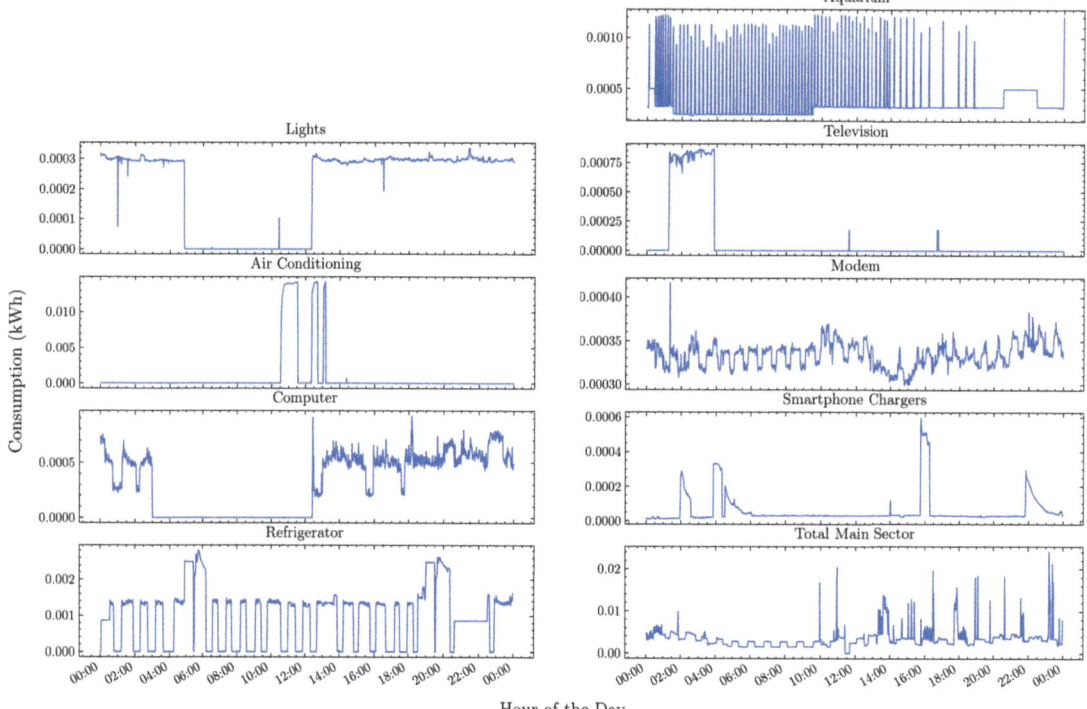

Figure 2. Daily load profile of different appliances during a weekday.

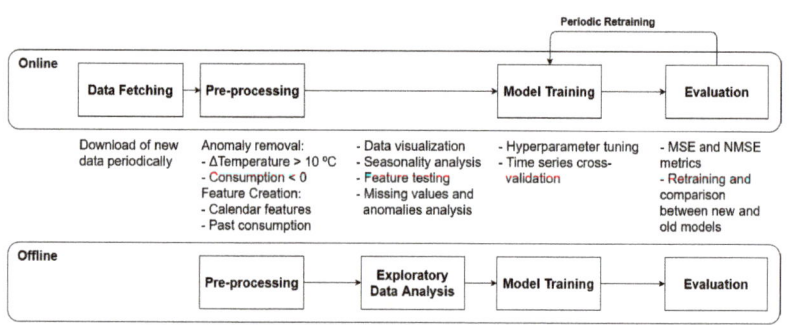

Figure 3. Online and offline pipelines for machine learning projects.

In offline environments, exploratory data analysis is then performed, in which data familiarization, anomaly detection, and distribution and correlation analysis between features occur in order to iteratively refine the previous pre-processing step. In the next step, the model is built by defining its hyper parameters, either manually or automatically through grid search, and trained according to available data.

Finally, in the model evaluation step, its accuracy is measured, and the hyper parameters that optimize the defined metric are selected. Thus, it is important to analyze and to choose which metrics will be the most appropriate and relevant to the problem.

3.3.1. Data Loading

Input data are received from cloud storage service, which stores the total energy consumption and by sectors, as well as the internal temperature of the residence, with a sampling frequency of one hour. Every hour, files are searched for in the cloud, and if the files not present in the local files are found, they are downloaded to the local directory.

3.3.2. Pre-Processing

During pre-processing, raw data are checked for missing hours and anomalies. To be considered an anomalous value, the consumption must be less than zero, and temperature must vary by more than 10°C from the previous value.

There were occurrences of temperature anomalies in which variations in relation to the previous hour exceeded 20 °C, as well as instants with missing temperature and consumption readings. In these cases, it was assigned as a reading error, and these values were discarded.

For defining features, three past hourly consumptions were added, referring to 25, 24, and 23 h ago in relation to the instant to be predicted, in addition to calendar-related attributes, such as the time of day, day of month, and month of the year to be predicted. The choice of these features for the final prediction model was made in the exploratory data analysis step.

However, new features can be added by modifying input files, such as adding the internal temperature of the residence, or also generated by modifying the source code of the pre-processing step, such as adding the first derivative of hourly energy consumption. In this manner, the other steps of the pipeline do not need further modifications.

3.3.3. Exploratory Data Analysis

In [3], energy consumption prediction models were developed by using Extreme Gradient Boosting (XGBoost), long short-term memory neural networks (LSTM), and support vector machines (SVM) architectures. The results showed that the XGBoost architecture obtained better accuracy in most of the monitored households, and this architecture was chosen for this study.

XGBoost is an open source ML library for regression and classification models using decision tree ensembles [54]. Its implementation allows training models in a parallelized and distributed fashion. The models also accept the existence of missing values in the input data in both training and prediction stages.

In order to analyze the gain of introducing new features, a base reference model was deployed, using only the last 24-hour consumption and calendar data: time, day of the week, day of the month, month, and year. This reference model was compared with other models with additional features in addition to those used in the reference model, as shown in Table 3. Cross-validation was used for each household, obtaining the mean squared error (MSE) and the adjusted error with a 2-hour window and norm 4 [32] from all households. Table 3 also shows the percentage reductions of MSE and the adjusted error relative to the reference model.

Table 3. Analysis of the addition of features on model accuracy.

Model	MSE	Adjusted Error
Reference Model	0.0479 (0%)	0.2535 (0%)
Reference + 1st Derivative	0.0476 (−0.62%)	0.2529 (−0.24%)
Reference + Indoor Temperature	0.0502 (+4.80%)	0.2568 (+1.30%)
Reference + Consumption from 25 and 23 h ago	0.0415 (−13.36%)	0.2043 (−19.40%)

The addition of the residence internal temperature as a feature of the model eventually reduced its accuracy, while the use of the first derivative of the energy consumption had little significant gains. Due to these results, these features were not considered in the final model.

Low weekly correlation was observed for all residences, with no great variation between weekdays and weekends, as shown in Figure 4 for one of the residences. Note that the consumption data refers to the year 2020, and this low variation may be related to the quarantine period due to the COVID-19 pandemic. This effect confirms what is presented by [55], in which consumption during weekends was higher than in weekdays for the residential sector in 2018 and 2019 but had similar consumptions for weekdays and weekends in 2020.

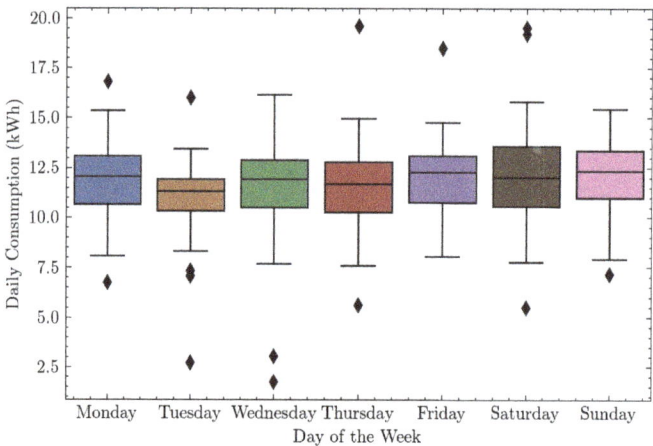

Figure 4. Box plot of the total consumption per day of the week for one of the households.

Energy consumption was higher during winter, as observed in Figure 5, due to the increased usage of air conditioning. A greater variance of consumption can be observed during the summer, although its median is similar to other seasons. This could be explained by greater air conditioning usage, as well as holidays and inhabitants absences.

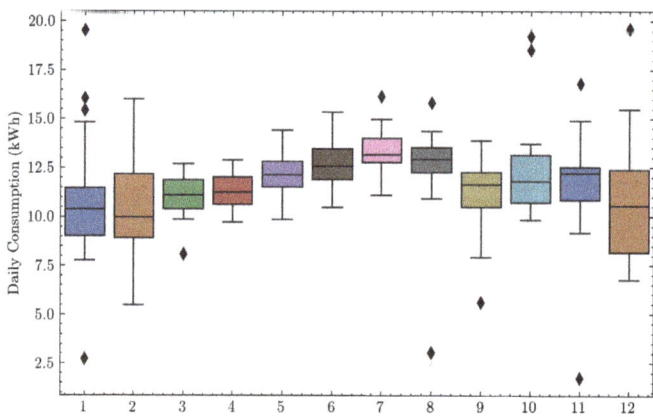

Figure 5. Box plot of the daily consumption for each month for one of the households.

Figure 6 shows the autocorrelation function of sampled energy consumption with hourly frequency for one of the monitored households. In the plot, a larger value on the ordinate axis indicates high correlation between the time series and the series lagged in time by k units, with k represented by the abscissa axis. One can observe autocorrelation peaks for 24-h lags, evidencing daily seasonality.

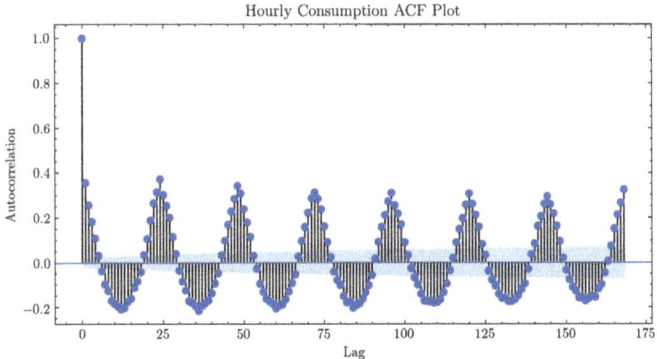

Figure 6. Autocorrelation function of the hourly energy consumption (95% significance band).

Figure 7 shows the first decision tree of the model, which observed the relevance of the time of day in model prediction, while Figure 8 shows the importance of the features for one of the households calculated as the number of times each feature appeared in XGBoost's decision trees. The possibility of obtaining information related to the internal structure of the model is important as it allows debugging the operation and investigating performance drops or instability.

The models are also evaluated with extreme or even invalid inputs, assessing their robustness. The inputs tested are as follows: consumption equal to zero, negative, infinite, and with missing values.

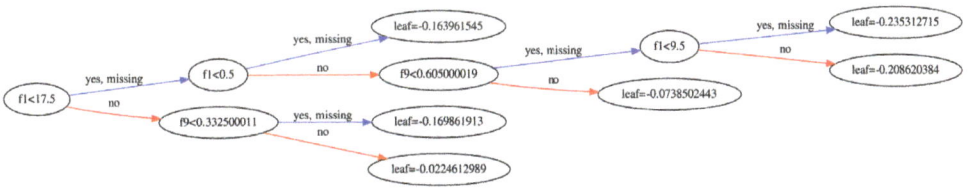

Figure 7. First decision tree for XGBoost prediction model.

3.3.4. Model Training and Prediction

The model uses the XGBoost library to predict the hourly consumption of the next 24 h, and it is trained with consumption features of the last 23, 24, and 25 h, the time of day, day of the week, day of the month, day of the year, and month.

In order to perform XGBoost hyper parameter tuning, a grid search is performed with cross validation with partitions of four subsets for each household, varying tree size, learning rate, and objective function to be minimized. After training the models for each combination of hyper parameters, the one with the smallest mean square error is chosen. The random seed used by XGBoost is fixed automatically, ensuring the reproducibility of results. Both data and code are versioned via Git and DVC version control systems.

Figure 9 shows an example of energy consumption prediction for one of the households performed during the month of July, and it is possible to observe the daily seasonality of energy consumption.

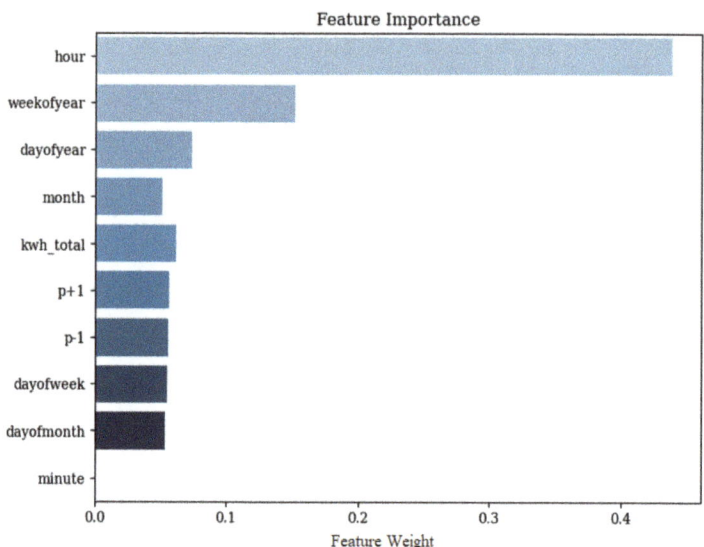

Figure 8. Importance of features for the prediction model of one of the project residences.

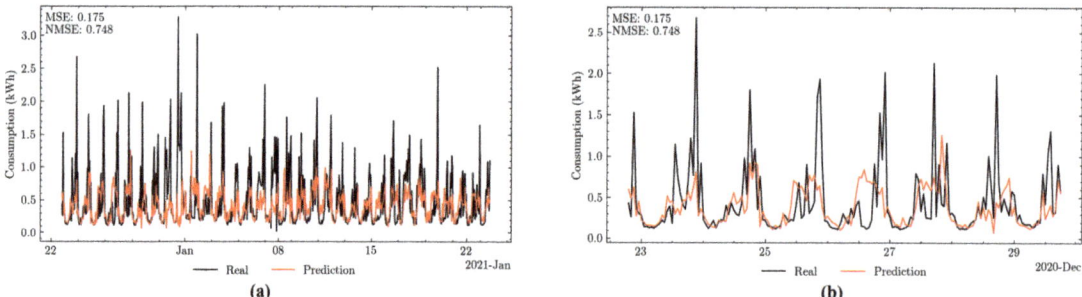

Figure 9. (a) Prediction (red) and actual value (black) of energy consumption for the months of December 2020 and January 2021. (b) Zoom in on the first week of the test data.

3.3.5. Inference

The system in online environment was deployed as a Flask application on Apache2 server hosted on Amazon Elastic Compute Cloud (EC2), performing retraining periodically every 24 h and permitting the reception of calls in REST API format for the consumption forecast of households monitored.

The API can be used by other systems to query users' consumption forecast. Figure 10 shows an example of an application, whereby a website was developed in Dash platform [56] to perform consumption forecasts in user-customizable time periods.

The calls made to the API and training time are monitored and saved in log files. When an anomalous value is encountered, as defined in Section 3.3.2 (negative consumption or temperature variation greater than 10 °C), an alert is added to the log files.

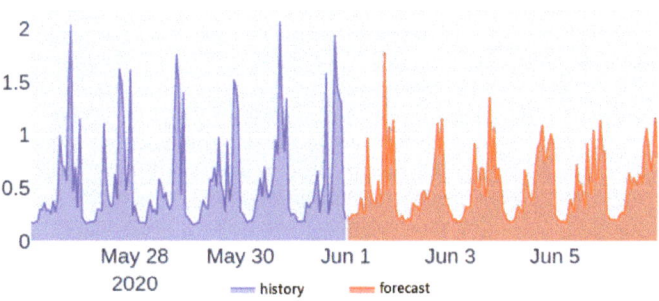

Figure 10. Website for visualizing consumption forecasts.

3.3.6. Evaluation

In order to analyze the effects of time granularity, different house appliances were evaluated using multiple time granularities. The prediction result for each combination of appliance and granularity is compared by using both error metrics and ACF for seasonality analysis.

Accuracy evaluation in a static environment is performed using the method proposed in [41]. In this method, multiple models are trained, each based on training data from different instants, to reflect the arrival of new data in an online environment.

Daily training seasonality is considered, with data partitioning 80% for training and 20% for testing. The hyper parameters are set by means of grid search. The adjusted error is used to compare the updated model with the previous one, and the one with the lowest error is used.

4. Results

4.1. Mlops Tests

Table 4 shows the tests performed by the system automatically following the metrics defined in [36] and whether they were performed autonomously (A), manually (M), not performed (-), or are not applicable (N/A).

Data Tests 4 and 5 are not applicable to the project in the current status as there is no personal data collection that allows identifying them for privacy concerns of users. Model 2 test does not apply because they are not currently monitored online metrics. Test Infrastructure 6 is not applicable due to the insufficient number of users to launch new versions (rollouts), nor is Monitoring Test 3 because there is no difference between offline and online training data.

Table 4. Tests related to data.

Category	Test 1	Test 2	Test 3	Test 4	Test 5	Test 6	Test 7
Data	A	M	M	N/A	N/A	A	M
Model	A	N/A	A	A	M	M	-
Infrastructure	A	-	-	A	M	N/A	-
Monitoring	-	A	N/A	A	M	A	-

Due to the relatively low complexity of the pipeline and low retraining cost, no tests regarding integration (Infrastructure 3 test) and rollback (test Infrastructure 7) were performed. Since the XGBoost library already performs a large series of unit tests to ensure correct code execution for training and predicting models, the verification of the model specification was considered as outside the scope of the project (Infrastructure 2).

As there is a low number of users at the moment, the project has not yet addressed issues of social inclusion of the system (test Model 7). When new users are invited to participate, representativeness of the Brazilian population will be important so as not to bias the system.

So far, there have been no changes in the structure of the input data; thus, monitoring changes (Monitoring test 1) are not currently performed, although in future steps if new features obtained from external sources, such as the weather forecast, are entered, this test will be of greater importance.

Since the model forecasts consumption for the next 24 h, real-time monitoring of the quality of forecasts made (Monitoring test 7) was not performed, as its accuracy can only be measured 24 h after the forecast.

Exploratory data analysis proved to be extremely important, satisfying several tests (Data 2, Model 5, Infrastructure 5, and Monitoring 5), which, despite performed manually, can be reused in the future for additions to the pipeline running automatically.

4.2. Use of Digital Twin Data to Improve Forecasting Accuracy

As mentioned in Section 2, using error metrics to choose the most adequate time frequency for prediction model training might generate biased results depending on which metric is chosen.

In order to analyze if these results, from entire residence consumption, also applied to appliance level consumption, forecasting models were trained for nine different appliances—lights, air conditioning, computer, refrigerator, aquarium, television, modem, smartphone chargers, and total main sector.

Each appliance was trained with data from five different time granularities—1 min, 15 min, 1 h, 6 h, and 1 day. Thus, a total of 45 models were trained. The forecasts were evaluated by using MSE and NMSE metrics. Figure 11 shows the mean results for each metric. It can be observed that better results are achieved for higher time series frequencies when normalized metrics are used, while better results for lower frequencies are achieved with non-normalized metrics, confirming what was observed in [8,23] with total residence consumption data.

A possible solution for time granularity selection for appliance level forecasting is using Digital Twin house metadata to categorize appliances by ACF plots and analyzing their prevailing seasonalities. This solution allows the scalability and customization of forecasts according to specific digital twin models and improved quality and value for the user, as defined by [35].

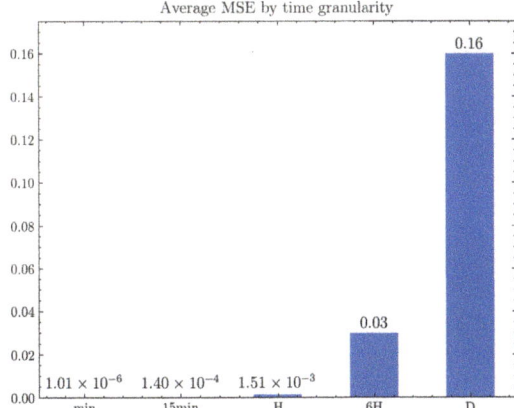 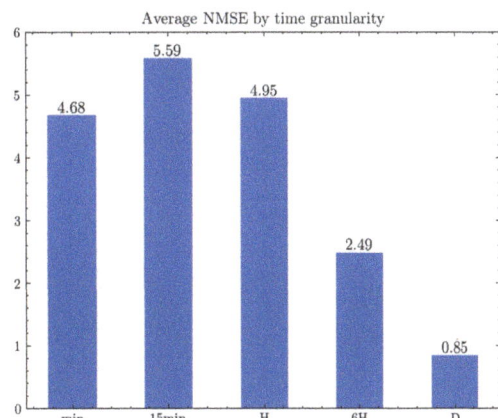

Figure 11. Average MSE and NMSE for 1 min, 15 min, 1 h, 6 h, and 1 day time period granularities.

Figure 12 shows different seasonalities for light energy consumption. Figure 13 shows their respective forecasts in different frequencies. From these results, choosing an adequate frequency is important for improving consistency, quality, and value, as defined by [35], as well as for avoiding information loss [19].

The minute and hourly frequency predictions show little temporal and shape dissimilarity when compared with daily data. There is daily seasonality present in the data, as observed in the ACF plot, which can be used to select the most adequate frequency for forecasting. Thus, in order to assist and automate this decision, the appliance classes retrieved from the digital twin model can be used in conjunction with ACF plots to select time frequencies to optimize information value for users by consumption forecasts.

4.3. Digital Twin Ontology

Figure 14 presents the smart home digital twin. It comprises persons, home, facility, room classes, and subclasses. The instances are related to the household used in the proof of concept. The four individuals live in Household ABC, which is an instance of Home. The Home class has the Room subclass, which is related to Household ABC instance. There are four household energy consumption sectors, all related to Household ABC. Each person may have a relation of private or shared room, and a device has a relation installed in some room. All these relations are illustrated in Figure 15. Additionally, each device has a data property describing its MQTT Topic, which is the Publish-Subscriber protocol used in smart home implementation.

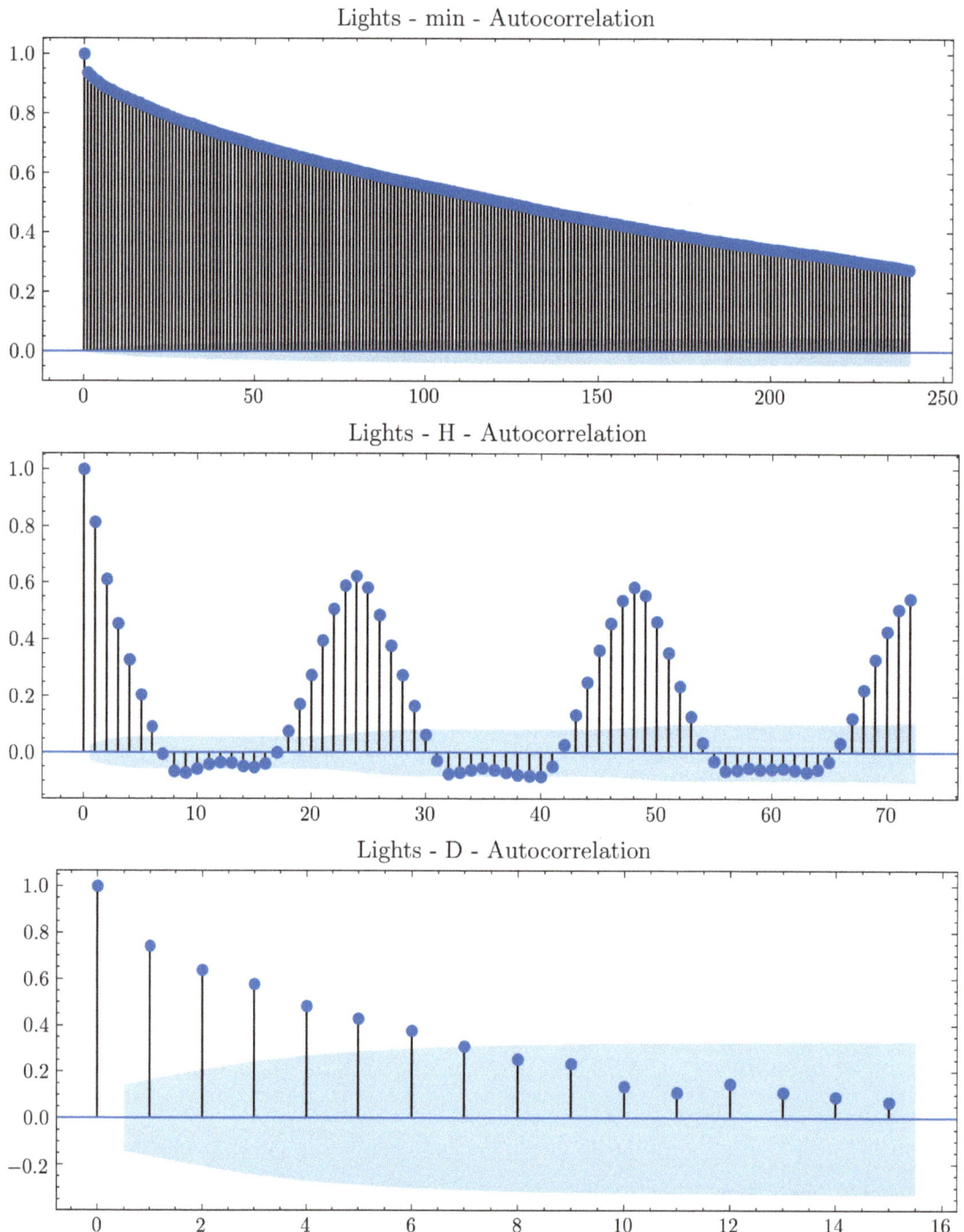

Figure 12. Lights ACF plot for minute, hourly, and daily frequency data.

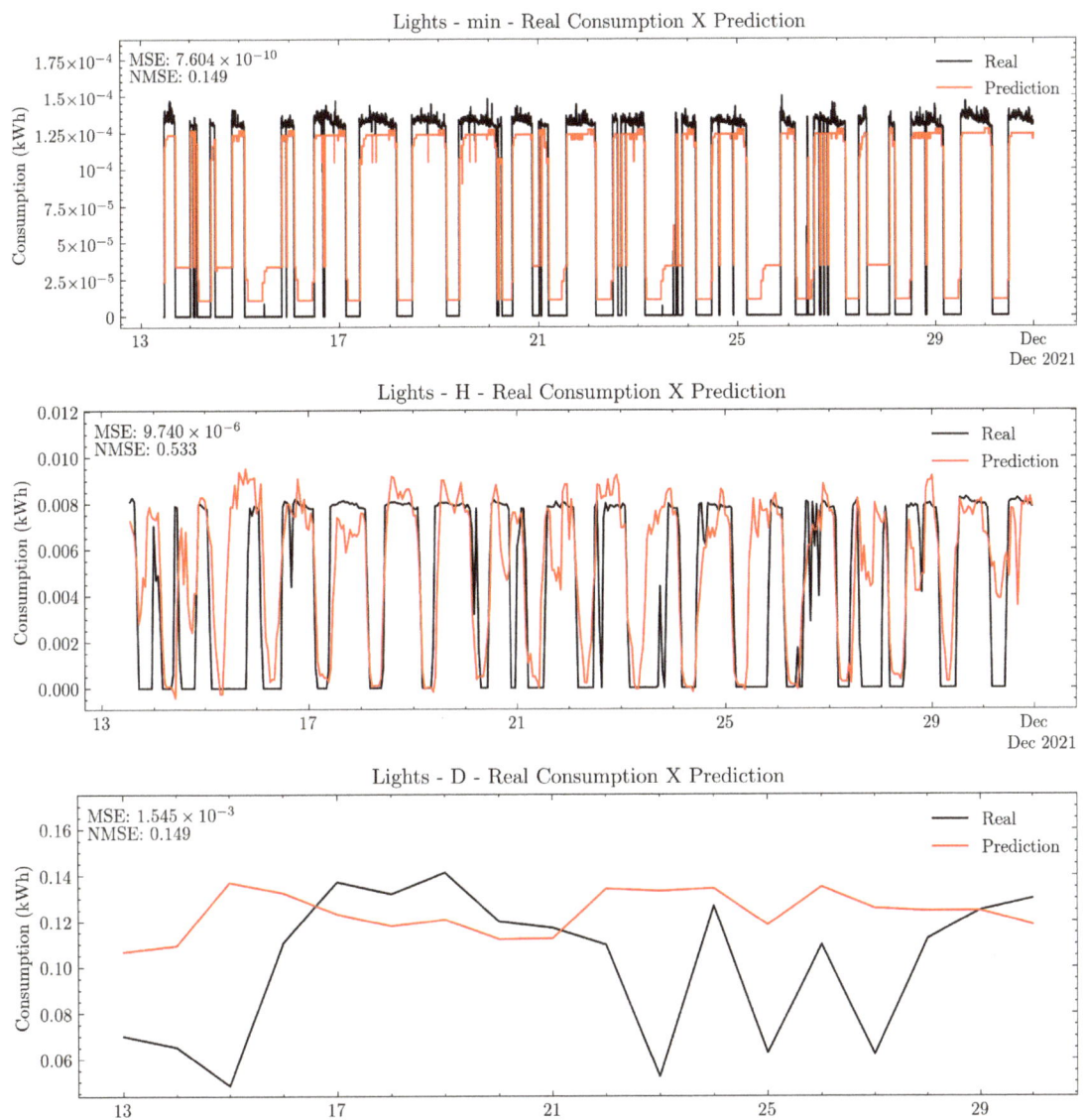

Figure 13. Lights forecast for minute, hourly, and daily frequency data.

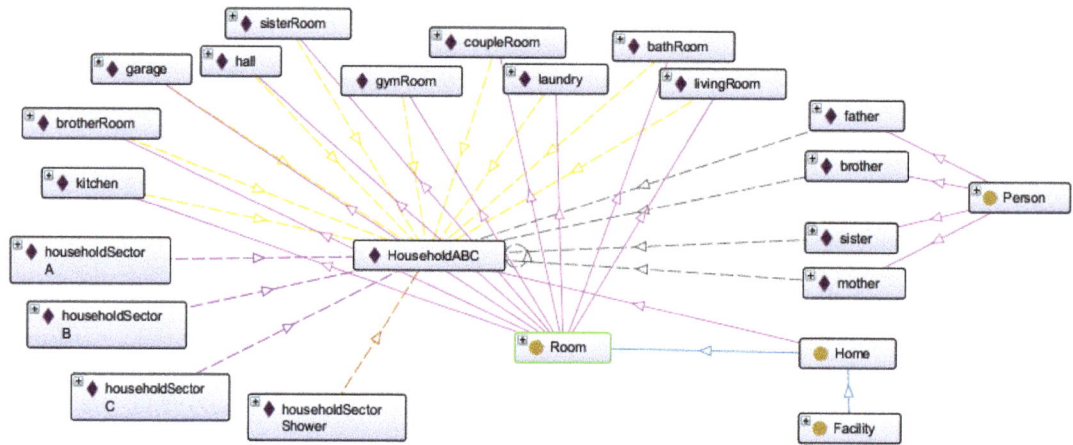

Figure 14. Household ABC digital twin ontology.

Figure 15. Digital twin smart home ontology relations and data property.

The brother person perspective is depicted in Figure 16. One may observe that the brother has a private room relation with his room. Brother room is a room of Household ABC, and home office and light bulb devices are installed in this room. A conversational agent may use this knowledge to recognize the speaker as the brother, and process the command "turn my light off" to infer that it must switch off the LightBrotherRoom and not another light bulb present in another room, thus saving a conversation iteration for increased usability. The automation command may be issued to the smart home backend based on the MQTT Topic of the LightBrotherRoom device.

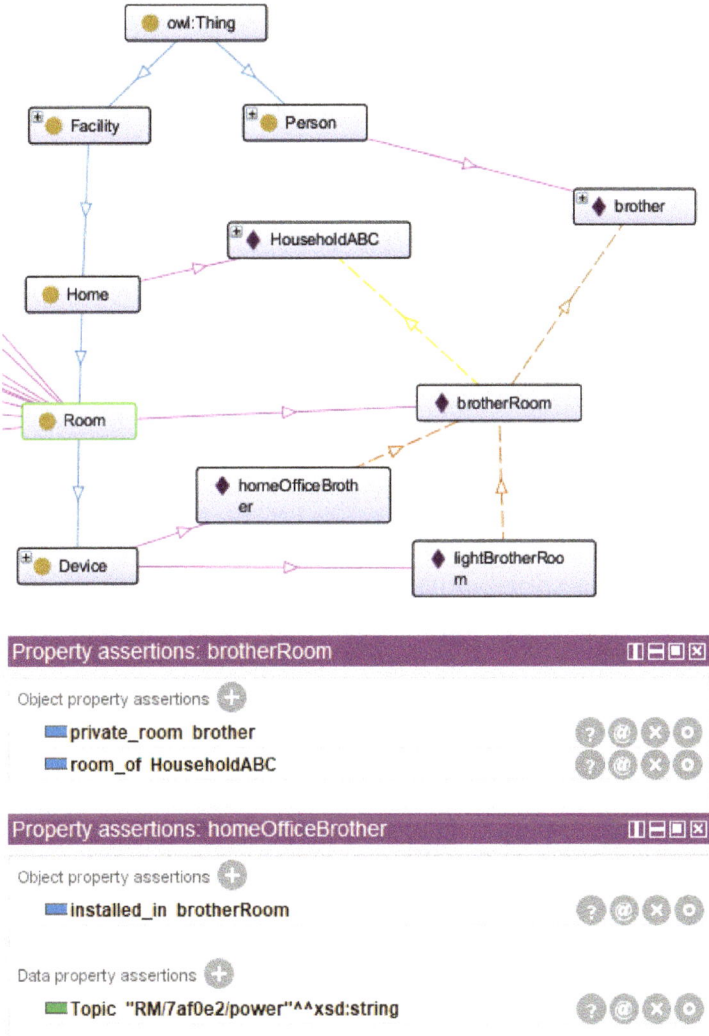

Figure 16. Brother perspective in digital twin smart home ontology.

The kitchen perspective shown in Figure 17 may be useful for a smart home automation and energy management system that must know all the devices installed in the kitchen. Based on smart plugs with device-level monitoring, household-level monitoring may be performed based on smart home digital twin ontology.

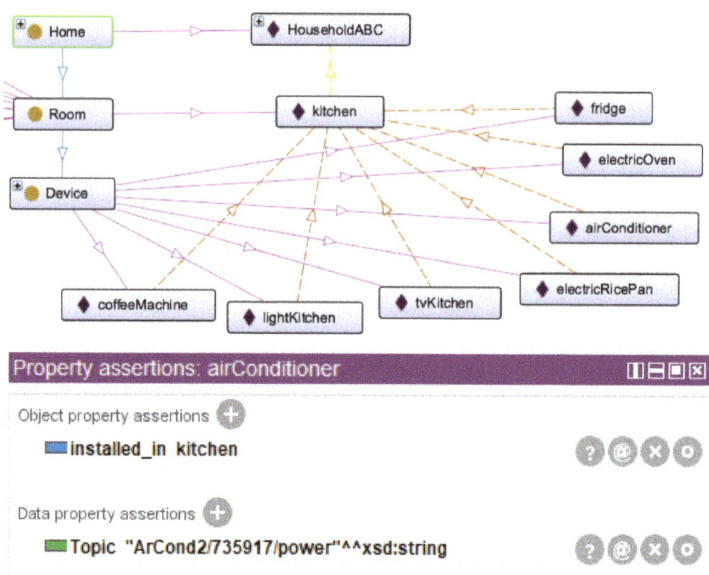

Figure 17. Kitchen perspective in digital twin smart home ontology.

5. Discussion

5.1. Comparison with Related Work

As shown in Table 5, most works used offline experiments, disregarding MLOps challenges to energy consumption forecasting. While [24] used hierarchical models to increase its resilience to missing data, his focus was mainly on data and model related challenges and did not address infrastructure and monitoring issues. A wider range of time granularities and forecasting horizons were used in comparison with related studies on residential energy consumption, and the period available for experiments is similar to other studies.

Table 5. Residential consumption forecasting papers comparison.

Paper	# of Houses	Appliance Data	Models Used	Metrics	Time Granularity	Forecast Horizon	Experiment Period
[6]	25	no	SVR, ANN	NRMSE	30 min	1–24 h	18 months
[8]	27	no	Gradient boosting	RMSE	30 min	1–24 h	35 months
[23]	7	yes	Multiple	MAPE	15–60 min	15–1440 min	9 months
[21]	1	yes	LSTM	MAPE	30 min	30 min	24 months
[24]	20	no	Hierarchical	NMAE, NQS	hourly	24 h	12 months
[25]	93	no	ANN	R2, MAPE, SDE	hourly	1–24 h	17 months
Wiseful	4	yes	Gradient Boosting	MSE, NMSE	1–1440 min	15–1440 min	19 months

One of the main difficulties in quantitatively comparing our results with related work is related to the different datasets and metrics used, as discussed in Section 2. Even if all studies used the same metrics, unless a universal dataset is used, comparing them quantitatively is unfeasible. The main focus of our study was to analyze the development approaches used to tackle MLOps challenges, as well as to optimize appliance level load forecasting.

Considering appliance consumption forecasting, it was possible to generate accurate predictions by choosing custom time frequencies for each appliance class using digital twin appliance metadata and ACF analysis for seasonality, as opposed to the approaches in [8,23] in which they used accuracy metrics.

5.2. Known Limitations and Future Work

The proposed solution supports more granular personalized recommendations than the approach based on collaborative filtering, as each household and its users are modeled in a specialized ontology [13,14]. However, a future study direction is to test, with real users, how smart home ontology supports engaging conversations and making personalized suggestions.

Human IoT is a concept explored in this project and added as an essential aspect of the platform. Distinct from the original objective of using IoT to collect data in real-time from the physical world [57], another key element is allowing the user to complement digital twin information regarding house physical dimensions; family people; energy monthly cost; electrical appliances with details, such as vendor, age, and technologies; and other pieces of information that, if combined with curate attitude, might produce valuable information. In this study, ontology was constructed with a manual method, but automating this process with user inputs in natural language may be a promising future research direction.

One opportunity is to integrate the gamification elements and other motivational factors to extend the gamified management platform proposed in [15] with a conversational interface based on a smart home digital twin ontology. Another opportunity is to use our smart home digital twin to investigate MLOps aspects when deploying reinforcement learning models as the ones presented in [17], in addition to the prediction models presented herein and found in the literature [16].

Finally, one of the most relevant challenges is to secure the digital twin [11], which is considered out of the scope of this article.

5.3. Development Considerations

One of the main fears at the beginning of the project is related to the large number of functional changes arising from the start of the project, and the assumption that the tests implemented at this stage would quickly become obsolete. However, this fear proved unfounded, since the simple definition of the tests not only verified the correct execution of the code but also guided the process of development, following Test-Driven Development (TDD) [58].

Data versioning proved to be important in experimentation during exploratory data analysis, ensuring the reproducibility of experiments performed in previous versions. During the development process, it was necessary to balance delivering results and running tests so that the definition of priorities was extremely important in the course of the project. The choice of priorities was calculated according to the probability of related problems to occur, considering the impact of these problems on the system.

Another consideration for running the tests was the modularization of pipeline steps. By accurately defining its expected features, inputs, and outputs, it makes it easier to change the source code and experiment with new settings and it is easier to observe how specific changes impact the results.

6. Conclusions

Research Question 1 was properly addressed by the digital twin household ontology model that includes topological and behavioral aspects of the residence and relevant metadata that may be used with the energy consumption forecasts by other systems (e.g., smart home with energy management system with solar photovoltaic panels and batteries).

Research Question 2 was also addressed in MLOps experiments that show how the proposed solution is robust to missing data and supports multiple time granularity of 1 to 1440 min and MSE and NMSE metrics.

A smart home digital twin integrated with MLOps is proposed to effectively predict energy consumption at a device level. Our approach may be useful for tackling the challenges of deploying machine learning prediction models in online environments, considering the specific scenario of energy consumption forecast. Household metadata is modeled in an ontology to support facilitated integration of real-time monitoring and prediction information with new interfaces, such as personalized conversational agents and dashboards.

The approach was validated by using a residential study case with 31 IoT smart meter and smart plug devices with 19-month data (measurements performed each second). Our results show that choosing custom time frequencies for each appliance class and ACF analysis allowed generating accurate predictions, actively tackling MLOps challenges in the energy forecast scenario.

Author Contributions: Conceptualization, T.Y.F. and R.A.; methodology, R.A.; software, T.Y.F., V.T.H., R.B.J., F.H.H. and K.A.K.; validation, R.A. and W.V.R.; investigation, T.Y.F., R.A. and V.T.H.; resources, V.T.H. and F.H.H.; data curation, T.Y.F.; writing—original draft preparation, T.Y.F., V.T.H. and R.A.; writing—review and editing, V.T.H.; supervision, R.A.; project administration, V.T.H.; funding acquisition, V.T.H. and R.A. All authors have read and agreed to the published version of the manuscript.

Funding: This research was funded by FAPESP under grant number 20/05763-6 and PPGEE ("Programa de Pós Gradução em Engenharia Elétrica") from the Polytechnic School of the University of São Paulo.

Institutional Review Board Statement: Not applicable

Informed Consent Statement: Not applicable.

Conflicts of Interest: The authors declare no conflicts of interest. The funders had no role in the design of the study; in the collection, analyses, or interpretation of data; in the writing of the manuscript; or in the decision to publish the results.

References

1. Iqbal, S.; Sarfraz, M.; Ayyub, M.; Tariq, M.; Chakrabortty, R.K.; Ryan, M.J.; Alamri, B. A Comprehensive Review on Residential Demand Side Management Strategies in Smart Grid Environment. *Sustainability* **2021**, *13*, 7170. [CrossRef]
2. Cruz, C.; Palomar, E.; Bravo, I.; Aleixandre, M. Behavioural patterns in aggregated demand response developments for communities targeting renewables. *Sustain. Cities Soc.* **2021**, *72*, 103001. [CrossRef]
3. Hayashi, V.T.; Arakaki, R.; Fujii, T.Y.; Khalil, K.A.; Hayashi, F.H. B2B B2C Architecture for Smart Meters using IoT and Machine Learning: A Brazilian Case Study. In Proceedings of the 2020 International Conference on Smart Grids and Energy Systems (SGES), Perth, Australia, 23–26 November 2020. [CrossRef]
4. Hayashi, V.; Fujii, T.; Arakaki, R.; Amaral, H.; Souza, A. Boa Energia: Base de Dados Pública de Consumo Residencial com Qualidade de Dados. In *Anais de XXXVIII Simpósio Brasileiro de Telecomunicações e Processamento de Sinais*; Sociedade Brasileira de Telecomunicações: Rio de Janeiro, Brazil, 2020.
5. Carrie Armel, K.; Gupta, A.; Shrimali, G.; Albert, A. Is disaggregation the holy grail of energy efficiency? The case of electricity. *Energy Policy* **2013**, *52*, 213–234. [CrossRef]
6. Humeau, S.; Wijaya, T.K.; Vasirani, M.; Aberer, K. Electricity load forecasting for residential customers: Exploiting aggregation and correlation between households. In Proceedings of the 2013 Sustainable Internet and ICT for Sustainability (SustainIT), Palermo, Italy, 30–31 October 2013.
7. Martins, P.B.d.M.; Pinto, R.G.D.; Bittencourt, S.P. Load Disaggregation of Industrial Machinery Power Consumption Monitoring Using Factorial Hidden Markov Models. In Proceedings of the International Workshop on Non-Intrusive Load Monitoring (NILM), Austin, TX, USA, 7–8 March 2018; p. 6.
8. Lusis, P.; Khalilpour, K.R.; Andrew, L.; Liebman, A. Short-term residential load forecasting: Impact of calendar effects and forecast granularity. *Appl. Energy* **2017**, *205*, 654–669. [CrossRef]
9. Alanezi, A.; P Hallinan, K.; Elhashmi, R. Using Smart-WiFi Thermostat Data to Improve Prediction of Residential Energy Consumption and Estimation of Savings. *Energies* **2021**, *14*, 187. [CrossRef]
10. Villa, S.; Sassanelli, C. The Data-Driven Multi-Step Approach for Dynamic Estimation of Buildings' Interior Temperature. *Energies* **2020**, *13*, 6654. [CrossRef]
11. Kaur, M.J.; Mishra, V.P.; Maheshwari, P. The convergence of digital twin, IoT, and machine learning: Transforming data into action. In *Digital Twin Technologies and Smart Cities*; Springer: Berlin, Germany, 2020; pp. 3–17.

12. Fujii, T.Y.; Ruggiero, W.V.; do Amaral, H.L.; Hayashi, V.T.; Arakaki, R.; Khalil, K.A. Desafios para Aplicação de MLOps na Previsão do Consumo Energético. In Proceedings of the 2021 14th IEEE International Conference on Industry Applications (INDUSCON), São Paulo, Brazil , 15–18 August 2021; pp. 455–462.
13. Alsalemi, A.; Sardianos, C.; Bensaali, F.; Varlamis, I.; Amira, A.; Dimitrakopoulos, G. The role of micro-moments: A survey of habitual behavior change and recommender systems for energy saving. *IEEE Syst. J.* **2019**, *13*, 3376–3387. [CrossRef]
14. Alsalemi, A.; Himeur, Y.; Bensaali, F.; Amira, A.; Sardianos, C.; Varlamis, I.; Dimitrakopoulos, G. Achieving domestic energy efficiency using micro-moments and intelligent recommendations. *IEEE Access* **2020**, *8*, 15047–15055. [CrossRef]
15. Zehir, M.A.; Ortac, K.B.; Gul, H.; Batman, A.; Aydin, Z.; Portela, J.C.; Soares, F.J.; Bagriyanik, M.; Kucuk, U.; Ozdemir, A. Development and field demonstration of a gamified residential demand management platform compatible with smart meters and building automation systems. *Energies* **2019**, *12*, 913. [CrossRef]
16. Zhang, Y.; Srivastava, A.K.; Cook, D. Machine learning algorithm for activity-aware demand response considering energy savings and comfort requirements. *IET Smart Grid* **2020**, *3*, 730–737. [CrossRef]
17. Fathy, Y.; Jaber, M.; Nadeem, Z. Digital Twin-Driven Decision Making and Planning for Energy Consumption. *J. Sens. Actuator Netw.* **2021**, *10*, 37. [CrossRef]
18. Hyndman, R.J.; Athanasopoulos, G. *Forecasting: Principles and Practice*; OTexts: Melbourne, Australia, 2021.
19. Rossana, R.J.; Seater, J.J. Temporal Aggregation and Economic Time Series. *J. Bus. Econ. Stat.* **1995**, *13*, 441–445.
20. Sprenger, J.; Weinberger, N. Simpson's Paradox. In *The Stanford Encyclopedia of Philosophy*, 2021th ed.; Zalta, E.N., Ed.; Metaphysics Research Lab, Stanford University: Palo Alto, CA, USA, 2021.
21. Kong, W.; Dong, Z.Y.; Hill, D.J.; Luo, F.; Xu, Y. Short-term residential load forecasting based on resident behaviour learning. *IEEE Trans. Power Syst.* **2018**, *33*, 2017–2018. [CrossRef]
22. Ben Taieb, S.; Hyndman, R.J. A gradient boosting approach to the Kaggle load forecasting competition. *Int. J. Forecast.* **2014**, *30*, 382–394. [CrossRef]
23. Veit, A.; Goebel, C.; Tidke, R.; Doblander, C.; Jacobsen, H.A. Household electricity demand forecasting—Benchmarking state-of-the-art methods. In Proceedings of the e-Energy 2014—Proceedings of the 5th ACM International Conference on Future Energy Systems, Cambridge, UK, 11–13 June 2014; pp. 233–234.
24. Gerossier, A.; Girard, R.; Bocquet, A.; Kariniotakis, G. Robust day-ahead forecasting of household electricity demand and operational challenges. *Energies* **2018**, *11*, 3503. [CrossRef]
25. The daily and hourly energy consumption and load forecasting using artificial neural network method: A case study using a set of 93 households in Portugal. *Energy Procedia* **2014**, *62*, 220–229. [CrossRef]
26. Kong, W.; Dong, Z.Y.; Jia, Y.; Hill, D.J.; Xu, Y.; Zhang, Y. Short-Term Residential Load Forecasting Based on LSTM Recurrent Neural Network. *IEEE Trans. Smart Grid* **2019**, *10*, 841–851. [CrossRef]
27. Serrenho, T., Bertoldi, P. *Smart Home and Appliances: State of the Art*; Technical Report; Publications Office of the European Union: Luxembourg, 2019.
28. Amaral, H.L.; Maginador, J.A.; Ayres, R.M.; De Souza, A.N.; Gastaldello, D.S. Integration of consumption forecasting in smart meters and smart home management systems. In Proceedings of the SBSE 2018—7th Brazilian Electrical Systems Symposium, Niteroi, Brazil, 12–16 May 2018; pp. 1–6.
29. Wang, Y.; Chen, Q.; Hong, T.; Kang, C. Review of Smart Meter Data Analytics: Applications, Methodologies, and Challenges. *IEEE Trans. Smart Grid* **2019**, *10*, 3125–3148. [CrossRef]
30. Tushar, W.; Saha, T.K.; Yuen, C.; Liddell, P.; Bean, R.; Poor, H.V. Peer-to-Peer Energy Trading With Sustainable User Participation: A Game Theoretic Approach. *IEEE Access* **2018**, *6*, 62932–62943. [CrossRef]
31. Pratt, A.; Krishnamurthy, D.; Ruth, M.; Wu, H.; Lunacek, M.; Vaynshenk, P. Transactive Home Energy Management Systems: The Impact of Their Proliferation on the Electric Grid. *IEEE Electrif. Mag.* **2016**, *4*, 8–14. [CrossRef]
32. Haben, S.; Ward, J.; Vukadinovic Greetham, D.; Singleton, C.; Grindrod, P. A new error measure for forecasts of household-level, high resolution electrical energy consumption. *Int. J. Forecast.* **2014**, *30*, 246–256. [CrossRef]
33. Stefan, A.; Athitsos, V.; Das, G. The move-split-merge metric for time series. *IEEE Trans. Knowl. Data Eng.* **2013**, *25*, 1425–1438. [CrossRef]
34. Guen, V.L.; Thome, N. Shape and Time Distortion Loss for Training Deep Time Series Forecasting Models. *arXiv* **2019**, arXiv:1909.09020.
35. Murphy, A.H. What Is a Good Forecast? An Essay on the Nature of Goodness in Weather Forecasting. *Weather Forecast.* **1993**, *8*, 281–293. [CrossRef]
36. Breck, E.; Cai, S.; Nielsen, E.; Salib, M.; Sculley, D. The ML test score: A rubric for ML production readiness and technical debt reduction. In Proceedings of the 2017 IEEE International Conference on Big Data (Big Data), Boston, MA, USA, 11–14 December 2017; Volume 47, pp. 1123–1132.
37. Sugimura, P.; Hartl, F. Building a reproducible machine learning pipeline. *arXiv* **2018**, arXiv:1810.04570.
38. Senapathi, M.; Buchan, J.; Osman, H. DevOps Capabilities, Practices, and Challenges. In Proceedings of the 22nd International Conference on Evaluation and Assessment in Software Engineering 2018, Christchurch, New Zealand, 28–29 June 2018; ACM: New York, NY, USA, 2018; pp. 57–67.
39. Sculley, D.; Holt, G.; Golovin, D.; Davydov, E.; Phillips, T.; Ebner, D.; Chaudhary, V.; Young, M.; Crespo, J.F.; Dennison, D. Hidden technical debt in machine learning systems. *Adv. Neural Inf. Process. Syst.* **2015**, *2015*, 2503–2511.

40. Google Cloud. *MLOps: Continuous Delivery and Automation Pipelines in Machine Learning*; Google LLC: Mountain View, CA, USA, 2020.
41. Guajardo, J.A.; Weber, R.; Miranda, J. A model updating strategy for predicting time series with seasonal patterns. *Appl. Soft Comput. J.* **2010**, *10*, 276–283. [CrossRef]
42. Kent, L.; Snider, C.; Hicks, B. Early stage digital-physical twinning to engage citizens with city planning and design. In Proceedings of the 2019 IEEE Conference on Virtual Reality and 3D User Interfaces (VR), Osaka, Japan, 23–27 March 2019; pp. 1014–1015.
43. Koreshoff, T.L.; Leong, T.W.; Robertson, T. Approaching a human-centred internet of things. In Proceedings of the 25th Australian Computer-Human Interaction Conference: Augmentation, Application, Innovation, Collaboration, Adelaide, Australia, 19–25 November 2013; pp. 363–366.
44. Chen, S.; Xu, H.; Liu, D.; Hu, B.; Wang, H. A vision of IoT: Applications, challenges, and opportunities with china perspective. *IEEE Internet Things J.* **2014**, *1*, 349–359. [CrossRef]
45. Eyre, J.; Freeman, C. Immersive Applications of Industrial Digital Twins. *Ind. Track EuroVR* **2018**, 11–20. Available online: https://publications.vtt.fi/pdf/technology/2018/T339.pdf (accessed on 29 October 2021).
46. Bezborodova, O.; Bodin, O.; Gerasimov, A.; Kramm, M.; Rahmatullov, R.; Ubiennykh, A. «*Digital Twin» Technology in Medical Information Systems*; Journal of Physics: Conference Series; IOP Publishing: Bristol, UK, 2020; Volume 1515, p. 052022.
47. Raes, L.; Michiels, P.; Adolphi, T.; Tampere, C.; Dalianis, T.; Mcaleer, S.; Kogut, P. DUET: A Framework for Building Secure and Trusted Digital Twins of Smart Cities. *IEEE Internet Comput.* **2021**. [CrossRef]
48. Kuller, M.; Kohlmorgen, F.; Karaoğlan, N.; Niemeyer, M.; Kunold, I.; Wöhrle, H. Conceptual design of a digital twin based on semantic web technologies in the smart home context. In Proceedings of the 2020 IEEE 3rd International Conference and Workshop in Óbuda on Electrical and Power Engineering (CANDO-EPE), Budapest, Hungary, 18–19 November 2020; pp. 000167–000172.
49. Maryasin, O. Home Automation System Ontology for Digital Building Twin. In Proceedings of the 2019 XXI International Conference Complex Systems: Control and Modeling Problems (CSCMP), Samara, Russia, 3–6 September 2019; pp. 70–74.
50. Raggett, D. The web of things: Challenges and opportunities. *Computer* **2015**, *48*, 26–32. [CrossRef]
51. Goddard, M. The EU General Data Protection Regulation (GDPR): European regulation that has a global impact. *Int. J. Mark. Res.* **2017**, *59*, 703–705. [CrossRef]
52. Pinheiro, P.P. *Proteção de Dados Pessoais: Comentários à Lei n. 13.709/2018-LGPD*; Saraiva Educação SA: São Paulo, Brazil, 2020.
53. Arakaki, R.; Hayashi, V.T.; Ruggiero, W.V. Available and Fault Tolerant IoT System: Applying Quality Engineering Method. In Proceedings of the 2020 International Conference on Electrical, Communication, and Computer Engineering (ICECCE), Istanbul, Turkey, 12–13 June 2020; pp. 1–6.
54. Chen, T.; Guestrin, C. XGBoost: A scalable tree boosting system. In Proceedings of the ACM SIGKDD International Conference on Knowledge Discovery and Data Mining, San Francisco, CA, USA, 13–17 August 2016; pp. 785–794.
55. Burleyson, C.D.; Rahman, A.; Rice, J.S.; Smith, A.D.; Voisin, N. Multiscale effects masked the impact of the COVID-19 pandemic on electricity demand in the United States. *Appl. Energy* **2021**, *304*, 117711. [CrossRef]
56. Plotly Technologies Inc. *Dash*; Plotly Technologies Inc.: Montreal, QC, Canada, 2021.
57. Ashton, K. That 'internet of things' thing. *RFID J.* **2009**, *22*, 97–114.
58. Beck, K. *Test-Driven Development: By Example*; Addison-Wesley Professional: Upper Saddle River, NJ, USA, 2003.

Article

SQL and NoSQL Databases in the Context of Industry 4.0

Vitor Furlan de Oliveira *, Marcosiris Amorim de Oliveira Pessoa, Fabrício Junqueira and Paulo Eigi Miyagi

Escola Politécnica, Universidade de São Paulo, Sao Paulo 05508-030, Brazil; marcosiris@usp.br (M.A.d.O.P.); fabri@usp.br (F.J.); pemiyagi@usp.br (P.E.M.)
* Correspondence: vitor.furlan@usp.br; Tel.: +55-(11)-96033-7582

Abstract: The data-oriented paradigm has proven to be fundamental for the technological transformation process that characterizes Industry 4.0 (I4.0) so that big data and analytics is considered a technological pillar of this process. The goal of I4.0 is the implementation of the so-called Smart Factory, characterized by Intelligent Manufacturing Systems (IMS) that overcome traditional manufacturing systems in terms of efficiency, flexibility, level of integration, digitalization, and intelligence. The literature reports a series of system architecture proposals for IMS, which are primarily data driven. Many of these proposals treat data storage solutions as mere entities that support the architecture's functionalities. However, choosing which logical data model to use can significantly affect the performance of the IMS. This work identifies the advantages and disadvantages of relational (SQL) and non-relational (NoSQL) data models for I4.0, considering the nature of the data in this process. The characterization of data in the context of I4.0 is based on the five dimensions of big data and a standardized format for representing information of assets in the virtual world, the Asset Administration Shell. This work allows identifying appropriate transactional properties and logical data models according to the volume, variety, velocity, veracity, and value of the data. In this way, it is possible to describe the suitability of relational and NoSQL databases for different scenarios within I4.0.

Keywords: Industry 4.0; database; data models; big data and analytics; asset administration shell

Citation: de Oliveira, V.F.; Pessoa, M.A.d.O.; Junqueira, F.; Miyagi, P.E. SQL and NoSQL Databases in the Context of Industry 4.0. *Machines* **2022**, *10*, 20. https://doi.org/10.3390/machines10010020

Academic Editor: Xiang Li

Received: 19 November 2021
Accepted: 21 December 2021
Published: 27 December 2021

Publisher's Note: MDPI stays neutral with regard to jurisdictional claims in published maps and institutional affiliations.

Copyright: © 2021 by the authors. Licensee MDPI, Basel, Switzerland. This article is an open access article distributed under the terms and conditions of the Creative Commons Attribution (CC BY) license (https://creativecommons.org/licenses/by/4.0/).

1. Introduction

Industry 4.0 (I4.0) designates the technological transformation process in production systems, logistics, and business models observed since the last decade [1]. The integration of digital technologies has promoted changes in the development phase [2,3], flexibility of production [3,4], efficiency in the use of resources [5,6], and level of automation and digitalization of the organizations [7,8]. This new mode of production characterizes the so-called Intelligent Manufacturing Systems (IMS): more efficient, flexible, integrated, and digitized than the traditional manufacturing systems. In the context of I4.0, the companies where the IMS are present are referred to as Smart Factories [9–11].

Data emerged as a fundamental resource for the Smart Factory due to their characteristics such as low cost, apparent inexhaustibility, and the possibility of cost reduction and value creation [12]. The authors of [10] argue that Smart Factory status is achieved, among other factors, when artificial intelligence solutions use the data. The "smart products" resultant from IMS are objects capable of storing and making their data available to humans or machines [9]. Thus, the importance of the data-oriented paradigm in the context of I4.0 is clear [12].

New system architectures have been proposed to promote the integration of enabling digital technologies to use data for industrial innovations [13–15]. While there is concern about optimizing IMS architectures in many respects, the impact of databases on their performance is not always considered. It is possible to observe that, in many cases, databases are treated as mere entities that support the functions of architectures, even though they can significantly influence the performance of the IMS [16].

This paper proposes the identification of data models that better suit different scenarios in the context of I4.0. This phenomenon is characterized by its key enabling data-related technologies and methods so that a consistent description of the nature of the data in this context could be achieved. By identifying the advantages and limitations of relational and NoSQL data models for such data characteristics, it is possible to discuss the suitability of these models for different scenarios in the context of I4.0.

2. Materials and Methods

Presenting the context in which this paper is inserted is essential to justify the choice of materials and methods adopted in the work from which it derives. For this reason, this section begins with a contextualization of Industry 4.0. Understanding some of its main characteristics is important to justify the relevance of this paper to (i) demonstrate the gap that exists in system architectures for I4.0 concerning data storage solutions used by these systems and (ii) identify patterns, methods, and technologies whose relevance to I4.0 is such that, from them, it is possible to obtain a characterization of the data in this context. Thus, this information is combined to propose suggestions for data models in the context of I4.0.

Nowadays, there is a consensual understanding that manufacturing automation systems have been undergoing a continuous transformation of technological paradigms since the last decade [1]. Authors claim that these transformations, obtained from integrating a series of independent digital technologies and a certain degree of independence from each other, configure the Fourth Industrial Revolution [17]. Because of the global scale of these changes, several initiatives worldwide, such as the Plattform Industrie 4.0 (https://www.plattform-i40.de/PI40/Navigation/EN/Home/home.html accessed on 2 August 2021), the Industrial Internet Consortium (https://www.iiconsortium.org/ accessed on 2 August 2021), and the Standardization Council Industrie 4.0 (https://www.sci40.com/ accessed on 2 August 2021), seek to establish guidelines for this process of technological transformation. The need to have a guide (or multiple equivalent guides) for the technological transformation process associated with the Fourth Industrial Revolution is because, unlike the first three, the Fourth Industrial Revolution was identified as such already in its early stages. Thus, these initiatives become responsible for outlining the advancement of technological transformation in manufacturing, proposing a common understanding of the phenomenon, establishing standards, and so on.

Among the different technological aspects mentioned above, some are highlighted in this work and focus is given to the so-called I4.0, a term often used as a synonym for the Fourth Industrial Revolution. In Germany, the Plattform Industrie 4.0 was created, a consortium of various organizations, including industries, universities, and the German government, proposing to shape the digital transformation in manufacturing according to the precepts of I4.0. The meaning of the term "Industry 4.0" is the object of analysis by several researchers [18–20]. Instead of presenting a definition, the option is to describe the phenomenon in terms of its characteristics: I4.0 is characterized by Intelligent Manufacturing Systems (IMS) that quickly adapt to market demands and with effective interconnection between all entities involved in these processes. This phenomenon is the conception of the so-called Smart Factory, which aims at manufacturing based on intelligent services and processes [21].

The main materials used in this work were technical publications and academic works. Considering that I4.0 is the result of cooperation between academia, industry, and government organizations, it was impossible to use only literature review methods of academic publications. Characterizing a system's data is essential for choosing the database to be adopted in the architecture for this system. In this work, this characterization is made based on technologies, methods, and standards for data in I4.0. Other relevant features for database design are fundamentally application dependent and are beyond the scope of this paper which seeks to expand the coverage of its contributions. The following paragraphs describe the materials and methods adopted to characterize data in different scenarios in the context of I4.0 and identify which data models are suitable for different scenarios.

Ensuring interoperability among systems is one of the requirements for implementing the Smart Factory [20,22,23]. For this purpose, the entities that establish guidelines for the advancement of I4.0 proposed a standard format for digitally representing and managing elements involved in carrying out productive activities—the Asset Administration Shell (AAS)—whose concept, structure, metamodel, and perspectives for implementation will be presented in Section 3.2. Current works propose the implementation and use of the Asset Administration Shell in system architectures that seek to use data for different purposes. However, it is noted that less attention is paid to the design of the database to be used in these architectures. To confirm this statement, a systematic review of the literature was carried. The adopted procedures were the following:

- The paper databases used in the search were: Web of Science, IEEEXplore, Science Direct, Scopus, and Google Scholar;
- The following search string was defined to find papers: "Asset Administration Shell" AND "Database";
- It was observed that, among the selected databases, the only one to return a considerable number of papers was Google Scholar, which included papers from the other databases and, therefore, was the only one used. The application of search string returned 139 papers;
- The following keywords were defined for ranking the papers: "AAS"; "Asset Administration Shell"; "Database"; "DBMS" (database management system); "Implement*"; and "Storage";
- Each occurrence of any keyword in the title of the paper assigned 5 points to it (the Google Scholar platform does not allow exporting the abstract or keywords of the article). For instance, the paper entitled "Toward Industry 4.0 Components: Insights into and Implementation of Asset Administration Shells" contains the keywords implementation and "Asset Administration Shell" so it scored 10 points;
- Papers with a score greater than or equal to 5 were classified as accepted, and their content was analyzed;
- It was researched which of the papers classified as accepted cited the implementation data model and/or DBMS used.

Considering that I4.0 is a process of technological transformation, important database-based digital technologies and methods were identified. Those have such importance for this process that a description of the nature of the data in the context of I4.0 can be obtained from them. In addition to academic works, technical publications such as working papers from key organizations and entities for I4.0 were also considered in this process.

3. Basic Concepts

This section presents a theoretical framework composed of essential basic concepts for the work. Database-related topics include relational and NoSQL data models, transactional properties, and theorems regarding these properties. Moreover, the Asset Administration Shell concept, an artifact developed to represent Industry 4.0 components in the digital world, is presented.

3.1. Relational and NoSQL Databases

A logical and coherent collection of data with an intrinsic meaning forms a database [24]. A database stores and ensures the persistence and integrity of data that represent assets, in addition to allowing these data to be made available to interested users. A database is created and maintained through a database management system (DBMS), a computer program that helps maintain and use data sets that compose the databases [25]. These programs have the following advantages: they enable efficient and concurrent access to data; ensure data integrity and security; protect against failures and unauthorized access;

support multiple views of data; and, finally, they guarantee independence, that is, the isolation between data and applications through data abstraction [25,26].

Data abstraction is provided through data models. A data model is a set of concepts used to describe the structure of a database [24]. The logical data model describes data in such a level of abstraction that hides some details of the physical storage, which allows the end-user of the data to understand them. At the same time, as they are not so far from the low level, these concepts can be used directly to implement a database in a computer system. DBMSs are usually characterized by the logical data models they implement and, for this reason, this work focuses on this level of data abstraction.

3.1.1. Relational Data Model

The relational data model was, for many years, the default choice for database implementation [27]. It uses the concept of "relation" in a mathematical sense to represent data. Instead of presenting a formal mathematical definition of the term, which can be found in [28], it is presented how a relationship is perceived. Relations can be seen as tables of values. These tables have columns not necessarily distinct that consist of "attributes" used to characterize an element to be represented by the relation. Each line (formally called "tuple") of this table has values for the attributes. For each column, the values present in every tuple belong to a single domain with a well-defined name, data type, and format. Besides, only atomic values (each value in the domain is indivisible) are allowed.

A schema defines the structure of a relational database. From a schema, tables, their attributes, and relationships between them can be described to be used through a DBMS to create a database. The vast majority of DBMSs that implement a relational model use a standard language to perform queries—the Structured Query Language (SQL); the relational model is commonly called the SQL model. The same extends to the DBMSs and databases that implement it.

3.1.2. ACID and BASE Transactional Properties and CAP Theorem

Relational DBMS grants four properties to transactions to maintain data through concurrent access and system failures. These properties are atomicity (A), consistency (C), isolation (I), and durability (D), so they are often referred to as ACID properties. A brief description of each of them based on [25,26] is presented:

- Atomicity: The transaction must be executed in its entirety or not to be executed. If during the transaction, any failure occurs that prevents the transaction from being completed, any changes that it has performed in the database must be undone;
- Consistency: If a transaction runs entirely from start to finish, without interference from other transactions, it should take the database from one consistent state to another. A consistent database state satisfies the constraints specified in the schema as well as any other database constraints that must be maintained;
- Isolation: The execution of a transaction must not be interfered with by any other transaction running at the same time;
- Durability: Changes applied to the database by a committed transaction must persist in the database. These changes must not be lost due to any failure.

A distributed database is defined as "a collection of several logically interrelated databases distributed over a network of computers" [29]. There are three crucial reasons pointed out in the literature for the use of distributed databases: (1) the increase in the volume of data [27], which requires the ability to scale horizontally, that is, to distribute the systems across several nodes—instead of vertically scaling the hardware, adding more computing resources to the same machine, which would be more expensive and limited; (2) the need to better reflect the distributed organizational structure of companies [30]; and (3) the inherently distributed nature of a range of applications, including industrial ones [30]. There are three ways to implement a distributed database system [26,27]. It is noteworthy that specific systems implement hybrid versions, combining different forms of partitioning [27].

- Single server: There is no distribution. The database runs on a single machine that handles all operations. It is an example of a centralized AAS implementation;
- Partitioning: Different pieces of data on different machines. Aggregate models are ideal here, as they form a natural partitioning unit, making certain users access, most of the time, the same server so there is no need to gather information from different servers, which increases performance compared to a single server implementation;
- Replication: Data can be replicated in a master–slave schema where the master processes updates and replicates this data to other nodes or in a peer-to-peer schema where all nodes can process updates and propagate them.

Three other properties are desirable for database systems: availability (A), partition tolerance (P), and consistency (C), which, in this case, has a slight difference from the concept presented before. The analysis of these properties is fundamental in distributed database systems. The CAP theorem correlates these three properties. According to it, these three properties, whose descriptions are presented here, cannot coexist simultaneously in a database system:

- Consistency (C): Ensuring that all nodes have identical copies of replicated data visible to applications. It is a little different from the consistency concept of ACID properties. In that case, consistency means not violating database restrictions. However, it can be considered that "having the same copy of a data replicated in all nodes that this data is replicated on" is a restriction, so the concepts start to resemble each other;
- Availability (A): Each write or read operation will be successfully processed (system available) or will fail (system unavailable). A "down" node is not said to be unavailable;
- Partition Tolerance (P): A partition tolerant system continues to operate if a network fails to connect nodes, resulting in one or more network partitions. In this case, nodes in a partition only communicate with each other.

According to the CAP theorem, only two of the three properties presented can be guaranteed simultaneously. It is worth noting that this choice is not binary, it is possible to relax specific properties so that it is possible to privilege others. However, ACID transactional properties make this flexibility unfeasible. As a kind of alternative, you can have a database that works basically all the time (basically available) and is not consistent all the time (flexible state), only when the writes are propagated to all nodes (eventually consistent) in a distributed system. Thus, the characteristics of this model, named BASE model, although not strictly defined, are presented as:

- Basically available (BA): the system must be available even if partial failures occur;
- Flexible State (S): the system may not have consistent data all the time;
- Eventually consistent (E): consistency will be achieved once all writes are propagated to all nodes.

3.1.3. NoSQL Data Models

NoSQL does not have a solid definition, but is possibly better understood as a movement that proposes non-relational database solutions that do not use the SQL language [27,31]. Thus, the term NoSQL (often interpreted as Not Only SQL), used in its technical sense, is applied to designate a family of DBMSs that have specific characteristics in common (at least for most DBMSs), the main one being the non-implementation of a relational data model [32]. These features can mean advantages or disadvantages for specific applications:

- They do not implement the relational data model: Self-description and the absence of a fixed schema allow for greater flexibility concerning the content stored in DBMS, being suitable for handling semi-structured data [26,32];
- They do not use the SQL language: The absence of a declarative query language, with a wide range of "features" that are sometimes unnecessary, requires more significant

effort for developers since the functions and operations have to be implemented through the language of programming [26];
- Absence of ACID transactions: Because aggregate-oriented data models generally do not guarantee ACID transactional properties, DBMSs that implement these models have greater efficiency in distributed systems [32,33]. Alternatively, these DBMSs use the BASE model of transactional properties;
- Horizontally scalable: The ability of NoSQL DBMS to scale out is linked to two main characteristics. (1) By not having ACID transactional properties (aggregate-oriented models), it allows for relaxing consistency, and thus balancing the consistency–latency trade-off in the way which is most suitable for the application, without giving up partition tolerance, as it was previously discussed. (2) The orientation to aggregates allows for a "natural" or intuitive data partitioning unit, as data from an aggregate are commonly accessed together and can be allocated on the same server, which makes the user of this data access, in the majority sometimes, the same server [27].

NoSQL DBMS are commonly differentiated based on how they store the data, that is, the data model employed for the storage. There are four main data models implemented in NoSQL DBMSs. The description of each of these models is presented here:

- Key-value: Key-value DBMSs are possibly the simplest NoSQL systems. These DBMSs store their data in a table without a rigid schema, where each line corresponds to a unique key and a set of self-described objects called value. These can take different formats, from the simplest as strings, passing through tables as in the relational model, reaching more elaborate formats such as JavaScript Object Notation (JSON) and eXtensible Markup Language (XML) documents. Thus, they can store structured, semi-structured, and unstructured data in one format (key, value). The key-value data model is often represented as a hash table. This data model is aggregate oriented, meaning that each value associated with a unique key can be understood as an aggregate of objects that can be retrieved in their entirety through the key. The content of these aggregates can be different for each key. The aggregate's opacity guarantees the possibility of storing any data in the aggregates; that is, the DBMS does not interpret the aggregate content, seeing it only as a set of bits that must always be associated with its unique key. This has the practical implication of generally not allowing partial retrievals on aggregate content. The operations implemented by key-value DBMSs are the insertion or update of a pair (key, value), the retrieval of a value from its key, or the deletion of a key;
- Documents: Document-oriented DBMSs are those in which data is stored in document format. They can be understood as a key-value DBMS in which the only allowed formats for the values are documents such as XML, JSON, or PDF. A fundamental difference between the document and key-value data models is that the former allows for partial aggregate retrievals as it stores self-described data format. In other words, aggregates are not opaque, they are not seen by the DBMS merely as a set of bits, and it is possible to define indexes on the contents of the aggregates that allow operations to be performed on specific items of this data set. As with the key-value data model, the content of each document does not follow a fixed schema. Document labels that guarantee the self-description of the data and enable partial recoveries also allow different keys to have documents with different content (attributes). Thus, it allows the storage of structured and semi-structured data. There are still DBMSs that allow the storage of unstructured data such as texts;
- Column family: In column family databases, data is stored similarly to key-value databases. However, the value can only be composed of a set of tables, each of which has a name (identifier) and forms a column family. In each of these tables, columns are self-described; they have a key (also called a qualifier) and its value, which is the data itself. Thus, a column family database is formed by a table without a rigid structure containing unique keys and a series of column family associated with each key in each row. Some considerations can be made about this model. The first is that keys do

not need to have the same column family. The second observation is that, for each key, each column family can only contain the columns of interest; that is, the column family does not need to be composed of the same columns for all the keys. The column family forms a data aggregate that is frequently accessed together and, because these columns contain their keys, the aggregates are not opaque to the DBMS, thus being possible to perform partial recoveries through the aggregates through the indexes of the columns;

- Graphs: In graph-oriented DBMSs, data is stored in a collection of nodes, which represent entities, and directed vertices, which represent relationships among these entities. The set of nodes and vertices form graphs in which the two elements that compose them can contain labels and attributes associated with them, which are the data itself. Regarding the characteristics of the data models presented so far, the flexibility in data representation due to the absence of a fixed schema is one of the few similarities between the graph data model and the other data models mentioned, as both vertices and nodes can contain attributes different from each other. Concerning differences, the graph model is not aggregate oriented, it usually has ACID transactional properties, it is best suited for single server (non-distribution) implementations, can represent small records with complex relationships to each other, and it is more efficient in identifying patterns. Unlike aggregate-oriented models, where partial recoveries can only be made on one aggregate at a time (when allowed), in the graph model they can be conducted for the graph as a whole.

3.2. Asset Administration Shell

Before presenting the Asset Administration Shell (AAS), it is necessary to introduce the concept of "asset". The IEC describes an asset as "a physical or logical object owned or held in custody by an organization, having a perceived or real value to the organization". Based on this definition, also adopted by the Plattform Industrie 4.0 [34], it is possible to recognize that an asset can be something physical (a machine, equipment, materials, products) or not (electronic documents, computer programs). Some less intuitive examples of assets are location, time, state of an asset, human beings, and relationships between assets [35]. In summary, the asset is everything that has value and importance for an organization.

It is known that I4.0 characterizes a digital transformation process. For this process to occur, the assets need to be digitized; their data must be taken to the virtual world [36,37]. To perform this mapping to ensure interoperability between systems and components [37] in IMS, the AAS was created. It corresponds to a standardized digital representation of the asset containing all its technical information and functionalities. The AAS provides a minimum, unique, and sufficient description of the asset in different perspectives relevant to each use case [34,38]. By standardizing the representation format and communication interfaces of assets in the digital world, AAS enables the exchange of information among I4.0 participants, ensuring interoperability between components [34,38]. In summary, the AAS corresponds to the virtual and standardized representation of an asset in the context of I4.0.

The combination of asset and AAS gives rise to Component I4.0 (I4.0C or I4.0 Component). The I4.0C combines the physical and real world, composed by the asset and its respective AAS. The combination of these two elements, with the second "involving" the first, allows services and functionalities to be offered inside (through AAS) and outside (through the asset) of the I4.0C network. These features and services are made possible by the unique identification and communication capability of an I4.0C. Here, it is worth noting that a single I4.0C can be associated with multiple assets depending on the considered granularity. In this way, such a structure can be replicated to different levels of granularity (for example, at different levels of hierarchy). The following subsections discuss details of the AAS structure, elements, metamodel, and implementation perspectives.

The elements that compose an AAS are divided into classes; each has its attributes used to describe the asset. Elements in the AAS can be understood as subclasses, which have the same attributes as a specific superclass from which they are derived but also contain attributes that differentiate them from other elements of the same superclass. The first way to divide the element classes of an AAS is to designate them as Identifiable and Referable. Identifiable Elements have a globally unique identifier. Referable, in turn, has an identifier that is not globally unique, being unique only within the context (defined by an Identifiable) in which it finds itself. The element classes contained within the Identifiable and Referable superclasses are called subclasses and can also, in turn, be superclasses; that is, they can be composed of other subclasses. There is an inheritance relationship of attributes in this hierarchy between classes: subclasses inherit attributes from superclasses.

The Identifiable Elements class can present additional domain-specific (owner) identifiers. The "Asset Administration Shell" and "Asset" subclasses have already been described from the Identifiable class. The subclass "Conceptual Description" defines the standardized semantic description of certain elements. Finally, the subclass "Submodel" allows an asset to be represented in its different perspectives. Each Submodel can describe an asset from an electrical, mechanical, thermal, control, and other perspective.

The "Referable Elements" class has more subclasses than the "Identifiable Elements" class, so only some of them are presented here in more detail. The description of all subclasses can be found in [34]. Among the subclasses of Referable elements, the subclass "Submodel Element" stands out in this work, and it is composed of elements suitable for the description and differentiation of assets in perspective specified by the Submodel. This class of "Submodel Elements" can be understood as a superclass in which some of the main subclasses are "Submodel Element Collection"—a collection that can be composed of all other classes with the same hierarchical level—and "Data Element". Data Elements, in turn, form another superclass with one of the important AAS element subclasses, the "Property", described in more detail in the next paragraph.

The "Properties" class contains elements that allow representing the characteristics of an asset given a perspective defined by the Submodel in which they are found. These elements are standardized by the IEC 61360 and can be found in the IEC Common Data Dictionary (CDD, common data dictionary) or eCl@ss repositories [34,39,40]. In the IEC CDD repository, a property has, in addition to its value itself, some additional data such as code, version and revision, identifier, and definition. The free digital version of the IEC CDD provides examples of properties for some specific domains. The complete list of AAS element classes, including those that do not qualify as Identifiable or Referable, can be found in [34].

Once the structure and some of the main components of the AAS are presented, it is possible to illustrate its metamodel. Figure 1 illustrates this metamodel with the components that were presented through a UML class diagram. The representation of AAS in the diagram also contains an example for the content of the AAS elements to highlight that its strict, coherent structure can be composed of data in different formats to contemplate the heterogeneity of assets represented through this artifact. A generic servo motor was considered as the asset to be described by the AAS. In Figure 1, the acronyms SM, SMC, Prop, and CD stand for Submodel, SubmodelElementCollection, Property, and ConceptDescription, respectively.

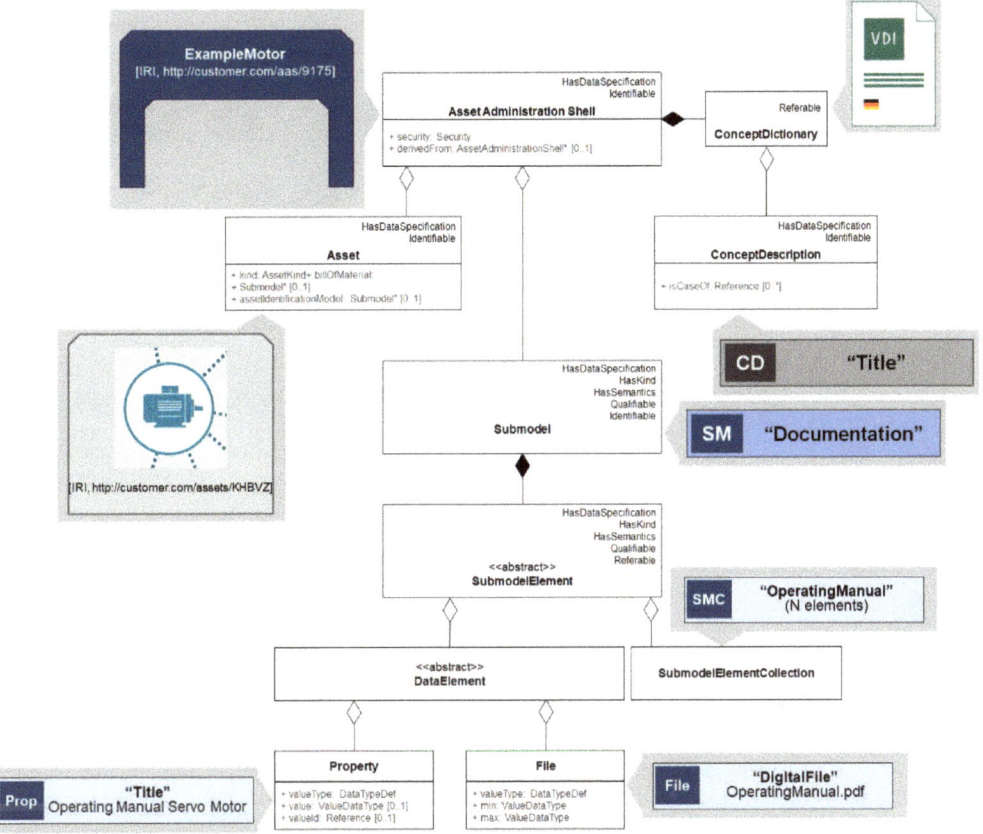

Figure 1. UML class diagram representing the metamodel of the AAS with main elements. Based on a generic servo motor as an asset, an example for the content of the AAS elements is also provided. Adapted from [34,41].

The standard solutions proposed for I4.0 need to be comprehensive enough not to limit the possibility of carrying out the Smart Factory in most different organizations. In this sense, no specific strategy for implementing the AAS is imposed by standardizing the digital format of representation and exchange of information. In [42], some of these possibilities are explored, taking into account different implementation perspectives. The different possibilities provide characteristics that can be advantages or disadvantages for specific applications. These characteristics include computational power, availability, performance and latency, security and reliability, maintenance, administration and management cost, failure identification, and recovery. Here, three perspectives of AAS implementation presented in [42] are described, along with the advantages and disadvantages.

The first issue to be discussed regarding AAS implementation is the computing platform. Three possibilities of implementation are presented, as illustrated in Figure 2. In the first one (Figure 2a), the AAS is embedded in the asset which, in turn, contains an execution environment for its digital representation. It is the case that an implementation based on an edge computing platform can be used as an example for such an implementation. In a second possibility (Figure 2b), the AAS can be physically separate from the asset but residing in the local IT infrastructure, connected to the asset through a local network. This case corresponds to a fog computing platform-based implementation. As a third possibility

(Figure 2c), the location of the AAS can be even further away from the asset in a cloud computing platform-based implementation. In this case, AAS and assets connect via external internet networks.

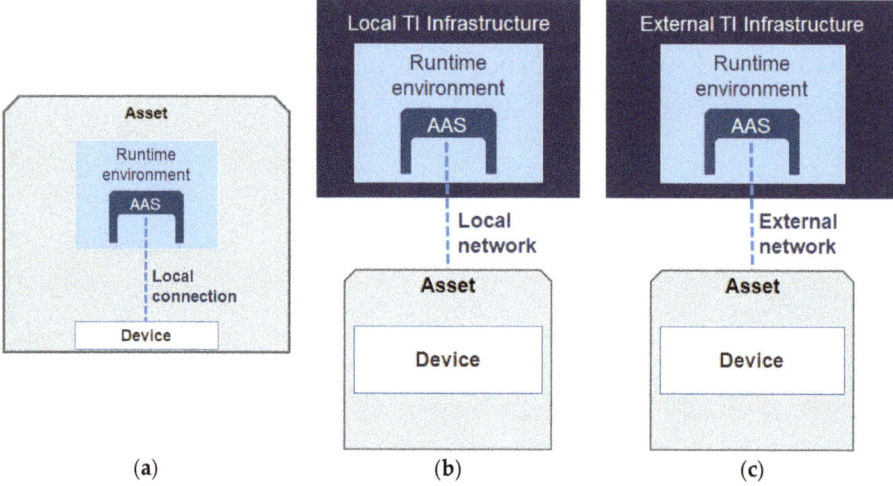

Figure 2. Computing platform perspective for implementation of the AAS: (**a**) edge, (**b**) fog, (**c**) cloud. Adapted from [42] with permission from the Federal Ministry for Economic Affairs and Climate Action. Scharnhorststraße 34–37, 10115 Berlin.

The second implementation perspective concerns the scalability of AAS. In a simplified way, scalability is related to the possibility of distributing data storage and processing across multiple nodes of a network. In this subsection, three possibilities for AAS distribution are considered, as illustrated in Figure 3: Figure 3a centralized, in which all information and services are allocated in a single node; Figure 3b loosely coupled distributed, where different nodes store information for the same asset (same identifier) and can be accessed individually; and Figure 3c distributed with aggregator node, which differs from the previous implementation by including an aggregator node, which gathers information from the nodes on which the AAS is distributed, forming a single data access point.

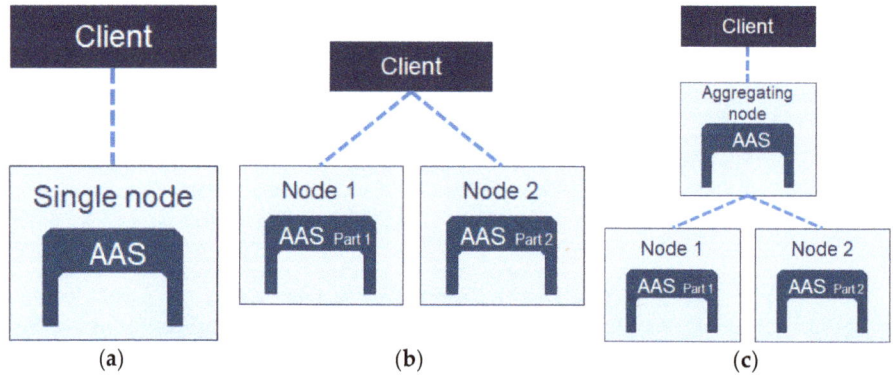

Figure 3. Distribution perspective for implementation of the AAS: (**a**) centralized, (**b**) distributed with loose-coupling, (**c**) distributed with aggregating node. Adapted from [42] with permission from the Federal Ministry for Economic Affairs and Climate Action. Scharnhorststraße 34–37, 10115 Berlin.

In Figure 2, the implementations differ in the distance between the AAS execution environment and the asset. However, no further consideration is made concerning the specific execution environment, which defines a form of AAS virtualization. This form of virtualization is an implementation perspective that implies advantages and disadvantages for the application. Three possibilities are presented and illustrated in Figure 4: the implementation based on Figure 4a operating system, Figure 4b hypervisor, and Figure 4c container. In the first case, the operating system is the AAS execution environment itself; that is, the execution environment consists of a process of a dedicated operating system or running inside another process. In the second case, the AAS execution environment is a virtual machine (VM). Multiple virtual machines with their own operating systems are allocated to a host (host) machine; they use its hardware and a hypervisor manages it. As in the previous case, AAS would still function as a complete operating system process but, in this case, this process would share hardware resources with other operating systems and applications. Finally, in the third possibility, the AAS execution environment consists of containers which run on top of an operating system. Unlike virtual machines, applications run in containers are run on the host machine's operating system, requiring only minimal resources such as applications and APIs needed to run the application, in this case, the AAS [43]. Two issues are related to the AAS execution environment, namely to its virtualization form: isolation and performance. These two characteristics form a trade-off, as pointed out by [44].

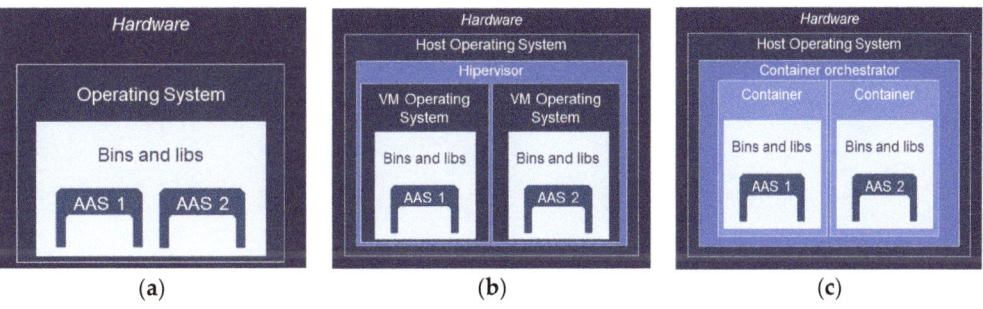

Figure 4. Virtualization perspective for implementation of the AAS: (**a**) operating system, (**b**) hypervisor, (**c**) container. Adapted from [42] with permission from the Federal Ministry for Economic Affairs and Climate Action. Scharnhorststraße 34–37, 10115 Berlin.

4. Results

This section presents results regarding the importance of technologies and methods gathered under the term "Big Data and Analytics" for I4.0 and correlates the characteristics of the data in this context with those of the relational and NoSQL databases.

4.1. Data Characteristics in the Context of I4.0

There is still no consensus on the technological pillars that support Industry 4.0 (I4.0). It is also observed that most authors put "Industry 4.0" and "Fourth Industrial Revolution" as synonyms, considering that there is no distinction between the phenomena, making it even more challenging to identify the technological pillars associated with each of the concepts. Table 1 presents the views of different authors about the key enabling technologies of the I4.0.

Despite this mentioned lack of consensus, it can be observed that there are certain convergences between authors and organizations about the enabling technologies of I4.0. It can be seen from Table 1 that the only technology identified as such in all the works consulted was big data and analytics. Thus, although there is no total consensus among the authors, the relevance of big data and analytics for the industry in the coming decades can be recognized. In brief, the term big data and analytics comprises (i) data sets "characterized by a high volume, velocity and variety to require specific technology and analytical methods for its transformation into value" [45], as well as the technologies and data analysis methods (big data analytics) themselves for this type of information asset, that is, specific to its characteristics [46,47].

Data generated, collected, transmitted, and possibly analyzed in real-time will be part of the Smart Factory [48]. The fact that big data and analytics systems are considered a pillar of I4.0 comes from the possibilities of improvements that can occur in a company based on its available data. The authors of [17,48] state that this type of system can be used to increase productivity and for better risk management, cost reduction, and aid in general decision making. For this reason, this set of technologies is considered fundamental for I4.0 [49]. Considering that big data and analytics is a key enabling technology of I4.0, the five dimensions of big data presented can be used to obtain a more general description of the nature of data in the context of I4.0. Still, based on the definition of big data presented, four dimensions characterize this information asset: the volume, speed, variety, and value of data. A fifth dimension—veracity—is still considered by some authors to encompass the reliability of the data [48,50]. These five dimensions, also called "5V", characterize big data and directly affect how data is stored, manipulated, and analyzed in IMS [51]; that is, they require technological solutions and specific methods for these functions. A description of these five dimensions of big data is presented in the next subsection with a discussion of how they are manifested in I4.0, taking into account the concept, structure, and implementation perspectives of the Asset Administration Shell and their impact on the design of databases for Intelligent Manufacturing Systems.

Table 1. Key enabling technologies for Industry 4.0 and/or the Fourth Industrial Revolution, according to different authors.

Technology	Rüssmann (2015) [52]	Bechtold et al. (2014) [53]	Lichtblau et al. (2015) [46]	Bauer et al. (2015) [54]	Petrillo et al. (2018) [55]	Wan, Cai, Zhou (2015) [56]
Big data and analytics	x	x	x	x	x	x
Advanced robotics	x	x		x	x	
Systems integration	x				x	x
Internet of things	x		x	x	x	x
Simulation	x				x	
Additive manufacturing	x	x	x	x	x	
Cloud computing	x	x	x		x	x
Virtual/augmented reality	x		x	x	x	
Cybersecurity	x				x	
Machine-to-machine		x		x		
Mobile technologies		x	x			
Location and detection technologies			x			
Human–machine interfaces			x	x		
Authentication and fraud detection			x			
Smart sensors			x			
Interaction with customers			x			
Community platforms		x				
Embedded projects						x
Self-guided vehicles						x
Social networks						x

4.2. Data Models in the Context of Industry 4.0

Section 3.2 allows for demonstrating that entities that seek to lead the evolution process of Industry 4.0 are clearly concerned with the description of data at a conceptual level. The creation of conceptual dictionaries, such as the aforementioned eCl@ss, IEC 61360, among others, defines the composition and semantics of the elements that make up the AAS. In terms of databases, it can be seen that efforts are directed towards building conceptual data models for I4.0. The same concern is not observed regarding the mapping of conceptual models at the logical level. The systematic review results, whose procedure was detailed in Section 2, demonstrates the gap that exists in the proposals for system architectures for I4.0 in terms of database design. Of the 139 papers resulting from the application of the search string on the Google Scholar platform, which reports the implementation of the AAS, 25 of them scored higher than five based on the criteria adopted and were analyzed. Only 32% (8 papers) at least inform the data model, or DBMS used, suggesting that studies in this area discuss this issue

Based on the results obtained from the systematic literature review, it is possible to identify the lack of rigor in the choice of logical data models that are used in databases for AAS implementations. Alternatively, several works discuss the applicability of data models in scenarios that can be observed in Industry 4.0 but consider a specific application [57–59] or do not propose a direct correlation with the phenomenon and its particularities [60,61]. Despite this finding, databases should not be understood as mere tools for data storage but as essential components of architectures, impacting their performance [16]. For this reason, database solution choices should not be made arbitrarily but based on criteria, application characteristics, users, and data. Given the reality exposed in the systematic review, i.e., the gaps in the implementation of database solutions in the context of I4.0, a correlation between the data characteristics described in the previous section—considering the five dimensions of big data—and the characteristics of relational and NoSQL data models can be introduced, discussing the adequacy of these models to the context of I4.0.

4.2.1. Operational and Analytical Databases

The subsections dealing with the dimensions of big data and analytics and the Asset Administration Shell make it possible to have a characterization of the data in the context of Industry 4.0. However, these characteristics manifest differently depending on application specifics. Therefore, in order to have a characterization of the data that allows for a deeper discussion about the suitability of data models for Industry 4.0, a brief differentiation between two types of databases, operational and analytical, is proposed. They differ in terms of the type of operations they perform most frequently, the volatility of the data stored, the number and type of users, the volume of data, their generation, and processing speed, among other characteristics, which are briefly described as follows:

- Operations: Operational databases are generally dedicated to online transaction processing (OLTP) applications, routine operations of an organization, which act on small fractions of the data, occur with great frequency, and must be processed efficiently, usually in real time [25]. Such operations are, for example, insertions, updates, deletions, and queries [26]. Analytical databases, in turn, are optimized for online analytical processing (OLAP), which allow you to extract value from the data through complex analytics. They are mainly dedicated to data recovery, involving grouping and joining operators, statistical functions, and complex Boolean conditions [26], applied to a large number of records and which, for this reason, usually occur in batches;
- Volume and volatility: Data analysis requires not only a comprehensive system perspective but often a considerable amount of historical data, such as time series. Thus, an analytical DB stores volumes of data much larger than its data sources— transactional DBs—in addition to ensuring the persistence of this data for much longer, while traditional DBs usually store current and non-historical data [62]. Thus, changes in the content of an analytical DB usually occur incrementally and in batches, while changes in transactional DBs occur continuously [26,62];
- Orientation and users: Analytical DBs are systems dedicated to facilitating the exploratory analysis of data, aiding decision-making and business processes [24]. For this reason, they are said to be subject oriented and generally dedicated to few users. Operational DBs, in turn, are purpose oriented and may have few to many users [62].

4.2.2. Volume

Volume is associated with the amount of data involved in big data applications. Although there is no consensus on a reference value for data volume for a database to be characterized as big data, this dimension is usually characterized by petabyte (PB) or exabyte (EB). According to [52], data volume in the modern industrial sector tends to grow by more than 1000 EB per year. In the context of I4.0, the digitization of the most diverse assets and the communication between them leads to an "unprecedented growth" in the volume of data, according to [56]. The authors of [49] state that big data cannot be manipulated in a single computer, requiring a distributed architecture for IMS. Thus, it is noted that the concern with the volume of data in I4.0 is reflected in the perspectives of implementing the AAS concerning its distribution.

The need for database distribution is a characteristic that imposes limitations on using the relational model for large volumes of data [63,64]. The main problem associated with using relational databases to implement distributed database systems is linked to the CAP theorem and, consequently, to ACID transactional properties. A single server system is a CA system: there is no partition tolerance because there is no partition on a single machine. Therefore, the two other properties—availability and consistency—are guaranteed. Most relational database systems are CA, and licenses for this type of DBMS are generally marketed to run on a single server [27].

On distributed systems, there is the possibility of partitioning the network. In this type of system, it is only possible to leave off partitioning tolerance if, in the event of a network partition failure, the system becomes completely inoperative, which is critically undesirable in some instances. Thus, it is generally not desirable to leave off the tolerance of network partitioning; that is, it is not desirable to have a distributed CA system. The other possibilities are leaving off consistency or availability. Therefore, the essence of the CAP theorem can be understood as: in a system that may be subject to partitioning, one must prioritize between consistency or availability. This turns out to be, in fact, a trade-off between consistency and latency: to have consistent transactions, a certain amount of time is needed for data changes to propagate to all copies, and the system can be available again [33].

Because transactions that adopt the ACID model are strongly consistent, it is impossible to balance the trade-off between consistency and latency in distributed databases that use this model of transactional properties [33]. For this reason, maintaining ACID properties generally implies higher latency [27,65] in a distributed database that implements the relational model. For high availability, data needs to be replicated across one or more nodes, so if one node fails, the data is available on another. Replication can increase availability and performance by reducing the overhead on nodes for reading operations. However, for "write" operations, where one wants to ensure that all nodes have an up-to-date copy of the data, one can experience a loss of performance (one must wait until the data is replicated across all nodes). On the other hand, in systems that adopt the BASE model, lower latency can be achieved, but inconsistencies can occur during a specific time interval (inconsistency window) since the different nodes can present different versions of the same record. Thus, it is understood that NoSQL systems, especially those oriented to aggregates, are beneficial for I4.0 in distributed database implementation scenarios, which usually contain large volumes of data, as they facilitate horizontal scalability.

In addition to performance, aggregate orientation is another reason why NoSQL data models better suit distributed database systems. Specific applications may contain data sets that are frequently accessed and manipulated together. In distributed systems, these sets can form a natural distribution unit [27] so that interested users are always directed to one or more specific nodes where they are stored. In databases, these sets are called aggregates: a rich structure formed by a set of data (objects) that can be stored as a unit, as they are often manipulated in this way. Elements of AAS as Submodel (set of Submodel Elements), Submodel Element Collection, and AAS itself (set of other elements) can be understood as aggregates. As such, aggregate-oriented NoSQL models are helpful for distributed implementations of AAS.

In terms of the implementation data model, the main problem of the relational model regarding the representation of aggregates is its rigid structure, which makes it impossible to treat data sets as a unit. The relational model allows representing the entities and relationships that are part of an aggregate. However, it does not allow to represent the aggregate itself; that is, it does not allow identifying which relationships constitute an aggregate nor the boundaries of this aggregate. For applications looking to process aggregates as a whole, the NoSQL key-value data model is beneficial because the aggregate is opaque. In case it is necessary to access parts of an aggregate, the document data model is more suitable than the key-value, as the aggregates are transparent in the case of the former; that is, partial operations can be performed on the data of the aggregates. For processing data from simpler formats such as numeric values, strings, etc., column family is also suitable for the purpose.

In summary, it can be argued that centralized implementations are suitable for smaller volumes of data, while distributed implementations are suitable for large volumes. There are relational databases that can be horizontally scalable; that is, they can be distributed [66] but the possible high unavailability of the system can make this distribution unfeasible. The trade-off between consistency and latency can be associated with two other dimensions of data—veracity and velocity—respectively, so that the speed dimension alone is not able to determine a more adequate data model.

4.2.3. Variety

Variety refers to the different formats of data. Big data applications can involve structured data, such as rigidly structured tables populated with scope-limited values; semi-structured as documents with a pre-defined template; and unstructured, such as multimedia content (image, audio) [67]. It is possible to argue that such heterogeneity may exist in the industrial context but it is possibly more "controlled". Despite this fact, variety is still a characteristic of the data in I4.0, considering that the AAS proposal presupposes a standardized format of representation and exchange that must be able to include assets of the most diverse natures. Thus, variety is a feature of the data in I4.0.

In the context of I4.0, this inability (or difficulty) can be verified in the attempt to create an AAS metamodel in a relational schema. Since the AAS must contemplate all I4.0 assets, it has a vast number of classes (entities) that represent each of its elements that would be translated into a large number of relations that could be even more significant if normalization procedures are applied. Associated with this complexity arising from mapping the AAS metamodel in the relational scheme, assets also have heterogeneity. When building a database composed of different assets, this heterogeneity can imply many null fields, which is undesirable.

Scenarios in which data heterogeneity is present may require flexibility in databases. A flexible data structure is not observed in the relational model, both in terms of the relationship scheme and the restrictions imposed by domains on possible values for attributes in the relationships. The flexibility of NoSQL systems makes them suitable for I4.0. Such flexibility enables the storage of semi-structured data, which best characterizes AAS. Technical reports from the Plattform Industrie 4.0 [34] and academic papers [68–70] present AAS implementations in XML and JSON format, which suggests that document-oriented NoSQL systems are advantageous, although it is not the only one capable of storing semi-structured data. This document encoding type is supported by essential communication technologies relevant to the I4.0, such as OPC UA [71] and HTTP. In summary, NoSQL data models adapt to the characteristic variety of data in I4.0, enabling the storage of heterogeneous records in the same DB. Thus, the flexibility of NoSQL models has its importance in the context of I4.0.

4.2.4. Velocity

This dimension can be understood as having two components—the velocity with which data is generated and the velocity with which it is processed [57]. In older big data applications, processing was commonly performed in batches, so the velocity at which data is generated and captured is critical to ensuring its reliability. Newer applications enable data processing in real time and in data streams so that, in addition to ensuring data reliability, the generation velocity must be consistent with the data processing velocity [52,54,57]. These two forms of processing are essential to guide the choice of database.

In addition to the high growth rate of data volume, which has already been mentioned and is more associated with data capture velocity, one can also discuss data processing speed and data analysis in IMS. Batch processing consists of processing a large volume of data at a time. The literature reports examples of this type of processing in an industrial environment [58,59]. Furthermore, Data Warehouse systems are typical examples of this way of processing and analysis. Applications of real-time processing and analysis in an industrial environment are also reported in the literature [60–62]. Comparing the AAS implementation platforms—edge, fog, and cloud—the last two are more suitable for batch processing since, in general, they have greater computational capacity than the first. However, this finding can be changed with the evolution of technology and the possibility of increasing data processing and storage capacity in devices closer and closer to the edges of networks. The asset-based implementation, that is, on edge devices, favors real-time processing due to low response latency.

Before introducing the discussion of data models suitable for batch and real-time processing, it is essential to introduce the Speed Consistency Volume principle, or SCV principle. While the CAP theorem presented in Section 3.1.2 concerns data storage, the SCV principle deals with data processing. The first attests that it is impossible to simultaneously guarantee consistency, availability, and tolerance to the partition. The second states that it is impossible to guarantee processing speed, consistency of results, and processing large volumes simultaneously. Based on [72], each of the elements that compose the SCV principle is described:

- Speed: deals with the speed at which data can be processed. To calculate the processing speed, the time spent to capture data should not be taken into account, considering only the actual processing time;
- Consistency: concerns the accuracy and precision of the results obtained from data processing. Inconsistent systems cannot use all available data to be processed, adopting sampling techniques, which leads to less precision and accuracy of results. On the other hand, systems with greater consistency use all available data in processing, obtaining more precise and accurate results;
- Volume: deals with the amount of data that can be processed. Large volumes of data require distributed processing, while smaller volumes can be processed centrally.

To analyze the "velocity" dimension, batch processing is initially considered. This type of processing is generally applied to analytical databases, which store large volumes of data and value the precision and accuracy of the results. Thus, from the point of view of the SCV principle, the properties that manifest themselves in this type of processing are volume and consistency. Analogously, considering the CAP theorem, the demands for consistency and tolerance to partition are manifested at the expense of speed. Thus, batch processing is often characterized by longer response times. Thus, data models that enable distribution and ensure data consistency are more suitable. Still regarding the velocity dimension in big data, the case of real-time processing is now analyzed. This type of processing is often used in operational databases. For those which store small volumes of data, there is not necessarily a distribution requirement, so the database system can be classified from the point of view of CAP theorem as a CA system, where the availability is low and consistency is ensured. In these cases, data models that implement ACID transactional properties are recommended. In cases where data have complex connections to each other, graph databases are particularly more efficient. For operational databases with small data volumes, the implementation of AAS based on edge computing platform is suitable for this type of processing, as being closer (or even embedded) to the asset, delays tend to be smaller.

Operational databases can also contain large volumes of data that cannot be left off, which imposes the need for storage and processing distribution. Thus, there are trade-offs between processing speed and consistency of results (SCV principle) and between availability and consistency (CAP theorem). In real-time processing, as delays are unwanted, trade-offs tend to prioritize speed and availability over consistency. It is known that ACID transactional properties do not allow the relaxation of consistency in favor of increased availability. Thus, aggregate-oriented NoSQL systems, which implement the BASE model of transactional properties, may be more suitable solutions for real-time processing. However, the level of consistency required by the application must be taken into account so that, by maximizing availability and processing speed, precision and accuracy requirements are not violated. Aggregate-oriented models are even more efficient; they guarantee higher processing speed if they do not need to perform operations on multiple aggregates simultaneously.

4.2.5. Veracity

Veracity is associated with the reliability of the captured data. The authors of [64] argue that the veracity dimension has three components: objectivity/subjectivity, which is more linked to the nature of the data source; deception, which refers to intentional errors in the content of the data or malicious modifications thereof; and implausibility (implausibility, irrationality) of the data, which refers to the quality of the data in terms of its validity, that is, its degree of confidence. Such concern is observed in the context of I4.0 as authors consider cybersecurity as a pillar of I4.0 (see Table 1), which presupposes protection against errors and intentional modifications to the data. It is also observed in the AAS implementation perspectives, where the virtualization strategy directly affects the isolation between applications [65] and, consequently, confidentiality and data integrity.

Some causes for veracity problems associated with implausibility, such as inconsistency, latency, and incompleteness, are pointed out by [67]. Therefore, it is observed that these causes and, consequently, the veracity is essential for the database design.

It is possible to associate the causes of implausibility with the properties of the CAP theorem and thus discuss the "truthfulness" dimension for different database systems. The inconsistency that affects the veracity of the data is directly linked to the consistency referred to in the CAP theorem. Latency is associated with the availability property, as seen in Section 3.1.2. The issue of incompleteness, in turn, is not directly associated with a property of the CAP theorem but with the transactional guarantee of atomicity, which establishes that a transaction must be performed entirely or not be performed at all. Thus, there is a foundation to discuss the impact of data models on veracity.

ACID transactional properties contribute to data veracity by enabling consistency and atomicity to operations. However, such properties imply high unavailability, which translates into delays in operations. Returning to the "speed" dimension of big data, if the data processing speed obtained through a system with ACID properties is consistent with the speed of data entry into the system, so there is no processing of outdated data, then these databases systems can be employed. Relational and graph-oriented DBMS generally adopt such properties.

Adopting the BASE model of transactional properties promotes an increase in availability at the expense of relaxation of consistency, which implies a decrease in the delay but rises the possibility of occurrence of inconsistencies. This does not mean that the BASE model is a bad choice when one wants to guarantee veracity based on completeness and consistency. The BASE model does not make it impossible to guarantee consistency but allows a balance of the trade-off between consistency and availability to better suit the application's need. Thus, one of the properties can be prioritized according to the characteristics of the application and the problems related to implausibility, whose susceptibility to the occurrence is greater. Thus, aggregate models can guarantee veracity, dealing with delay, incompleteness, and inconsistency, not simultaneously but balancing the problems according to the application demand.

This discussion is enriched with the distinction between analytical and operational databases. Analytical databases often have large volumes of data as they have less volatility. Therefore, they are generally implemented in distributed architectures, where the concurrency control problem is naturally less critical, especially when using aggregate-oriented data models, which allow users interested in a specific fraction of the data to always consult the network node that contains this fraction of the data, minimizing concurrent accesses. Additionally, analytical databases generally have fewer users, which further reduces the need for tight concurrency control. For these reasons, databases that implement the BASE model of transactional properties become a more suitable option. Operational databases have a greater number of users and, for this reason, they need to perform concurrency control more rigidly. For these cases, data models that implement the ACID model of transactional properties emerge as more viable options.

Finally, one can also consider in the discussion of veracity the differentiation between integration and application databases. The former store's data from multiple applications in a single database. This type of system has a much more complex structure than would be required by individual applications, as there is a need to coordinate and orchestrate applications, which differ above all in terms of performance requirements in terms of their operations. Application databases, in turn, are accessed and updated by a single application. This type of implementation allows databases to be encapsulated to applications, and the integration between them occurs through services so that application databases are fundamental for web applications and service-oriented architectures in general [27]. In an I4.0 context, it is possible to observe that the AAS implementation perspectives regarding its virtualization support both types of databases, especially concerning the degree of isolation between applications.

Integration databases generally implement the relational model, as the ACID properties precisely confer the desired concurrency control to coordinate the requests of different users/applications of the database [73]. However, for application databases, the relational model entails specific unnecessary and even undesirable characteristics: an application database usually requires a much smaller number of operations offered by the SQL language [27] and ACID transactional properties, which ensures concurrency control becomes unnecessary as only one application accesses the database [27].

4.2.6. Value

This dimension is associated with the value that can be extracted from the data through data analytics. Extracting value from data consists of converting the data into entities with a higher hierarchical level [57]. It involves a series of data analysis techniques, including machine learning, that requires a multidisciplinary approach, and, above all, it receives the name "value" because it offers prospects for improvement and cost reduction in terms of products and processes in the industry [52,63] so that it is possible to argue that there is a loss in not extracting value from the data. Extracting value from data in a significant data context presupposes the application of specific technologies and analytical methods. This dimension highlights the importance of data for the industrial sector as it brings the possibility of implementing improvements in organizations based on data.

Extracting value from data involves employing data analysis techniques in a multidisciplinary approach, which can translate into a naturally distributed organizational structure of a company or institution. Regarding the AAS implementation perspectives, its distribution in different network nodes, in fact, better reflects the structure of organizations today [41]. In these situations, an organization's subdivision is responsible for a fraction of the AAS or the whole AAS that concerns it. Because applying data analysis techniques in an organization to extract value from them requires the integration of different perspectives, it assumes that distributing AAS across different nodes also requires that these "AAS fragments" (or different AAS) be integrated to extract value from its data. Taking up the perspectives of AAS implementation regarding its forms of distribution, the distributed solution with the aggregator node can be an adequate solution for data integration. This does not mean that the aggregator node needs to act as a master node of the network, which manages all routine transactions, but that it can act as a data integration node and as a source of access by those members of the organization who intend to extract data value.

Although the extraction of value from the data is more linked to data analysis than to its storage, here is an excerpt of this dimension regarding database management systems, considering the interdisciplinarity and distribution of data in organizations. Thus, the analysis of the adequacy of database systems for this "value" perspective is mainly achieved by taking into account the importance of AAS distribution and the need for integration to extract value from asset data.

Network nodes that only contain AAS data referring to an organizational unit of the institution are those that process more routine transactions, the so-called online transaction processing. The databases of these nodes are called operational or transactional databases. A network node that promotes data integration, in turn, processes transactions with an analytical purpose, online analytical processing, and provides data for algorithms and other subsystems, acting, in fact, as an integrator database. In an institution with a distributed organizational structure, the data models that enable the horizontal scalability of the database, that is, the NoSQL DBMSs oriented to aggregates, are suitable for implementing "transactional" nodes, that is, those that process routine data transactions' specific AAS that pertain to a unit of the organization. If AAS distribution is not performed through the DBMS itself but through application databases that communicate by means of service interfaces, then NoSQL data models are still applicable. The flexibility provided by these models allows each company's subdivision to adopt the data models that best suit their application.

An integrator node is usually built from the so-called multidimensional data model at the conceptual level of data abstraction [74,75]. This is where the value is extracted from the data utilizing (big) data analytics methods. The multidimensional model is generally mapped at the implementation level through a relational scheme, although there are literature works that seek to map the multidimensional model in NoSQL models [76–78]. In particular, the importance of this mapping for the graph model is highlighted: the integrator node usually performs processing in batches and, based on the discussions in the previous subsection, it was argued that the graph-oriented model is suitable for this type of processing.

5. Discussion

Big data dimensions and other data characteristics in I4.0 were addressed in the previous subsections to discuss the suitability of data models to different realities of I4.0. However, interrelationships among these characteristics are analyzed for the database design, as it is possible to observe that one dimension can affect the others regarding the data model to be used. The dimensions "volume" and "velocity", for example, are correlated according to the SCV principle. When dealing with the "veracity" dimension, the impact of the BASE and ACID models on the veracity of the data was discussed. However, using one or another model of transactional properties also affects the distribution of data associated with the "value" dimension. The variety dimension, which concerns the possibility of storing data with more complex structures and, therefore, presupposes the use of more flexible data models such as those oriented to aggregates, for example, also implies the speed of processing multiple records, which is inferior in this type of data model.

Two qualitative analyses are presented to synthesize the results of the last section. The first of them is represented in Table 2, in which the dimensions "volume", "velocity", and "veracity" are associated with the two models of transactional properties, that is, BASE and ACID. As seen earlier, the first model is generally implemented in aggregate-oriented NoSQL databases, while the second is implemented in relational and graph databases.

Table 2 also includes the type of database—analytical or operational—where the scenario is more likely to be observed.

Table 2. Most suitable model of transactional properties according to the volume, velocity, and veracity of data.

Volume	Velocity	Veracity	Database Type	Suitable Model of Transactional Properties
Low	Low	Low	Operational	BASE
		High	Operational	Both
	High	Low	Operational	BASE
		High	Operational	ACID
High	Low	Low	Analytical	BASE
		High	Analytical	Both
	High	Low	Operational	BASE
		High	Analytical	BASE

The recommendations for each of the lines of Table 2 are presented here.

- The first scenario characterized by low volume, velocity, and veracity is more likely to be observed in operational DBs even though low velocity is usually not desirable. Considering the correlation between veracity and consistency presented in Section 4.2.5, the BASE model allows flexibility of consistency and, consequently, of veracity. This is the determinant factor for this recommendation;
- The second scenario with low volume and high veracity is again more likely to be observed in operational DBs despite the low velocity. As there is no need for distribution due to the low volume nor high-velocity requirements, ACID and BASE can ensure high veracity;
- The third scenario with low volume, high velocity, and low veracity can represent an operational DB. As there is a demand for high velocity at the expense of veracity, the BASE model is more suitable as it allows for relaxation of consistency in favor of availability;
- The scenario with low volume and high velocity and veracity well represent an operational DB as well as an analytical DB in its early stages. Since the database design needs to take into account the evolution of the DB, this analysis is made considering the former. Based on the CAP theorem and the SCV principle, the requirements of high speed and veracity imply the need for centralization of the database so that it is not subject to partition. Since the volume of data considered is small, there is no problem regarding distribution. For a CA system such as this, the ACID model is more suitable;
- Despite the low veracity, the fifth scenario may better represent an analytical DB than an operational DB. Although this type of DB requires high veracity (consistency), the distribution and the lack of strict concurrency control make the BASE model more suitable;
- The sixth scenario represents an analytical DB well. The requirement of high veracity (consistency) at the expense of speed can be initially associated with the ACID model. However, the distribution and no need for strict concurrency control present in an analytical DB mean that the BASE model can also be used;
- The scenario with high volume and velocity and low veracity illustrates an operational DB well. As there is a demand for high speed at the expense of veracity, the BASE model is more suitable as it allows for relaxation of consistency in favor of availability, especially in a distributed system;
- Regarding the eighth scenario, even though it is ideal for both an operational and an analytical DB, based on the CAP theorem and the SCV principle, it is not possible

to guarantee the three properties simultaneously neither with the BASE model nor with ACID. However, it is important to recognize that the very nature of the analytical distributed DB without the need for concurrency control contributes to high veracity. Thus, in an analytical database, to ensure distribution of a large volume of data and high processing speed, the BASE model can be used;

Table 2 does not refer to one or more data models specific to each scenario. The "variety" dimension, in addition to data linkage complexity and the flexibility of access, can be taken into account so that, based on a model of transactional properties, the choice for a data model can be made. Table 3, inspired by [51], synthesizes these three characteristics also qualitatively, suggesting the most appropriate data model with ACID transactional properties. Likewise, Table 4 suggests the most suitable data models with BASE transactional properties according to the veracity dimension, access flexibility, and data linkage complexity.

Table 3. Most suitable data model with ACID properties according to veracity, access flexibility, and data linkage complexity.

Variety	Access Flexibility	Data Linkage Complexity	Suitable Logical Data Model
Low	Low	Low	Relational
		High	Graph
	High	Low	Relational
		High	Graph
High	Low	Low	Graph
		High	Graph
	High	Low	Graph
		High	Graph

Table 4. Most suitable data model with BASE properties according to veracity, access flexibility, and data linkage complexity.

Variety	Access Flexibility	Data Linkage Complexity	Suitable Logical Data Model
Low	Low	Low	Key-value
		High	Document
	High	Low	Column family
		High	Document
High	Low	Low	Column family
		High	Document
	High	Low	Column family
		High	Document

Initially, database recommendations that implement the ACID model of transactional properties are discussed. For scenarios where data have high complexity in connections, the graph model is strongly recommended. This type of data model is also ideal in scenarios with high variety, where the rigid structure of the relational model is a disadvantage. Both allow flexible access to data and, therefore, in scenarios with less variety and complexity of connections, the relational model emerges as a viable option.

Some considerations about the recommendations of data models which implement the BASE transactional properties are presented: Key-value databases can easily store data with high complexity and variety but, in these cases, the complexity of handling the data is transferred to the application that deals with the data since the aggregate is opaque. That is the reason why the key-value data model is only recommended here for the scenario with low variety and linkage complexity.

Column family databases are strongly recommended for analytical databases. The way related data are organized in groups (the column families) optimizes not only operations for retrieving records (especially for similar data), as the rows are indexed, but also aggregate functions such as statistical operations, as the column families are also associated with primary keys. This data model can provide high access flexibility, but the structure of the database needs to be previously known.

Document databases are strongly recommended for storing and handling unstructured and semi-structured data. That is why they are suggested over column family databases for scenarios with high variety and data linkage complexity. They also provide higher access flexibility in comparison to column family as the aggregate is transparent; metadata is encapsulated into the document.

It is possible to observe that, when considering the specifications of a given application along the dimensions, the choice for a database generates trade-offs in terms of the requirements that can be met. In specific applications, conflicting characteristics from a database standpoint can be equally important. For this reason, it is common to find applications, especially in service-oriented architectures, in which multiple databases are used to meet the different application specifications satisfactorily. Works in this area are referred to as polyglot persistence [16,27], in which each database is responsible for managing data from a part of the application.

Finally, it is important to highlight that there are other factors linked to the specific characteristics of applications that can significantly affect the performance of databases. Optimization solutions are also constantly being developed [79,80]. Prominence is given to the class of databases called NewSQL, which seek to optimize the scalability of traditional relational databases such as that of NoSQL systems. The mapping of the conceptual model to the logical model itself can impact the performance of the database, as pointed out in [81]. All these factors may eventually modify the recommendations presented in this work.

6. Conclusions

This work presents different contributions regarding the database in the context of Industry 4.0 (I4.0). Given the importance of understanding the characteristics of the data for the design of a database, this article provided a comprehensive description of the data in the present context, identifying, for this purpose, the technologies, methods, and standards related to data that show fundamentals for the I4.0.

Systems architectures that organize the fundamental technologies and methods to provide functionalities of Intelligent Manufacturing Systems (IMS) can be provided. An indispensable element for these architectures is the database. Regarding the design of this element, this paper seeks to corroborate the assertion that, among the works that propose architectures for IMS, including adopting its standardizations such as Asset Administration Shell (AAS), few demonstrate evident concern and justification about the choice of data models to be used and how databases can influence the performance of these architectures. Subsequently, based on the characterization of the data in an I4.0 context, analysis was made of how the characteristics of relational and NoSQL data models fit into the dimensions of the data—volume, velocity, variety, veracity, and value. These analyses were summarized in Tables 2–4, in which hypothetical scenarios were built based on four of the five dimensions of data and other characteristics such as flexibility of access and complexity of data connections. The transactional guarantee models (ACID and BASE) and the data models (relational and NoSQL) that best fit each scenario were suggested.

The results presented in this paper adopted a qualitative comparison between data models. Works found in the literature propose comparisons between the performance of relational and NoSQL databases based on quantitative metrics [32,57,82]. However, the dimensions dealt with in this article can be analyzed quantitatively. The velocity dimension is widely used for performance comparisons across databases. This dimension can be measured in terms of the time it takes for database instantiation, reading, writing, removing, and searching operations to be performed on the database, as conducted by [83]. The volume dimension, strongly associated with the distributed databases, can also be evaluated quantitatively. In [84], the performance of databases is compared in terms of the number of operations performed per second, but having as parameters in the comparisons the amount of data stored and the number of nodes in the cluster where the data is distributed. Quantitative metrics for evaluating flexibility are presented in [85]. Data linkage and structure complexity can be quantitatively accessed by the metrics defined in [86]. Thus, future work can explore the dimensions by which the data were characterized in this paper and quantitatively assess the performance of the data models for the scenarios presented.

Furthermore, the previous section briefly mentions the concept of polyglot persistence, in which multiple databases are used in the architecture for the IMS as a whole or its subsystems. This work considered the use of a single relational or NoSQL data model for each scenario and then it was pointed out which would be a possible, most adequate choice. Future work can explore the combination of different data models for each scenario and discuss the possible improvements this combination would have, as well as the cost of managing more than one database for each application.

Author Contributions: Conceptualization, V.F.d.O. and F.J.; data curation, V.F.d.O. and F.J.; formal analysis, V.F.d.O. and P.E.M.; funding acquisition, F.J., M.A.d.O.P. and P.E.M.; investigation, V.F.d.O. and F.J.; methodology, V.F.d.O., F.J. and P.E.M.; project administration, V.F.d.O., F.J., M.A.d.O.P. and P.E.M.; resources, V.F.d.O., F.J., M.A.d.O.P. and P.E.M.; software, V.F.d.O.; supervision, F.J., M.A.d.O.P. and P.E.M.; validation, V.F.d.O., F.J., M.A.d.O.P. and P.E.M.; visualization, V.F.d.O., F.J., M.A.d.O.P. and P.E.M.; writing—original draft preparation, V.F.d.O. and F.J.; writing—review and editing, V.F.d.O., F.J., M.A.d.O.P. and P.E.M. All authors have read and agreed to the published version of the manuscript.

Funding: This research was supported by the Coordenação de Aperfeiçoamento de Pessoal de Nível Superior (CAPES), grant number 88887.508600/2020-00; Fundação de Amparo à Pesquisa do Estado de São Paulo (FAPESP, São Paulo Research Foundation), grant number 2020/09850-0; and Conselho Nacional de Desenvolvimento Científico e Tecnológico (CNPq, National Council for Scientific and Technological Development), grant numbers 303210/2017-6 and 431170/2018-5.

Institutional Review Board Statement: Not applicable.

Informed Consent Statement: Not applicable.

Data Availability Statement: The database from the systematic literature review is available upon request from the corresponding author of this paper.

Conflicts of Interest: The authors declare no conflict of interest. The funders had no role in the design of the study; in the collection, analyses, or interpretation of data; in the writing of the manuscript; or in the decision to publish the results.

References

1. Nakayama, R.S.; de Mesquita Spínola, M.; Silva, J.R. Towards I4.0: A Comprehensive Analysis of Evolution from I3.0. *Comput. Ind. Eng.* **2020**, *144*, 106453. [CrossRef]
2. Tyrrell, A. Management Approaches for Industry 4.0: A Human Resource Management Perspective. In Proceedings of the 2016 IEEE Congress on Evolutionary Computation (CEC), Vancouver, BC, Canada, 24–29 July 2016; pp. 5309–5316.
3. Lasi, H.; Fettke, P.; Kemper, H.G.; Feld, T.; Hoffmann, M. Industry 4.0. *Bus. Inf. Syst. Eng.* **2014**, *6*, 239–242. [CrossRef]
4. Fragapane, G.; Ivanov, D.; Peron, M.; Sgarbossa, F.; Strandhagen, J.O. Increasing Flexibility and Productivity in Industry 4.0 Production Networks with Autonomous Mobile Robots and Smart Intralogistics. *Ann. Oper. Res.* **2020**. [CrossRef]
5. Yazdi, P.G.; Azizi, A.; Hashemipour, M. An Empirical Investigation of the Relationship between Overall Equipment Efficiency (OEE) and Manufacturing Sustainability in Industry 4.0 with Time Study Approach. *Sustainability* **2018**, *10*, 3031. [CrossRef]
6. Mohamed, N.; Al-Jaroodi, J.; Lazarova-Molnar, S. Leveraging the Capabilities of Industry 4.0 for Improving Energy Efficiency in Smart Factories. *IEEE Access* **2019**, *7*, 18008–18020. [CrossRef]
7. Brozzi, R.; D'Amico, R.D.; Pasetti Monizza, G.; Marcher, C.; Riedl, M.; Matt, D. Design of Self-Assessment Tools to Measure Industry 4.0 Readiness. A Methodological Approach for Craftsmanship SMEs. *IFIP Adv. Inf. Commun. Technol.* **2018**, *540*, 566–578. [CrossRef]
8. Morkovkin, D.E.; Gibadullin, A.A.; Kolosova, E.V.; Semkina, N.S.; Fasehzoda, I.S. Modern Transformation of the Production Base in the Conditions of Industry 4.0: Problems and Prospects. *J. Phys. Conf. Ser.* **2020**, *1515*, 032014. [CrossRef]
9. Hozdić, E. Smart Factory for Industry 4.0: A Review. *Int. J. Mod. Manuf. Technol.* **2015**, *7*, 28–35.

10. Shi, Z.; Xie, Y. Smart Factory in Industry 4.0. *Syst. Res. Behav. Sci.* **2020**, *37*, 607–617. [CrossRef]
11. Wang, S.; Wan, J.; Li, D.; Zhang, C. Implementing Smart Factory of Industrie 4.0: An Outlook. *Int. J. Distrib. Sens. Netw.* **2016**, *12*, 3159805. [CrossRef]
12. Klingenberg, C.O.; Borges, M.A.V.; Antunes, J.A.V. Industry 4.0 as a Data-Driven Paradigm: A Systematic Literature Review on Technologies. *J. Manuf. Technol. Manag.* **2021**, *32*, 570–592. [CrossRef]
13. Adolphs, P.; Bedenbender, H.; Dirzus, D.; Ehlich, M.; Epple, U.; Hankel, M.; Heidel, R.; Hoffmeister, M.; Huhle, H.; Kärcher, B.; et al. *Reference Architecture Model Industrie 4.0 (RAMI4.0)*; ZVEI—German Electrical and Electronic Manufacturers' Association: Berlin, Germany, 2015.
14. IVI—Industrial Value Chain Initiative. *Industrial Value Chain Reference Architecture (IVRA)*; Chiyoda: Chuo-ku, Japan, 2016.
15. Lin, S.-W.; Murphy, B.; Clauser, E.; Loewen, U.; Neubert, R.; Bachmann, G.; Pai, M.; Hankel, M. Architecture Alignment and Interoperability. *Plattf. Ind. 4.0* **2017**, *19*, 2–15.
16. De Oliveira, V.F.; Pinheiro, E.; Daniel, J.F.L.; Guerra, E.M.; Junqueira, F.; Santos Fo, D.J.; Miyagi, P.E. Infraestrutura de Dados Para Sistemas de Manufatura Inteligente. In Proceedings of the 14th IEEE International Conference on Industry Applications, São Paulo, Brazil, 15–18 August 2021; pp. 516–523.
17. Schwab, K. *The Fourth Industrial Revolution*; World Economic Forum: Geneva, Switzerland, 2017; ISBN 9781944835019.
18. Castelo-Branco, I.; Cruz-Jesus, F.; Oliveira, T. Assessing Industry 4.0 Readiness in Manufacturing: Evidence for the European Union. *Comput. Ind.* **2019**, *107*, 22–32. [CrossRef]
19. Lu, Y. Industry 4.0: A Survey on Technologies, Applications and Open Research Issues. *J. Ind. Inf. Integr.* **2017**, *6*, 1–10. [CrossRef]
20. Hermann, M.; Pentek, T.; Otto, B. *Design Principles for Industrie 4.0*; Technology University Dortmund: Dortmund, Germany, 2015; Volume 15.
21. Kagermann, H.; Riemensperger, F.; Hoke, D.; Schuh, G.; Scheer, A.-W. Recommendations for the Strategic Initiative Web-Based Services for Businesses. *Acatech Rep.* **2014**, *112*, 5–6.
22. Da Xu, L.; Xu, E.L.; Li, L. Industry 4.0: State of the Art and Future Trends. *Int. J. Prod. Res.* **2018**, *56*, 2941–2962. [CrossRef]
23. Bangemann, T.; Bauer, C.; Bedenbender, H.; Braune, A.; Diedrich, C.; Diesner, M.; Epple, U.; Elmas, F.; Friedrich, J.; Göbe, F.; et al. *Status Report—Industrie 4.0 Service Architecture—Basic Concepts for Interoperability*; ZVEI—German Electrical and Electronic Manufacturers' Association: Frankfurt, Germany, 2016; Volume 1.
24. Elmasri, R.; Navathe, S.B. *Sistemas de Banco de Dados*, 4th ed.; Pearson: London, UK, 2005; ISBN 9788578110796.
25. Ramakrishnan, R.; Gehrke, J. *Sistemas de Gerenciamento de Banco de Dados*, 3rd ed.; McGraw Hill: New York, NY, USA, 2008; Volume 14, ISBN 9788577803828.
26. Elmasri, R.; Navathe, S.B. *Fundamentals of Database Systems*, 7th ed.; Pearson: Hoboken, NJ, USA, 2000; ISBN 9780133970777.
27. Sadalage, P.J.; Fowler, M. *NoSQL Distilled: A Brief Guide to the Emerging World of Polyglot Persistence*; Pearson: Hoboken, NJ, USA, 2013; ISBN 9780321826626.
28. Codd, E.F. A Relational Model of Data for Large Shared Data Banks. *Commun. ACM* **1983**, *26*, 64–69. [CrossRef]
29. Özsu, M.T.; Valduriez, P. Distributed and Parallel Database Systems. *ACM Comput. Surv.* **1996**, *28*, 125–128. [CrossRef]
30. Özsu, M.T.; Valduriez, P. *Principles of Distributed Database Systems*, 3rd ed.; Springer: London, UK, 2011; ISBN 9781441988331.
31. NOSQL Databases. Available online: https://www.christof-strauch.de/nosqldbs.pdf (accessed on 2 August 2021).
32. Moniruzzaman, A.B.M.; Hossain, S.A. NoSQL Database: New Era of Databases for Big Data Analytics—Classification, Characteristics and Comparison. *Int. J. Database Theory Appl.* **2013**, *6*, 1–14. [CrossRef]
33. Abadi, D.J. Consistency Tradeoffs in Modern Distributed Database System Design: CAP Is Only Part of the Story. *Computer* **2012**, *45*, 37–42. [CrossRef]
34. Bader, S.; Barnstedt, E.; Bedenbender, H.; Billman, M.; Boss, B.; Braunmandl, A. *Details of the Asset Administration Shell Part 1—The Exchange of Information between Partners in the Value Chain of Industrie 4.0*; Federal Ministry for Economic Affairs and Energy (BMWi): Berlin, Germany, 2019.
35. Bedenbender, H.; Billmann, M.; Epple, U.; Hadlich, T.; Hankel, M.; Heidel, H.; Hillermeier, O.; Hoffmeister, M.; Huhle, H.; Jochem, M.; et al. *Examples of the Asset Administration Shell for Industrie 4.0 Components—Basic Part*; German Electrical and Electronic Manufacturers' Association: Frankfurt, Germany, 2017.
36. Gastaldi, L.; Appio, F.P.; Corso, M.; Pistorio, A. Managing the Exploration-Exploitation Paradox in Healthcare: Three Complementary Paths to Leverage on the Digital Transformation. *Bus. Process Manag. J.* **2018**, *24*, 1200–1234. [CrossRef]
37. Inigo, M.A.; Porto, A.; Kremer, B.; Perez, A.; Larrinaga, F.; Cuenca, J. Towards an Asset Administration Shell Scenario: A Use Case for Interoperability and Standardization in Industry 4.0. In Proceedings of the IEEE/IFIP Network Operations and Management Symposium, NOMS 2020, Budapest, Hungary, 20–24 April 2020. [CrossRef]
38. Gayko, J. The Reference Architectural Model Rami 4.0 and the Standardization Council as an Element of Success for Industry 4.0. 2018. Available online: https://www.din.de/resource/blob/271306/340011c12b8592df728bee3815ef6ec2/06-smart-manufacturing-jens-gayko-data.pdf (accessed on 2 August 2021).
39. Adolphs, P.; Auer, S.; Bedenbender, H.; Billmann, M.; Hankel, M.; Heidel, R.; Hoffmeister, M.; Huhle, H.; Jochem, M.; Kiele-Dunsche, M.; et al. *Structure of the Asset Administration Shell: Continuation of the Development of the Reference Model for the Industrie 4.0 Component*; Federal Ministry for Economic Affairs and Energy (BMWi): Berlin, Germany, 2016.
40. Ye, X.; Hong, S.H. Toward Industry 4.0 Components: Insights into and Implementation of Asset Administration Shells. *IEEE Ind. Electron. Mag.* **2019**, *13*, 13–25. [CrossRef]

41. Boss, B.; Malakuti, S.; Lin, S.-W.; Usländer, T.; Clauer, E.; Hoffmeister, M.; Stokanovic, L. *Digital Twin and Asset Administration Shell Concepts and Application in the Industrial Internet and Industrie 4.0*; Industrial Internet Consortium: Boston, MA, USA, 2020.
42. Bedenbender, H.; Bentkus, A.; Epple, U.; Hadlich, T.; Heidel, R.; Hillermeier, O.; Hoffmeister, M.; Huhle, H.; Kiele-Dunsche, M.; Koziolek, H.; et al. *Industrie 4.0 Plug-and-Produce for Adaptable Factories: Example Use Case Definition, Models, and Implementation*; Federal Ministry for Economic Affairs and Energy (BMWi): Berlin, Germany, 2017.
43. Gerend, J. Contêineres vs. Máquinas Virtuais. Available online: https://docs.microsoft.com/pt-br/virtualization/windowscontainers/about/containers-vs-vm (accessed on 15 June 2021).
44. Xavier, M.G.; Neves, M.V.; Rossi, F.D.; Ferreto, T.C.; Lange, T.; De Rose, C.A.F. Performance Evaluation of Container-Based Virtualization for High Performance Computing Environments. In Proceedings of the IEEE 2013 21st Euromicro International Conference on Parallel, Distributed and Network-Based Processing (PDP 2013), Belfast, UK, 27 February–1 March 2013; pp. 233–240. [CrossRef]
45. De Mauro, A.; Greco, M.; Grimaldi, M. A Formal Definition of Big Data Based on Its Essential Features. *Libr. Rev.* **2016**, *65*, 122–135. [CrossRef]
46. Lichtblau, K.; Stich, V.; Bertenrath, R.; Blum, M.; Bleider, M.; Millack, A.; Schmitt, K.; SChmitz, E.; Schröter, M. *Industry 4.0 Readiness*; IMPULS—Institute for Mechanical Engineering, Plant Engineering, and Information Technology: Frankfurt, Germany, 2015.
47. Russom, P. *Big Data Analytics*; TWDI—Transforming Data with Intelligence: Renton, WA, USA, 2011.
48. Yin, S.; Kaynak, O. Big Data for Modern Industry: Challenges and Trends. *Proc. IEEE* **2015**, *103*, 143–146. [CrossRef]
49. Kagermann, H.; Wahlster, W.; Helbig, J. Securing the Future of German Manufacturing Industry: Recommendations for Implementing the Strategic Initiative INDUSTRIE 4.0. *Final Rep. Ind. 4.0 Work. Gr.* **2013**, 1–84. Available online: https://www.din.de/blob/76902/e8cac883f42bf28536e7e8165993f1fd/recommendations-for-implementing-industry-4-0-data.pdf (accessed on 2 August 2021).
50. Gil, D.; Song, I.Y. Modeling and Management of Big Data: Challenges and Opportunities. *Futur. Gener. Comput. Syst.* **2016**, *63*, 96–99. [CrossRef]
51. NIST Big Data Public Working Group. *NIST Big Data Interoperability Framework: Volume 6—Reference Architecture*; NIST—National Institute of Standards and Technology: Gaithersburg, MD, USA, 2019; Volume 6.
52. Russmann, M.; Lorenz, M.; Gerbert, P.; Waldner, M.; Justus, J.; Engel, P.; Harnisch, M. Industry 4.0: The Future of Productivity and Growth in Manufacturing Industries. *Bost. Consult. Gr.* **2015**, 1–20. Available online: https://www.bcg.com/pt-br/publications/2015/engineered_products_project_business_industry_4_future_productivity_growth_manufacturing_industries (accessed on 2 August 2021).
53. Bechtold, J.; Kern, A.; Lauenstein, C.; Bernhofer, L. *Industry 4.0—The Capgemini Consulting View*; Capgmenini Consulting: Paris, France, 2014.
54. Bauer, H.; Baur, C.; Camplone, G.; George, K.; Ghislanzoni, G.; Huhn, W.; Kayser, D.; Löffler, M.; Tschiesner, A.; Zielke, A.E.; et al. *Industry 4.0: How to Navigate Digitization of the Manufacturing Sector*; McKinsey Digital: New York, NY, USA, 2015.
55. Petrillo, A.; De Felice, F.; Cioffi, R.; Zomparelli, F. Fourth Industrial Revolution: Current Practices, Challenges, and Opportunities. In *Digital Transformation in Smart Manufacturing*; Intechopen: London, UK, 2018; pp. 1–20.
56. Wan, J.; Cai, H.; Zhou, K. Industrie 4.0: Enabling Technologies. In Proceedings of the International Conference on Intelligent Computing and Internet of Things, ICIT 2015, Harbin, China, 17–18 January 2015; pp. 135–140. [CrossRef]
57. Fatima, H.; Wasnik, K. Comparison of SQL, NoSQL and NewSQL Databases for Internet of Things. In Proceedings of the 2016 IEEE Bombay Section Symposium (IBSS), Baramati, India, 21–22 December 2016. [CrossRef]
58. Rautmare, S.; Bhalerao, D.M. MySQL and NoSQL Database Comparison for IoT Application. In Proceedings of the 2016 IEEE International Conference on Advances in Computer Applications, Coimbatore, Tamilnadu, India, 24 October 2016; pp. 235–238.
59. Di Martino, S.; Fiadone, L.; Peron, A.; Vitale, V.N.; Riccabone, A. Industrial Internet of Things: Persistence for Time Series with NoSQL Databases. In Proceedings of the IEEE 28th International Conference on Enabling Technologies: Infrastructure for Collaborative Enterprises, Naples, Italy, 12–14 June 2019; pp. 340–345.
60. Bonnet, L.; Laurent, A.; Sala, M.; Laurent, B.; Sicard, N. Reduce, You Say: What NoSQL Can Do for Data Aggregation and BI in Large Repositories. In Proceedings of the International Workshop on Database and Expert Systems Applications, Toulouse, France, 29 August–2 September 2011; pp. 483–488.
61. Guzzi, P.H.; Veltri, P.; Cannataro, M. Experimental Evaluation of Nosql Databases. *Int. J. Database Manag. Syst.* **2014**, *6*, 656–660. [CrossRef]
62. Teorey, T.; Lightstone, S.; Nadeau, T.; Jagadish, H.V. *Database Modeling and Design*; Elsevier: Burlington, MA, USA, 2005.
63. Fowler, M. Reporting Database. Available online: https://martinfowler.com/bliki/ReportingDatabase.html (accessed on 15 June 2021).
64. Han, J.; Haihong, E.; Le, G.; Du, J. Survey on NoSQL Database. In Proceedings of the 2011 6th International Conference on Pervasive Computing and Applications, Port Elizabeth, South Africa; 2011; pp. 363–366. [CrossRef]
65. Khine, P.P.; Wang, Z. A Review of Polyglot Persistence in the Big Data World. *Information* **2019**, *10*, 141. [CrossRef]
66. Cattell, R. Scalable SQL and NoSQL Data Stores. *SIGMOD Rec.* **2010**, *39*, 12–27. [CrossRef]
67. Lee, E.A. *Cyber Physical Systems: Design Challenges*; Electrical Engineering and Computer Sciences, University of California at Berkeley: Berkeley, CA, USA, 2017.

68. Cavalieri, S.; Salafia, M.G. A Model for Predictive Maintenance Based on Asset Administration Shell. *Sensors* **2020**, *20*, 6028. [CrossRef] [PubMed]
69. Lang, D.; Grunau, S.; Wisniewski, L.; Jasperneite, J. Utilization of the Asset Administration Shell to Support Humans during the Maintenance Process. In Proceedings of the 2019 IEEE 17th International Conference on Industrial Informatics, Helsinki, Finland, 22–25 July 2019; pp. 768–773. [CrossRef]
70. Ye, X.; Jiang, J.; Lee, C.; Kim, N.; Yu, M.; Hong, S.H. Toward the Plug-and-Produce Capability for Industry 4.0. *IEEE Ind. Electron. Mag.* **2020**, *14*, 146–157. [CrossRef]
71. OPC Foundation. OPC 10000-6: OPC Unified Architecture—Part 6: Mappings 2017. Available online: https://reference.opcfoundation.org/v104/Core/docs/Part6/ (accessed on 2 August 2021).
72. Erl, T.; Khattak, W.; Buhler, P. *Big Data Fundamentals: Concepts, Drivers & Techniques*; Prentice Hall: Hoboken, NJ, USA, 2016; ISBN 0975442201.
73. Stonebraker, M. SQL Databases v. NoSQL Databases. *Commun. ACM* **2010**, *53*, 10–11. [CrossRef]
74. Inmon, W.H. *Building the Data Warehouse*, 3rd ed.; John Wiley & Sons: Hoboken, NJ, USA, 2002; ISBN 0-471-08130-2.
75. Thomsen, E. *OLAP Solutions: Building Multidimensional Information Systems*, 2nd ed.; John Wiley & Sons: Hoboken, NJ, USA, 2002; ISBN 0471400300.
76. Bicevska, Z.; Oditis, I. Towards NoSQL-Based Data Warehouse Solutions. *Procedia Comput. Sci.* **2016**, *104*, 104–111. [CrossRef]
77. Chevalier, M.; El Malki, M.; Kopliku, A.; Teste, O.; Tournier, R. Implementing Multidimensional Dasta Warehouse into NoSQL. In Proceedings of the 17th International Conference on Enterprise Information Systems, Barcelona, Spain, 27–30 April 2015.
78. Yangui, R.; Nabli, A.; Gargouri, F. Automatic Transformation of Data Warehouse Schema to NoSQL Data Base: Comparative Study. *Procedia Comput. Sci.* **2016**, *96*, 255–264. [CrossRef]
79. Mathew, A.B. Data Allocation Optimization for Query Processing in Graph Databases Using Lucene. *Comput. Electr. Eng.* **2018**, *70*, 1019–1033. [CrossRef]
80. Tan, Z.; Babu, S. Tempo: Robust and Self-Tuning Resource Management in Multi-Tenant Parallel Databases. In Proceedings of the 42th International Conference on Very Large Databases, New Delhi, India, 5–9 September 2016; Volume 9, pp. 720–731.
81. Atzeni, P.; Bugiotti, F.; Cabibbo, L.; Torlone, R. Data Modeling in the NoSQL World. *Comput. Stand. Interfaces* **2020**, *67*, 103149. [CrossRef]
82. Ali, W.; Shafique, M.U.; Majeed, M.A.; Raza, A. Comparison between SQL and NoSQL Databases and Their Relationship with Big Data Analytics. *Asian J. Res. Comput. Sci.* **2019**, *4*, 1–10. [CrossRef]
83. Li, Y.; Manoharan, S. A Performance Comparison of SQL and NoSQL Databases A Performance Comparison of SQL and NoSQL Databases. In Proceedings of the IEEE Pacific Rim Conference on Communications, Computers and Signal Processing, Victoria, BC, Canada, 27–29 August 2013; pp. 15–19.
84. Swaminathan, S.N.; Elmasri, R. Quantitative Analysis of Scalable NoSQL Databases. In Proceedings of the International Congress on Big Data, San Francisco, CA, USA, 27 June–2 July 2016; pp. 323–326.
85. Aravanis, A.I.; Voulkidis, A.; Salom, J.; Townley, J.; Georgiadou, V.; Oleksiak, A.; Porto, M.R.; Roudet, F.; Zahariadis, T. Metrics for Assessing Flexibility and Sustainability of next Generation Data Centers. In Proceedings of the 2015 IEEE Globecom Workshops, San Diego, CA, USA, 6–10 December 2015.
86. Nagpal, S.; Gosain, A.; Sabharwal, S. Complexity Metric for Multidimensional Models for Data Warehouse. In Proceedings of the CUBE International Information Technology Conference, Pune, India, 3–5 September 2012; pp. 360–365.

Article

Industrial Upper-Limb Exoskeleton Characterization: Paving the Way to New Standards for Benchmarking

Vitor Neves Hartmann, Décio de Moura Rinaldi, Camila Taira and Arturo Forner-Cordero *

Departamento de Engenharia Mecatrônica e Sistemas Mecânicos, Escola Politécnica da Universidade de São Paulo, São Paulo 05508-030, Brazil; vitor.hartmann@usp.br (V.N.H.); decio.rinaldi@usp.br (D.d.M.R.); catai@usp.br (C.T.)
* Correspondence: aforner@usp.br

Abstract: Exoskeletons have been introduced in industrial environments to prevent overload or repetitive stress injuries in workers. However, due to the lack of public detailed information about most of the commercial exoskeletons, it is necessary to further assess their load capacity and evolution over time, as their performance may change with use. We present the design and construction of a controlled device to measure the torque of industrial exoskeletons, along with the results of static and dynamic testing of an exoskeleton model. A step motor in the test bench moves the exoskeleton arm in a pre-defined path at a prescribed speed. The force measured with a beam load cell located at the interface between the exoskeleton arm and the test bench is used to derive the torque. The proposed test bench can be easily modified to allow different exoskeleton models to be tested under the same conditions.

Keywords: exoskeletons; test bench; industry; benchmarking

1. Introduction

Work-related musculoskeletal disorders (WMSDs) are a major concern in working environments [1], and they are often associated with repetitive movements and prolonged unfavorable postures during working shifts. WMSDs are the most prevalent disease among Brazilians, and to make matters worse, the number of cases escalated 184% within a period of 10 years, from 2007 to 2016, according to the Health Department in Brazil, 2018 [2].

Among WMSDs, shoulder injuries are one of the most severe, accounting for lost working days three times more than those of back injuries [1].

In Brazil, almost 39,000 workers were on sick leaves in 2019 due to WMSDs [3], and shoulder injuries are again among the leading causes [2].

Several tasks in the industry leave workers vulnerable to various injuries risks, which poses threats to factory workers' health as well as to their employers, and eventually, it may become an economic burden to the society due to therapies, leave period, and lost working days [1].

Handling, supporting, and lifting heavy materials may cause overexertion injuries, therefore contributing to WMSDs [4]. Working in awkward postures and repetitive motions are among the top 10 causes of disabling injuries also in the USA [5] apart from bringing massive costs on medical care and lost-wage expenses [1].

Assembly line workers and machine operators are frequently exposed to many unfavorable positions in their daily tasks, such as in handling heavy tools, drilling, countersinking, riveting, bucking, swaging [6] and paneling, electrical work, welding, grinding, picking, pruning, painting, inspection, and overhead assembly, all of which require sustained work at chest level to overhead level [7].

Exoskeletons are wearable devices that augment power, as well as assist and enable physical activity through mechanical interaction with the body by increasing strength and endurance, amplifying and reinforcing performance. In the case of industrial exoskeletons, mainly lower back and upper limbs work frequently objects of research [8–10].

Exoskeleton technology is an emerging technology used in different fields, for example, for military purposes [11], in the medical field for rehabilitation or assistance [12–14], or for industrial applications [15–20]. Even though there is a growing trend in automation and mechanization of a variety of processes in the industry, human–robot collaboration has been targeted in order to keep human flexibility, allowing high levels of dexterity and fine handling, much needed in dynamic manufacturing, for example [9,21]. Moreover, full automation is unfeasible and costly, and a wide range of products and tasks in the modern industry require human thinking skills, such as being able to observe, decide, and take actions [9], and this variation in demand allows us to outperform robots [16].

Upper-limb exoskeletons are designed to help sustain working posture, supporting the upper limbs during overhead tasks. To this end, they generate joint torques, for instance, in the shoulder, aiming at reducing shoulder overload and preventing injury [17]. Most of the commercially available passive exoskeletons can provide different levels of assistance by adjusting springs stiffness.

Although industrial exoskeletons have been extensively tested with subjects [18,19,22], there are no standardized procedures to test exoskeletons independently of the user that provide direct quantitative measurements. In this respect, it is difficult to assess different exoskeletons under the same conditions to obtain benchmarking metrics. Moreover, there are still gaps in the literature about design requirements and assessment of the exoskeletons in terms of the level of assistance provided to the users. In addition, the lack of validation standards also hinders their adoption in the industry, thus making their comparison difficult [23].

To disseminate knowledge and to better explore outcomes from different studies on such trendy technology, standardization is necessary. It provides minimum acceptable requirements for quality and safety for manufacturing products, because of materials that were previously tested, which also helps to minimize costs due to this testing and selection processes. Furthermore, complying with standards involves meeting legal demands, which facilitates commercialization and their adoption in global markets [24].

In a previous study [25], we presented the design of a test bench to measure the actuating torque of shoulder exoskeleton models, in order to compare these values between different manufacturers under controlled conditions. In this study, we present the final design and construction of this test bench and apply it to assess the load capacity of a commercially available exoskeleton model. We propose a set of tests that involve both static and dynamic measurements, and we present the design and construction of a controlled device to measure the torque of industrial exoskeletons, along with the results of static and dynamic testing of this model.

Therefore, the goal of this manuscript is to present a device and procedure involving static and dynamic tests to assess exoskeletons. The capabilities of the device and the validity of the procedure are demonstrated with one exoskeleton model. In this way, this study contributes to filling the literature gap in terms of exoskeleton assessment and paves the way to benchmarking and standardization. This is a step toward making exoskeleton technology viable for the market to ultimately contribute to reduce and prevent injuries at workstations.

2. Materials and Methods

The objective was to measure exoskeleton torque as a function of arm angles, under the same conditions for all exoskeletons under study. This was achieved by means of a beam load cell located at a known distance in the test bench. The transmitted force was then directly converted into torque, and angle positions were determined by the number of steps sent to the step motor. In order to demonstrate the utility of the system, we tested the commercially available exoskeleton MATE (v1.0 Comau, Turin, Italy). Figure 1 shows the exoskeleton mounted on the test bench.

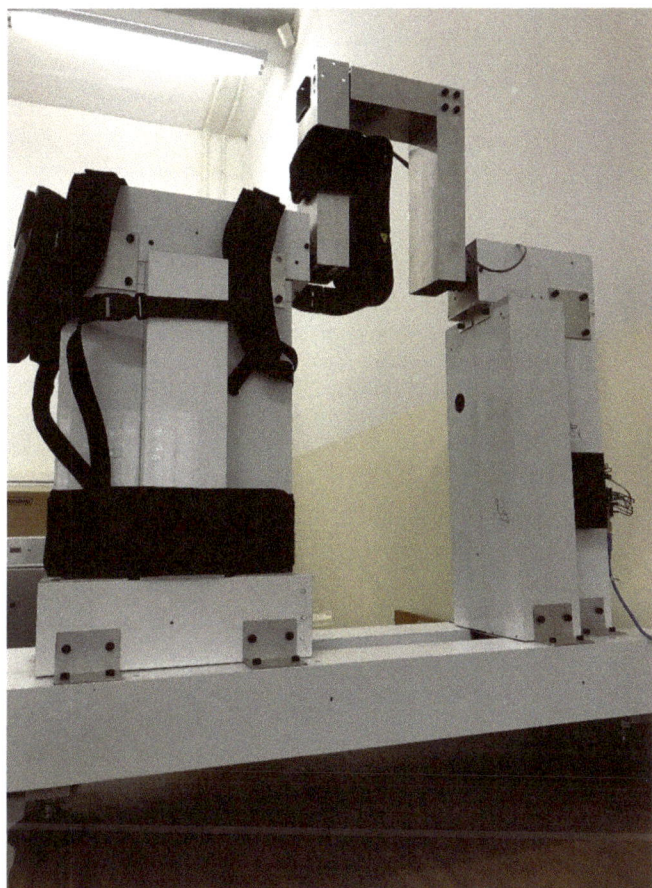

Figure 1. Exoskeleton mounted on the test bench.

The structure of the test bench was made of structural aluminum, and moving parts were built with steel. It was composed of two towers, one for the actuation and the other for the actuated exoskeleton. A step motor controlled position and a 10 kg beam load cell provided information about forces and, thus, torque. The machine was controlled by an Arduino Uno R3 board connected to a personal computer. A dedicated power source fed the motor driver, as shown in Figure 2. For more information about the test bench, see [25].

Validation of the test bench was carried out against results measured with the device. The masses listed in Section 2.1 were directly used to calculate the resulting torque at different positions of the exoskeleton actuator. The same exoskeleton was then fixed on the test bench and used as a reference for comparison between the directly measured torque and the torque calculated by the device. It is a procedure similar to those used for interlaboratory comparisons.

The following subsections describe how several conditions were tested and considered in the tests, as summarized in Table 1, in order to validate the test bench with an exoskeleton.

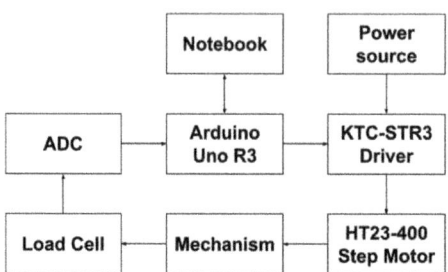

Figure 2. Diagram describing the actuation, control, and data acquisition elements. The Notebook has a custom-made program to select the testing procedure that would be fed to the microcontroller (Model Uno Rev. 3, Arduino Uno, Arduino SpA, Italy) to control the driver (KTC-STR3, Kalatec, São Paulo, Brazil) that drove the step motor (HT23-400, Kalatec, São Paulo, Brazil). The forces acting on the beam of the mechanism by the exoskeleton actuation were measured with a load cell. This measurement was converted to a digital signal (ADC: Analog-to-Digital Converter) and fed to the Arduino Uno microcontroller.

Table 1. Considerations and their effects on the tests.

Condition	Effect
Beam load cell calibration	Compensate for both directions.
Beam load cell linearity	Linear in both directions.
Beam load cell hysteresis	Negligible.
Beam load cell drift	Negligible.
Table deflection	Negligible.
Vertical vibration	Negligible.
Horizontal vibration	Negligible.
Vertical resonance	Limit test speed.
Horizontal resonance	Limit test speed.
Effect of masses	Negligible.

2.1. Masses

A set of already existing nine test masses were used to calibrate the beam load cell. The values were chosen so that the complete range (from 0 kg to 3 kg) could be achieved in steps of 250 g, as follows:

- M1 = 175 g;
- M2 = 109 g;
- M3 = 172 g;
- M4 = 169 g;
- M5 = 137 g;
- M6 = 103 g;
- M10 = 856 g;
- M11 = 998 g;
- M12 = 468 g.

A metal basket was also used to allow (un)load of the main masses, with weight Mb = 41 g.

All masses were weighted against a common scale with a resolution of 1 g.

2.2. Beam Load Cell Calibration

A simple flexion beam load cell was used in the tests, with a capacity of 10 kg. It must be calibrated before use, and as it was measured in both directions, both should be calibrated. The calibration curves were different in each case and were assumed different but close to each other.

Mass M11 = 998 g was used to calibrate initially the beam load cell. After that, the maximum mass value of 3013 g was used to generate a second calibration factor. Zero was added as a third point, and the mean value between them was defined as the calibration factor. The same procedure was used for both working sides of the beam load cell. The calibration factors obtained were as follows:

- Positive: 233,000;
- Negative: 45,000.

The initial assumption was confirmed partially because they were indeed different but not close to each other. Their values were corrected post-acquisition.

2.3. Beam Load Cell Linearity

Although linearity was assumed, we used a set of 13 points to assess linearity, from zero to the maximum expected value of 3 kg at 158 mm from the rotation axis, in incrementing steps of ~50 g. Table 2 shows the masses and values used.

Table 2. Beam load cell linearity test points.

Point (g)	Masses	Exact Value (g)
0	None	0
250	2, 5	246
500	1, 3, 4	516
750	2, 3, 12	749
1000	11	998
1250	2, 5, 11	1244
1500	1, 3, 4, 11	1514
1750	2, 3, 11, 12	1747
2000	5, 10, 11	1990
2250	4, 5, 6, 10, 11	2263
2500	1, 3, 4, 5, 10, 11	2506
2750	2, 3, 4, 10, 11, 12	2773
3000	2, 3, 4, 5, 6, 10, 11, 12	3013

In both positive and negative cases, beam load cell linearity was confirmed, with R^2 of 0.99. Figure 3 shows the results for positive deformations, and Figure 4 shows the results for negative ones. Thus, the initial assessment was confirmed, and the beam load cell could be considered linear in all desired ranges.

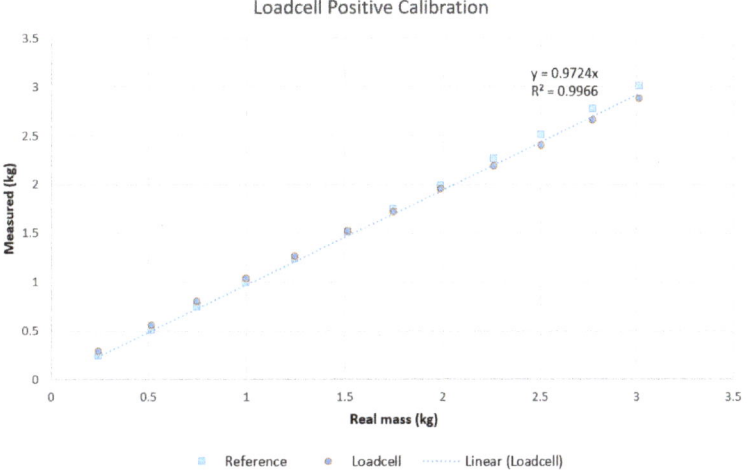

Figure 3. Beam load cell positive linearity.

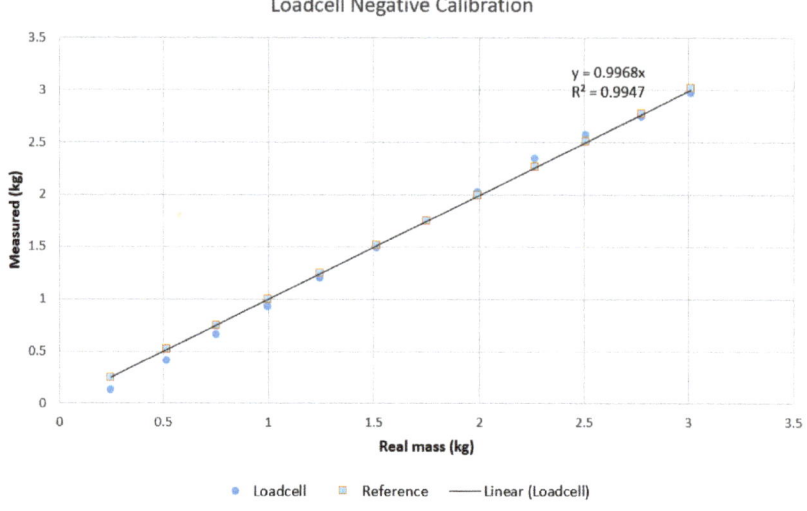

Figure 4. Beam load cell negative linearity.

2.4. Beam Load Cell Hysteresis

In order to assess hysteresis, we applied a sequence of load/unload/load with each one of the following sequence of masses: M6, M4, M1, M12, M11, M10, M3, M5, and M2.

Figure 5 shows that the hysteresis was negligible.

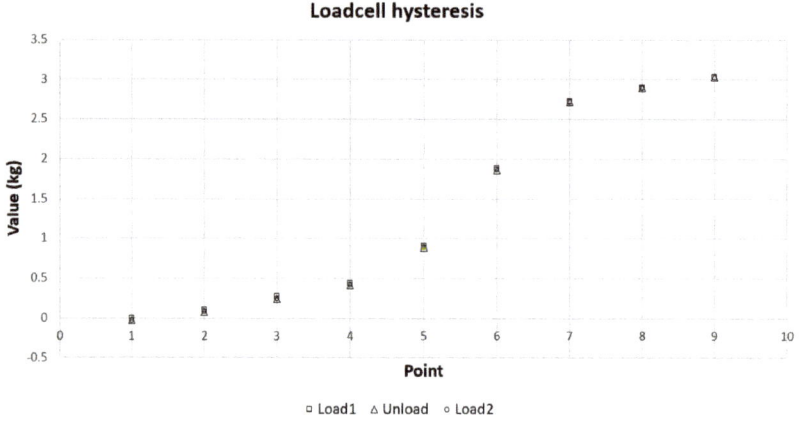

Figure 5. Beam load cell hysteresis behavior.

2.5. Beam Load Cell Drift

Drift was assumed to be negligible for the duration of each run. It was measured twice: first, at the "power on" of the Arduino board and at the establishment of communication and, second, after 3 h of uninterrupted work. In both cases, drift was measured during the subsequent 2 h period. Figure 6 shows the results obtained.

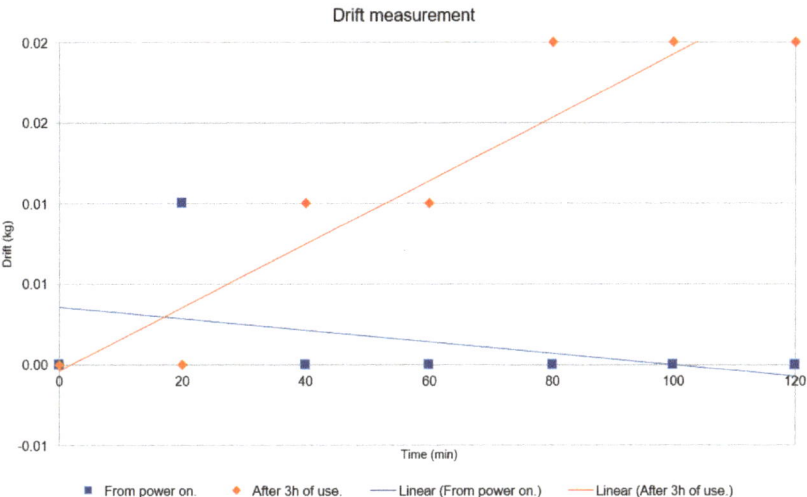

Figure 6. Beam load cell drift at the start and after 3 h of use.

From "power on", measurements remained stable for the period of 2 h and could be considered zero. After 3 h of use, the measured values kept rising at a rate of 12 g/h.

Thus, considering that the range of the expected force values ranged from −1 kg up to 5 kg and that the total time of each run reached a maximum of 1 min, the initial assumption was confirmed, and drift was considered negligible for the duration of each run.

2.6. Beam Load Cell Interferences

Interference due to different sources was assumed significant. The following five cases were analyzed:

- Effect of deflection of the table on which the workbench was laid;
- Effect of vertical vibration of the workbench;
- Effect of horizontal vibration of the workbench;
- Effect of vertical vibration of the moving arm;
- Effect of horizontal vibration of the moving arm.

For each of them, a series of measurements were performed, and the maximum value was taken as the worst possible error. In all cases, the moving arms were placed in position zero, which pointed downward. Each test was carried out as follows:

1. Effect of deflection of the table on which the workbench was laid: The reason for this test was that sometimes, it was necessary for the user to use the table as support for an activity. In this case, a mass of ~70 kg was placed during the measurements on the center of the largest side of the table. No change was detected in the measurement.
2. Effect of vertical vibration of the workbench: The reason for this test was to check if small vibrations due to mass operation on the table could affect the measurement. In this case, a mass of ~1 kg (mass 11) was released from a height of 10 mm on the top surface of the workbench, precisely over the actuator (not a direct hit, however). No change was detected in the measurement.
3. Effect of horizontal vibration of the workbench: The reason for this test was to check if small vibrations due to mass operation on the table could affect the measurement. In this case, with the help of a wire, a mass of ~1 kg (mass 11) was released from a height of 10 mm on the side surface of the workbench, precisely to the side of the actuator (not a direct hit, however). No change was detected in the measurement.
4. Effect of vertical vibration of the moving arm: The reason for this test was to check if vibrations due to resonance on the arm could affect the measurement. In this case, a

mass of ~1 kg (mass 11) was released from a height of 10 mm on the top surface of the actuated arm, precisely over the axis (not a direct hit, however). A change of 0.05 kg was detected in the measurement.

5. Effect of horizontal vibration of the moving arm: The reason for this test was to check if vibrations due to resonance on the arm could affect the measurement. In this case, with help of a wire, a mass of ~1 kg (mass 11) was released from a height of 10 mm on the side surface of the actuated arm, precisely on the side of the axis (not a direct hit, however). A change of 0.03 kg was detected in the measurement.

Table 3 summarizes the results of the tests. During normal operation, the initial assumption was refuted, and no interference was expected in the tests. However, in the case in which resonance was identified, data either required additional filtering or had to be acquired under different conditions. In this particular case, speed was limited to 18°/s due to resonance at higher speeds.

Table 3. Beam load cell interference test results.

Condition	Reference (kg)	Worst Case (kg)
Table deflection	0.00	0.00
Vertical vibration	0.00	0.00
Horizontal vibration	0.00	0.00
Vertical resonance	0.00	0.05
Horizontal resonance	0.00	0.03

2.7. Testing Procedures

Two types of tests were performed: static and dynamic. In both tests, angular zero was defined as the position in which the arms were pointing downward, in agreement with the anatomical reference position. Angles increased as the arms were raised, corresponding to an increase in shoulder flexion up to a maximum of 180°. Angles decreased in the direction of shoulder extension up to −20°. At these extreme values, there was no exoskeleton actuation. Beam load cell zero was taken at the starting position of 180°, at which there was no actuation of the exoskeleton under analysis. The final position was set at −20°. Figure 7 shows a kinematic diagram of the test bench with the exoskeleton.

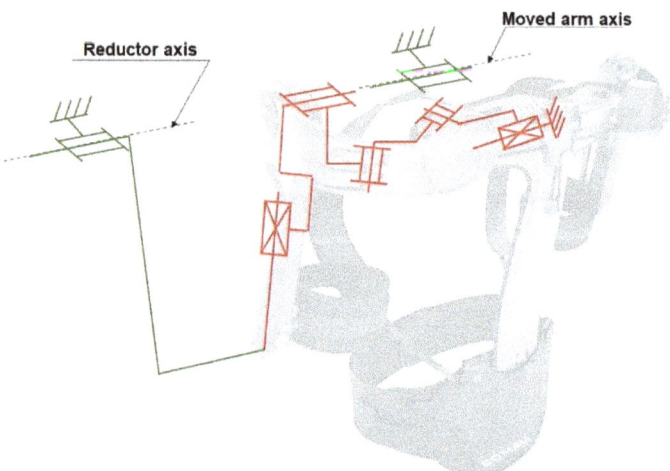

Figure 7. Kinematic diagram of the test bench with the exoskeleton.

In the static tests, a set of 10 consecutive measurements were taken from 180° up to −20°, in steps of 8°. At every step, the machine remained still for 500 ms for mechanical stabilization. To allow for compensation of the position and masses of the test bench, three tests without exoskeleton were carried out to measure the background forces. The background force at a certain angle was the average of these 30 measurements, after outlier removal. This value was subtracted from the actual measurements.

In the exoskeleton MATE (v1.0, Comau, Turin, Italy) used in this study, similar to most of the commercially available exoskeletons, it is possible to select different values of spring stiffness that provide different supporting torques. For the exoskeleton measurements, three runs were performed for each one of the seven spring levels available in MATE. The torque adjustment in MATE was performed with a mechanical switch with seven positions that indicate the level of strength. Level 1 is the weakest (lowest stiffness), and Level 7 is the strongest (highest stiffness). The value of the point of interest was the average of these 30 measurements, after removing the outliers.

In dynamic tests, three cycles of consecutive measurements were taken from 180° up to −20°, in steps of 2°. The velocity of movement was 18°/s, to avoid vibration due to both step transitions at lower speeds and resonance at higher speeds. A single measurement was taken from each angle during a single run.

Background measurement was taken as a set of three runs without any exoskeleton attached to the test bench. The value of the point of interest was the average of the three measurements, after outlier removal.

Three runs were carried out for each level of the spring. The value of the point of interest was the average of the three measurements, after outlier removal.

3. Results

In the next sections, static and dynamic results are presented.

3.1. Static Tests

The torques obtained in the static measurements for each spring level are presented in Figure 8.

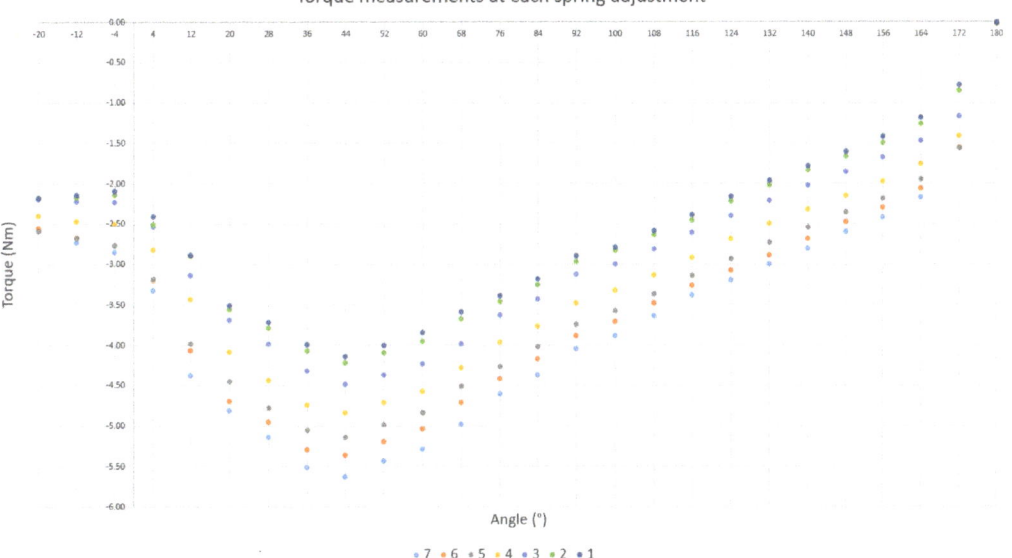

Figure 8. Computed static torques with the testbench configuration (at one extreme of the shoulder translation joint) for each of the seven spring levels in MATE. Level 7 is the strongest one, and level 1 is the weakest.

All spring levels started at zero in position 180°. Actuation increased up to a maximum of around 45°, and then, they decreased again until position 0°. From 0° to −20°, the torque remained constant between −2 Nm and −3 Nm. This behavior between 0° and −20° did not correspond to the expected traction forces, which is discussed in the next section.

The maximum torque in position 45° was 5.6 Nm.

3.2. Dynamic Tests

The torques obtained during the dynamic tests are shown in Figure 9.

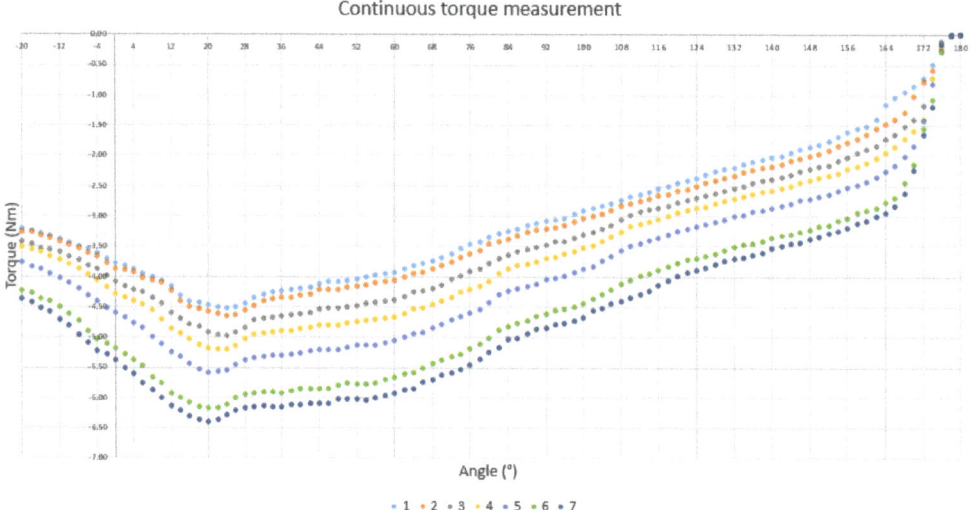

Figure 9. Computed dynamic torques with the testbench configuration (at one extreme of the shoulder translation joint) for each of the seven spring levels in MATE. Level 7 is the strongest one, and level 1 is the weakest.

All spring levels started at zero in position 180°. Actuation increased up to a maximum between 20° and 30°, decreasing until position −20°. The behavior between 0° and −20° did not correspond to the expected traction forces and, as in the case of the static measurements, it is discussed in the next section.

4. Discussion

The results showed that it is possible to use our test bench to measure the performance of industrial exoskeletons under repeatable conditions. The test bench is open access, it mostly requires commercial off-the-shelf components, and the parts that need some custom fabrication are simple, making it low cost (less than 800 USD), with easy implementation.

In the tests, it was found that dynamic measurements were increased in torque when compared with static measurements. That was expected since, in dynamic measurements, inertia and other forces also contribute to the actual measured values. This is relevant because it is not possible to compare two tests carried out under different dynamic conditions, while our device can provide the same testing conditions.

There were small variations in the torques measured in both tests. It is possible that those variations are due to the sliding of the biceps strap during movement. This underscores one of the contributions of this study: the torques effectively applied on the user's arm were not necessarily the torques measured directly in the exoskeleton because of the relative movement of the equipment to the trunk and the biceps.

The main advantage of using a standardized test that considers effectively transmitted torques for the exoskeletons is the possibility of comparison under the same conditions. In this way, it is possible to obtain the characteristics of each exoskeleton and the specific angular positions in which it provides the higher torque levels. This information is critical to adapt the exoskeleton to the task and the user. For instance, if the task requires a certain shoulder angle, and the exoskeleton is not sufficiently assisting the user, it is possible to increase the torque level. It is also possible that the exoskeleton is providing an excessive torque in a certain position that demands extra user effort to maintain that position. Finally, the exoskeleton angle-torque curves provided by our test bench can be very useful to plan the introduction of exoskeletons in precision tasks: if the torque changes abruptly between two angular positions, this would be perceived by the user as a perturbation that may hinder precision performance.

One of the main aspects in the design of a test bench for industrial exoskeletons is the generalization of the attachment, because there are several exoskeletons in the market, and each one has its unique characteristics. In this study, we used MATE (v1.0, Comau, Turin, Italy) to validate the test bench. This exoskeleton fitted very easily for the test bench, which was designed to simulate a human arm. However, it is possible that other exoskeleton models may need further adaptations of the test bench.

It must be noted that the results for the static and dynamic tests show a compression force between $0°$ and $-20°$. However, when using MATE, it is possible to feel that the spring changed its actuation direction, which should have been detected by the test bench by a transition to positive values. The reason for this behavior within the machine is believed to be due to the way the exoskeleton was fixed for testing. In the human body, that range of movement is followed by a slight torsion of the trunk and slight sliding of the exoskeleton, which was not reproduced in the machine. Thus, from $0°$ to $-20°$, the exoskeleton started to offer resistance to the movement, which was registered as a compression force in that region, not as a traction force, as expected.

It is important to keep in mind that the goal of this research was to evaluate the test bench, and not the exoskeleton itself. Thus, there might be differences between the values presented here and other studies about the same exoskeleton, due to differences in the testing conditions. In this work, the shoulder translation joint was at one extreme of its range of motion. This underscores the need for testing devices that provide homogeneous testing conditions for benchmarking.

Actuation can be further improved to allow better dynamic response, and additional sensors, e.g., inertial or pressure sensors, could provide more information about the tests, such as the vibration during exoskeleton use or the pressure at the contact surfaces between the exoskeleton and the user.

It is also possible to include additional fixation points to measure a variety of exoskeletons with different sizes while avoiding extreme joint positions along with improving the beam load cell g remodeling of the connections with the machine.

Finally, we presented the design and validation of a test bench for industrial exoskeletons. This device allows for the implementation of standardized tests that can be useful to assess objectively one exoskeleton allowing the comparison between different exoskeletons. Moreover, the results from the test bench combined with measurements with subjects, considering physiology, biomechanics, and user's perception, would allow designing and/or choosing the best exoskeleton for a certain user and a given task.

Author Contributions: Conceptualization, A.F.-C., V.N.H. and D.d.M.R.; methodology, V.N.H. and D.d.M.R.; software, V.N.H.; validation, V.N.H. and D.d.M.R.; formal analysis, A.F.-C.; investigation, A.F.-C.; resources, A.F.-C.; data curation, A.F.-C.; writing—original draft preparation, C.T.; writing—review and editing, C.T.; visualization, C.T.; supervision, A.F.-C.; project administration, A.F.-C.; funding acquisition, A.F.-C. All authors have read and agreed to the published version of the manuscript.

Funding: This research received no external funding.

Institutional Review Board Statement: Not applicable.

Informed Consent Statement: Not applicable.

Data Availability Statement: The data presented in this study are available on request from the corresponding author.

Acknowledgments: The authors thank Pedro Parik Americano and Victor Bartholomeu for the technical support with the electronics, and João Pedro Pinho, for providing the Myotrace for these measurements. We would like to thank Mercedes-Benz do Brasil for lending the exoskeletons for the tests.

Conflicts of Interest: The authors declare no conflict of interest.

References

1. Lowe, B.D.; Dick, R.B. Workplace Exercise for Control of Occupational Neck/Shoulder Disorders a Review of Prospective Studies. *Environ. Health Insights* **2015**, *8*, 75–95. [CrossRef]
2. Maciel, V. LER e DORT São as Doenças Que Mais Acometem os Trabalhadores, Aponta Estudo—Português (Brasil). Available online: https://www.gov.br/saude/pt-br/assuntos/noticias/ler-e-dort-sao-as-doencas-que-mais-acometem-os-trabalhadores-aponta-estudo (accessed on 5 November 2021).
3. SCS Fundacentro. Quase 39 Mil Trabalhadores São Afastados por LER/Dort em 2019—Português (Brasil). Available online: https://www.gov.br/fundacentro/pt-br/assuntos/noticias/noticias/2020/3/a (accessed on 5 November 2021).
4. Bernard, B.P. Musculoskeletal Disorders and Workplace Factors. *Natl. Inst. Occup. Saf. Health (NIOSH)* **1997**, *104*, 97B141.
5. Liberty Mutual. 2020 Workplace Safety Index: The Top 10 Causes of Disabling Injuries—Liberty Mutual Business Insurance. Available online: https://business.libertymutual.com/insights/2020-workplace-safety-index-the-top-10-causes-of-disabling-injuries/ (accessed on 5 November 2021).
6. Weston, E.B.; Alizadeh, M.; Knapik, G.G.; Wang, X.; Marras, W.S. Biomechanical evaluation of exoskeleton use on loading of the lumbar spine. *Appl. Ergon.* **2018**, *68*, 101–108. [CrossRef] [PubMed]
7. SuitX. ShoulderX | suitX. Available online: https://www.suitx.com/shoulderx (accessed on 11 June 2019).
8. Lowe, B.D.; Billotte, W.G.; Peterson, D.R. ASTM F48 Formation and Standards for Industrial Exoskeletons and Exosuits. *IISE Trans. Occup. Ergon. Hum. Factors* **2019**, *7*, 230–236. [CrossRef] [PubMed]
9. de Looze, M.P.; Krause, F.; O'Sullivan, L.W. The Potential and Acceptance of Exoskeletons in Industry. In *Wearable Robotics: Challenges and Trends*; Volume 16: Biosystems & Biorobotics; González-Vargas, J., Ibáñez, J., Contreras-Vidal, J., van der Kooij, H., Pons, J., Eds.; Springer: Cham, Switzerland, 2017. [CrossRef]
10. Howard, J.; Murashov, V.V.; Lowe, B.D.; Lu, M.L. Industrial exoskeletons: Need for intervention effectiveness research. *Am. J. Ind. Med.* **2020**, *63*, 201–208. [CrossRef] [PubMed]
11. Crowell, H.P.; Park, J.-H.; Haynes, C.A.; Neugebauer, J.M.; Boynton, A.C. Design, Evaluation, and Research Challenges Relevant to Exoskeletons and Exosuits: A 26-Year Perspective From the U.S. Army Research Laboratory. *IISE Trans. Occup. Ergon. Hum. Factors* **2019**, *7*, 199–212. [CrossRef]
12. Van Dijsseldonk, R.B.; van Nes, I.J.W.; Geurts, A.C.H.; Keijsers, N.L.W. Exoskeleton home and community use in people with complete spinal cord injury. *Sci. Rep.* **2020**, *10*, 15600. [CrossRef] [PubMed]
13. Daunoraviciene, K.; Adomaviciene, A.; Grigonyte, A.; Griškevičius, J.; Juocevicius, A. Effects of robot-assisted training on upper limb functional recovery during the rehabilitation of poststroke patients. *Technol. Health Care* **2018**, *26*, 533–542. [CrossRef] [PubMed]
14. Varghese, R.J.; Freer, D.; Deligianni, F.; Liu, J.; Yang, G.Z. *Wearable Robotics for Upper-Limb Rehabilitation and Assistance*; Elsevier Inc.: London, UK, 2018.
15. Der Vorm, V.; Sullivan, J.O.; Nugent, R.; De Looze, M. Considerations for Developing Safety Standards for Industrial Exoskeletons New Hybrid Production Systems in Advanced Factory Environments Based on New Human-Robot Interactive Cooperation. 2015. Available online: www.robo-mate.eu (accessed on 11 June 2019).
16. Fox, S.; Aranko, O.; Heilala, J.; Vahala, P. Exoskeletons: Comprehensive, comparative and critical analyses of their potential to improve manufacturing performance. *J. Manuf. Technol. Manag.* **2020**, *31*, 1261–1280. [CrossRef]
17. de Vries, A.W.; de Looze, M.P. The Effect of Arm Support Exoskeletons in Realistic Work Activities: A Review Study. *J. Ergon.* **2019**, *9*, 255. [CrossRef]
18. Alabdulkarim, S.; Nussbaum, M.A. Influences of different exoskeleton designs and tool mass on physical demands and performance in a simulated overhead drilling task. *Appl. Ergon.* **2019**, *74*, 55–66. [CrossRef] [PubMed]
19. Pinho, J.P.; Taira, C.; Parik-Americano, P.; Suplino, L.O.; Bartholomeu, V.P.; Hartmann, V.N.; Umemura, G.S.; Forner-Cordero, A. A comparison between three commercially available exoskeletons in the automotive industry: An electromyographic pilot study. In Proceedings of the IEEE/RAS-EMBS International Conference on Biomedical Robotics and Biomechatronics (BioRob), New York, NY, USA, 29 November–1 December 2020; pp. 246–251. [CrossRef]
20. Pinho, J.P.; Parik Americano, P.; Taira, C.; Pereira, W.; Caparroz, E.; Forner-Cordero, A. Shoulder muscles electromyographic responses in automotive workers wearing a commercial exoskeleton. In Proceedings of the Annual International Conference of the IEEE Engineering in Medicine and Biology Society (EMBC), Montreal, QC, Canada, 20–24 July 2020; pp. 4917–4920. [CrossRef]
21. MacDougall, W. Industrie 4.0: Smart Manufacturing for the Future. In *Germany Trade & Invest*; Gesellschaft für Außenwirtschaft und Standortmarketing mbH.: Berlin, Germany, 2014; p. 40.

22. Manna, S.K.; Dubey, V.N. Comparative study of actuation systems for portable upper limb exoskeletons. *Med. Eng. Phys.* **2018**, *60*, 1–13. [CrossRef] [PubMed]
23. Pesenti, M.; Antonietti, A.; Gandolla, M.; Pedrocchi, A. Towards a functional performance validation standard for industrial low-back exoskeletons: State of the art review. *Sensors* **2021**, *21*, 808. [CrossRef] [PubMed]
24. O'Sullivan, L.; Nugent, R.; van der Vorm, J. Standards for the Safety of Exoskeletons Used by Industrial Workers Performing Manual Handling Activities: A Contribution from the Robo-Mate Project to their Future Development. *Procedia Manuf.* **2015**, *3*, 1418–1425. [CrossRef]
25. Hartmann, V.N.; de Rinaldi, D.M.; Taira, C.; Pinho, J.P.; Forner-Cordero, A. Design of a torque measurement unit for upper limbs industrial exoskeletons. In Proceedings of the 2021 4th IEEE/IAS International Conference on Industry Applications (INDUSCON), São Paulo, Brazil, 16–18 August 2021; pp. 1005–1010. [CrossRef]

MDPI
St. Alban-Anlage 66
4052 Basel
Switzerland
Tel. +41 61 683 77 34
Fax +41 61 302 89 18
www.mdpi.com

Machines Editorial Office
E-mail: machines@mdpi.com
www.mdpi.com/journal/machines

www.ingramcontent.com/pod-product-compliance
Lightning Source LLC
LaVergne TN
LVHW070502100526
838202LV00014B/1776